Social Trust and Life Insurance

Social Trust and Life Insurance: The Impact of Genetic Test Results in the Republic of Ireland

By

Louise Morris

CAMBRIDGE
SCHOLARS
PUBLISHING

Social Trust and Life Insurance:
The Impact of Genetic Test Results in the Republic of Ireland,
by Louise Morris

This book first published 2011

Cambridge Scholars Publishing

12 Back Chapman Street, Newcastle upon Tyne, NE6 2XX, UK

British Library Cataloguing in Publication Data
A catalogue record for this book is available from the British Library

ISBN (10): 1-4438-3001-1, ISBN (13): 978-1-4438-3001-0

The author would like to thank the Irish Research Council
for the Humanities and Social Sciences for their funding
of the research contained within this book.

TABLE OF CONTENTS

LIST OF TABLES

LIST OF FIGURES

CHAPTER ONE

INTRODUCTION

1.1 Introduction

> "Within a few short decades, society rearranges itself, its world view, its basic values, its social and political structure, its arts, its key institutions… There is a new world."
> —Peter F. Drucker

The life insurance industry is regularly confronted with changes in its surrounding environment that may influence the business in a positive or a negative manner, depending upon the nature of such changes and their effect upon society. One such change that impinges upon both life insurance business and society is genetic testing. Developments in genetic knowledge have enabled the medical profession to test individuals for their genetic predisposition to inheritable disease. The results of genetic tests can reduce uncertainty surrounding certain future events, such as a person's life expectancy or possibility of developing a particular disease. The advantages of this reduction in uncertainty for society include several desirable medical possibilities.[1] However, among the myriad of uses of genetic testing, its use by life insurance companies has been received rather negatively by society, with strong opposition to the use of genetic test results in the underwriting of life insurance policies (Hall Mark *et al.*, 2005; Voss, 2000:8; RADAR, 1999).[2] Given that the result of a genetic

---

[1] There exists the possibility to pre-treat disease in order to prevent future illness and the potential to individually tailor drugs to ensure direct treatment of disease with the foresight of an individual's negative reactions to specific medications (Robertson, *et al.* 2002)

[2] The 1999 BSA (British Sociological Association) indicated that 75 per cent of the British public thought that insurance companies shouldn't be allowed to use genetic tests when determining premium rates (Voss, 2000:8).

A study completed in 2005, illustrates that 40 per cent of the study's population of 86,859 participants agreed with the statement: "Genetic testing is not a good idea

test constitutes the evaluation of a risk, this information is deemed usable by life insurance companies in determining the risk status of applicants. Reasons for the controversy surrounding this use of genetic test results are based upon perceptions of discrimination and fairness, whereby in today's society, no person's genetic inheritance is considered a priori better than any other person's, with the corollary that discrimination on genetic grounds is deprecated (MacDonald, 1999:83). However, the business of life insurance is based upon the ability of an insurer to discriminate between risks and to charge clients a premium that is representative of the risk which they bring forth to the life insurance fund. Indeed, insurers have been using family history information for decades in order to determine the mortality and morbidity of an individual. However, the fresh moral and ethical concerns introduced by the more accurate and reliable information provided by genetic tests, has lead to widespread debate surrounding what can be considered as acceptable and unacceptable forms of discrimination and what constitutes fairness.

Society's perception of insurers' use of genetic test results has received little attention in the media and political circles in the Republic of Ireland. During the conduct of the primary research of the present study, insurers were allowed to use the results of two genetic tests that indicate predisposition to Huntington's disease, in addition to other underwriting factors. Following some discussion between the insurance industry and the Government, new legislation imposed by the Government in June 2005, dictates that life insurance companies are no longer permitted to access applicants' genetic test results.[3] Although, this legislation is designed to protect Irish society from perceived risks relating to insurers' use of genetic test results, the attitudes of the Irish public regarding the risks associated with this situation were not assessed. Given that a good

---

because you might have trouble getting or keeping your insurance"; Hall, Mark A. *et al.* (2005)

Research by RADAR (the disability network; registered British charity) in 1999 demonstrated strong public opposition to insurers gaining access to genetic tests (90 per cent of British RADAR members surveyed

[3] The study was conducted over a three year period (2003-2006), while the primary research for the study was completed during approximately one of those three years. Hence, the present study must be considered as producing a snapshot of individuals' risk acceptance attitudes during that particular period. Changes in the policies governing life insurers' use of genetic test results have since been enforced, in particular the Disability Act, 2005, and this may affect applicability of the study. However, this picture can be used to compare the impact of the changes that have been incorporated into legislation since the research was carried out.

risk management decision "emerges from a decision-making process that elicits the views of those affected by the decision, to ensure that differing technical assessments, public values, knowledge and perceptions are considered" (The Presidential/Congressional Commission on Risk Assessment and Risk Management, 1997), without public consultation, the entities managing the risks related to the use of genetic test results in life insurance underwriting are incapable of making any appropriate decisions regarding the control of these risks. This study shall question if society perceives any risks from insurers using genetic test results, what these risks are, and if individuals accept insurers' use of these tests.

Insurers are generally concerned about the practical and financial consequences of an increasing number of people knowing from genetic tests that they have a genetic mutation which increases their susceptibility to illness that may be the cause of premature death. Hence, this study is primarily concerned with insurance policies specifically designed to cover the risk of premature death. There are a variety of other life insurance policies available on the market that may also be affected by genetic testing, such as critical illness, income protection and medical expenses insurance policies. It is important at this stage to underline a distinction between several of these policies. The medical expenses insurance market in Ireland is governed differently than other types of health insurance. There are only three suppliers of medical expenses insurance in the Republic of Ireland at present. These include VHI (Voluntary Health Insurance), BUPA Ireland (British United Provident Association) and Vivas Health. Genetic testing should not present the same risks to society from medical expenses insurers using genetic test results due to the underwriting system of this form of insurance. Medical details are requested by medical expenses insurers, however, individuals' health status details only restrict them from obtaining full cover for a certain period of time (known as the waiting period). The waiting period depends upon the age of an applicant upon enrolment. Given the less stringent approach to paying claims based upon health status, medical expenses insurers cannot be considered to present the same or equivalent risks as life insurers from their use of genetic test results.

Life insurance companies also provide types of health insurance, such as permanent health insurance, critical illness insurance and long-term care insurance. Each of these products tends to have stricter claiming conditions than medical expenses insurance. The underwriting criteria for permanent health insurance and long-term care insurance are more stringent than those for critical illness insurance, where it is possible to obtain a large level of cover without provision of a medical. These contracts expose society to more risks than medical expenses insurance, indeed, the level of risk could be comparable to that deemed to derive from death insurance. This is due to the nature of these policies whereby a failure in the health of an individual, often to a certain pre-described degree, necessitates insurers to pay out. Therefore, for the majority of these types of health insurance contract, insurers require a great detail of medical history from applicants and, in some cases, regular medical examinations. This information may alter the premium or level of cover provided to an individual. In this manner, the risks for society from insurers gaining access to genetic test results are similar to death insurance contracts and the risks from insurers being denied access to this information is also comparable. However, it is difficult to investigate these risks as there can be a substantial difference across insurers with regard to the requirements and criteria for payment of each type of health insurance contract. Given the difficulties in the convergence of all types of health insurance into one bracket, it is considered beyond the scope of the present study to explore perceptions of the risks related to insurers using genetic test results to underwrite such policies. The present study is, therefore, limited to the investigation of the risks and benefits perceived by society as relating to life insurers using genetic test results to underwrite policies associated with premature death.

Research in the area of risk perceptions and risk acceptance has been particularly intense when it comes to explaining perceptions of nuclear risks and the acceptance of hazardous waste facilities. In addition, trust in the entities managing such risks has been examined and the affect that trust has upon perceived risk analysed. Viklund (2003) suggests that the special importance of these particular risks when studying the relationship between trust and risk perception, is related to people's low level of knowledge about such risks. It is likely that Irish society has incomplete knowledge of the risks related to insurers using genetic test results, as this matter has received relatively little attention by the Irish media. Unlike nuclear waste and genetically modified foods, there has been little significant research on society's perceived risks and benefits of the use of genetic test results by the life insurance industry or trust in the entity(s) governing such usage. In addition, few studies examining the relationship between trust and risk perceptions have been conducted within Europe, instead, the majority of research has been performed in the United States (Viklund, 2003:727). Therefore, as well as gaining an understanding of the risks perceived by Irish society of insurers' use of genetic test results and the acceptance of such usage, this study intends to bridge the research gap that exists on the trust-risk relationship and examine the nature of the association between knowledge, trust, risk perception and benefit perception and their influence upon risk acceptance. To fully understand acceptance of the risks related to insurers using genetic test results, this study shall explore the nature of this relationship and identify and examine the factors and conditions that may influence this affiliation and consequently risk acceptance.

1.2 Theoretical Context of the Study

The study is based on the theory that there exists a relationship between knowledge, social trust, risk perception, benefit perception and risk acceptance.[4] For the purpose of the study, knowledge is defined in terms of a lay person's knowledge which is considered less formal and more intuitive and qualitative than expert knowledge, which is seen as objective, analytic and rational (Slovic, 1999:675). This knowledge represents a different kind of expertise that is a function of both the cognitive and the motivational system and of the conditions of the social, political and cultural environment (Jungermann and Slovic, 1993:87; Otway, 1992:220). Social trust focuses on public trust in those responsible for risk management, including politicians, authorities, and corporations and refers to people's willingness to rely on experts and institutions in the management of risks and technologies (Viklund, 2003:729; Earle & Cvetkovich, 1995). Social trust is assumed to be socially constructed and dependent upon a number of factors that constitute and contribute to trust (Weber and Carter, 2002). These factors indicate reasons for trust or lack of trust in an individual or an organisation and include competence, reliability, credibility, integrity, care, fairness, and openness (Poortinga and Pidgeon, 2003). Social trust in two entities, the Irish Government and the insurance industry, who were both involved in the management of

[4] Land and Hallman, 2005; Siegrist *et al.*, 2003:707; Viklund, 2003; Pisano and Woods, 2002; Sjoberg, 2002; Figueiredo and Drottz-Sjoberg, 2000:233-242; Wright, Pearman and Yardley, 2000:681-690; Siegrist, Cvetkovich, Roth, 2000:353-362; Siegrist and Cvetkovich, 2000; Slovic, 2000; Hunt, Frewer, and Shepherd, 1999:167-180; Grobe *et al.* 1999; Groothius and Miller, 1997:241-257; Jungermann *et al.*, 1996; Biel and Dahlstrand, 1995:25-36; Earle and Cvetkovich, 1995; Freudenburg, 1993; Flynn *et al*, 1992:417-429; Flynn *et al.*, 1992; Bord and O'Connor, 1992, and Pijawka and Mushkatel, 1991; Drottz-Sjoberg and Sjoberg, 1991:2007-2036 and Baird, 1986:425-435

insurers' use of genetic test results during the time which the study was conducted, are analysed. Risk perception and benefit perception are defined as perception of risks/benefits for 'family and I' and for 'society as a whole'. The term 'risk' implies a possible future negative and uncertain outcome of an event or situation caused by nature or human activities where insufficient knowledge was available to prevent such an outcome (Holzheu, 1993:241; Renn, 1992:57-58). The positive outcome of a situation or an event can be referred to as a gain, a profit or, in this case, as a benefit. Risk acceptance is defined as agreement or satisfaction with a risk situation and the current strategies designed to manage such circumstances (Sjoberg, 1999:130).

This research, to a certain extent, is replicating segments of previous studies, albeit with a different approach. The three studies that are of particular relevance to this research are those of Siegrist and Cvetkovich (2000), Siegrist (2000) and Poortinga & Pidgeon (2003). According to Siegrist and Cvetkovich (2000), where incomplete knowledge of a hazardous situation exists, people make decisions regarding their acceptance of risk based upon their trust of those managing the risk situation. Within this risk acceptance pattern, trust does not have a direct influence upon risk acceptance but results in a positive evaluation of the benefits and a negative evaluation of the risks which leads to the acceptance or tolerance of risk. Therefore, according to Siegrist (2000), trust has an indirect affect upon risk acceptance. Rather than examining a general evaluation of trust, as in Siegrist (2000), the present study shall adopt a multi-dimensional definition of trust as defined by Poortinga and Pidgeon (2003). This influential pattern shall be explored in relation to the risk acceptance of life insurers using genetic test results in their underwriting procedures. In addition, other factors and conditions that influence risk acceptance attitudes associated with this matter shall be examined.

This study adopts the central features of critical realism in that an attempt is made to preserve a 'scientific' attitude toward social analysis, as well as acknowledge the importance of people's meanings in the exploration of the relationship between trust and risk (Layder, 1993:16). This particular approach conveys the 'textured' or interwoven nature of the different dimensions of social reality as depicted by a realist philosophy, taking into account the perspectives of the people involved. To fully understand the complex social reality that influences public views, the study teases out the different, although interconnected and interdependent, dimensions that shape this reality. Layder (1993) outlines several dimensions of social reality that intertwine with each other to produce a multifaceted or stratified model of society. The dimensions include 'context', 'setting', situated activity, self, plus the overriding dimension of history. Layder (1993) also refers to the need to consider power; a dimension of social reality that pervades all other levels. For the purposes of this study, an attempt is made to disentangle four of the elements that make up social reality and examine them and their affect on the research questions separately. It is important to note at this stage that history is not examined as it is considered beyond the scope of the study. In addition, given the nature and objectives of this particular study, there is a stronger emphasis placed upon analysis of the self-dimension than the remaining three dimensions. This is not to say that these dimensions play a lesser role upon the social reality surrounding insurers use of genetic test results, it is merely that the self-dimension preoccupies a predominant position in this study. This inadequate examination of each of the dimensions somewhat contributes to a lack of full disclosure of the social reality surrounding the risk acceptance of insurers using genetic test results. However, further analysis of these dimensions and their affect upon trust, risk perceptions and risk acceptance may prove a valuable contribution to the findings for future research. By applying Layder's (1993) guide to examining social reality, all the dimensions that impact on acceptance of the perceived risks related to insurers' use of genetic test results are considered, thereby increasing our understanding of the factors that influence the risk acceptance pattern to be examined.

1.3 Objectives

The present study intends to explore the role that incomplete knowledge, social trust, risk perception and benefit perceptions have on society's acceptance of the perceived risks related to insurers using genetic test results. In doing so, the study shall examine society's knowledge of insurers' use of genetic test results in their underwriting processes and individuals' perceptions of the risks and benefits related to such usage, as each of these factors have been attributed to influencing individuals' risk acceptance attitudes. Social trust is also considered a rather eminent factor in affecting risk acceptance, hence, this item shall also be analysed with regard to the effect it may have upon society's acceptance of the risks related to this situation.

In addition, the factors and conditions that influence society's knowledge, social trust, risk perceptions, benefit perceptions and acceptance of the risks perceived from insurers using genetic test results are explored. The intention of doing so is to understand the reasoning behind individuals' trust and risk related evaluations. This ensures that an explanation regarding the acceptance of risks perceived by individuals in relation to insurers using the results of genetic tests is presented as close to its entirety as possible.

To achieve these objectives, the following sub-objectives are identified:

- Test a model adapted from Siegrist (2000) suggesting an influential link between incomplete knowledge, social trust, risk perception, benefit perception and risk acceptance in relation to the use of genetic test results by the life insurance industry in the Republic of Ireland, in order to understand if these factors influence society's risk acceptance.
- Examine whether or not trust is an important determinant in judgements of risks and benefits in the absence of knowledge.
- Verify if social trust in the insurance industry and the Government influences individuals' perceptions of the associated risks and benefits.
- Investigate any influence that risk and benefit perceptions may have on public acceptance of the use of genetic test results by life insurance companies.
- Explore differences in self-assessed knowledge, assumptions of social trust, risk perceptions, benefit perceptions and risk acceptance of individuals based on gender, age, economic status and life insurance ownership to understand if and how these factors may affect knowledge reports and trust evaluations, risk perception, benefit perception and risk acceptance judgements.
- Further develop an explanation of acceptance of the risks related to insurers using genetic test results by further investigating the factors and conditions that underlie self-reports of incomplete knowledge, trust in the Government and the insurance industry, perceived risks and perceived benefits related to insurers using genetic test results and, indeed, risk acceptance itself.

1.4 Methodology

To achieve the objectives of the present study, the use of genetic test results by life insurers and to examine the consequences of such usage and the policies that govern this practice are explored. In doing so, a review of previous literature on this area is considered in order to provide a fuller picture and a means of assessing the situation in the Republic of Ireland. Literature relating to the formation of a risk society is also investigated to gain an understanding of the risks perceived as a result of insurers using genetic test results. To provide a framework in which to study the factors that affect risks acceptance, previous research on the influence that knowledge, social trust, risk and benefit perceptions have upon risk acceptance is examined. In addition, several socio-demographic factors, which are considered in the literature to affect risk acceptance, are explored and analysed.

The study is both extensive and intensive by design and embraces both quantitative and qualitative elements, reflecting a critical realist approach to the research. Participants from two counties in the Republic of Ireland completed a questionnaire that evaluated their self-assessed knowledge, social trust, risk perception, benefit perception and risk acceptance of the use of genetic test results by life insurance companies. The survey measured sets of items related to knowledge, social trust, risk perception and benefit perception to obtain a fuller analysis of the relationship between these factors (see Appendix A.1). Statistical analysis of the survey using SPSS was conducted to reveal whether or not an influential relationship between incomplete knowledge, trust, risk perception, benefit perception and risk acceptance exists based upon the replies of respondents.[5] Survey analysis also indicated whether people differed in their responses across trust in the Irish Government and the insurance industry. In addition, the effect that socio-demographic factors and ownership of life insurance have upon knowledge, trust, risk perception, benefit perception and risk acceptance was explored.

This analysis was followed by semi-structured interviews based on a strategic sample, whereby four types of case were illustrated and explored based upon the findings of the survey analysis – extreme case, extremely

---

[5] Statistical Package for the Social Sciences: A software system for data management and analysis. SPSS may be used for many univariate and multivariate statistical analyses and has facilities for sorting and merging files and manipulating data. SPSS can deal automatically with complex files.

varied case, critical case and representative case (Danermark, 2002:170). The use of a comparative case study analysis allowed the attainment of information surrounding the significance of various factors and conditions that serve to influence the risk acceptance pattern deduced by the questionnaire. Given the relatively small number of cases comprising of few representatives, it was decided to conduct the analysis manually without the assistance of any specialised computer package. Interview analysis is contained in the Appendices for review (see Appendix D.1).

1.5 Significance of the Research

The research provides an analysis of the influential nature of the relationship between incomplete knowledge, trust, risk perception, benefit perception and risk acceptance in the Republic of Ireland and the factors and conditions that serve to influence this relationship. Unlike previous studies that examine this relationship, trust is measured through the use of previously defined dimensions of trust (Poortinga and Pidgeon, 2003), thereby, providing a fuller analysis of this pattern. In addition, given that most other studies conducted on this pattern of risk acceptance have been completed in the United States of America and, on different hazards, this study reveals valuable findings that might be considered for exploration in further studies.

The Government and the insurance industry have paid little attention to the perceptions of society regarding insurers' use of genetic test results. There has been no significant research on society's perceived risks and benefits of the use of genetic test results by the life insurance industry in the Republic of Ireland. This research underlines the importance of acknowledging that society feels affected by insurers using genetic test results and that these feelings should be accounted for in the decision-making process regarding the policies designed to mitigate the risks identified. This study should aid policy makers in their management of the risks posed by this commercial use of genetic testing.

The present study applies the risk theories of Beck (1992) and Giddens (1990; 1991 and 2000) to a partly new empirical field, making this study a somewhat different undertaking. The purpose of doing so is to understand if these theories are productive when it comes to understanding the perceived risks related to the use of genetic test results by the life insurance industry. By bringing such theoretical perspectives into a field where they have not been applied before, the theories could contribute to new ways of understanding and thinking about the management of the use of genetic test results by the life insurance industry. In addition, the findings from the present study could further develop our comprehension of risk theory and risks in late modernity.

1.6 Plan of the Study

Chapter 1 of this study has provided an overview of the research. Chapter 2 explains genetic testing, insurers' use of genetic test results and the policies designed to mitigate the risks perceived from such usage. The nature of the relationship between Beck's (1992) and Giddens' (1990; 1991 and 1999) theories of the risks faced by modern society and the alleged risks posed by insurers using genetic test results is identified, and the bearing that this relationship has on society discussed. Chapter 3 provides an overview of previous research findings on trust and risk perception and the relationship between incomplete knowledge, trust, risk perception, benefit perception and risk acceptance. Chapter 4 outlines the methodology for carrying out the primary research and findings from the survey and case study are revealed in chapters 5 and 6. Chapter 7 draws conclusions from the literature review and the primary research regarding the factors and conditions that influence acceptance of the perceived risks related to insurers using genetic test results. Based upon the findings of this study, implications for the field of risk research and the insurance are explored and criticisms of the research explained. Finally, Chapter 8 assesses achievement of the objectives of the study and discusses recommendations for future research in this area.

1.7 Summary

This chapter has detailed the nature and objectives of the research being undertaken and has outlined how the study is organised. It was revealed that there has been little significant research on society's perceived risks and benefits of the use of genetic test results by the life insurance industry, acceptance of these risks or trust in the entity(s) governing such usage. It was considered that it would be advantageous for entities managing the risks related to this matter to identify and take into account the perceptions that society has of the risks associated with insurers using genetic test results.

An introduction was given to the context of the study and it was found that although researchers have previously examined the relationship between trust, risk perception and risk acceptance, a critical realist approach to the study of this relationship has rarely been conducted. It was outlined how Layder's (1993) research map is used as a guide from which to explore the web of influence surrounding society's acceptance of the risks related to insurers using genetic test results and the generative mechanisms that underlie such acceptance. By following this approach, it should be possible to obtain a holistic view of the nature of this relationship and the factors and conditions that generate this risk acceptance pattern.

The methodology for the primary research was identified and discussed briefly. It was outlined how a survey and case study were used as research tools from which to explore the relationship between knowledge, trust, risk perception, benefit perception and risk acceptance and other factors that may influence this risk acceptance pattern. Finally, the value or significance to be derived from the completion of this study was presented. On theses foundations, the study can proceed with a detailed description of the research.

The following chapter shall explore the nature of genetic testing and its use by life insurance companies as a factor to be considered in their underwriting of applications. The reported consequences of insurers demanding genetic test results from individuals are identified and explored. In addition, the resemblance of the perceived consequences of insurers using genetic test results and the risks identified that contribute to a risk society is explained in the context of theories from Beck (1992) and Giddens (1990; 1991 and 2000). Finally, the policies that govern life insurers' use of genetic test results in the Republic of Ireland shall be examined.

CHAPTER TWO

LIFE INSURANCE AND GENETIC TESTING

2.1 Introduction

The previous chapter provided a brief overview of the background to the present study. The theoretical standpoint of the research was discussed and the objectives described. The purpose of this chapter is to explore insurers' use of genetic test results and the consequences that arise from such usage for insurers and society. To begin with, it is important to acknowledge the development of genetic test results as a source of hereditary information. The potential for rapid advances in our understanding of the genetic basis for health and disease has been heralded by the publication of the sequence and analysis of the human genome in February 2001. DNA (deoxyribonucleic acid) based tests are among the first commercial applications of the new genetic discoveries.[1] Genetic testing involves the examination of an individual's DNA in order to identify certain characteristics about that individual, such as gender or allergic reaction. In addition, genetic tests can be used to look for mutated genes and gene products that could signal an aberrant gene. Such tests can ascertain if a person is a carrier of a particular gene, which may determine whether the person is predisposed to a particular disease or is presymptomatic for a specific illness. From the onset, these new developments were publicly and politically acclaimed as a positive step toward a new era in pre and post medical healthcare. However, within a short space of time, genetic testing is increasingly perceived as a very

[1] DNA is an extremely long macromolecule that is the main component of chromosomes and is the material that transfers genetic characteristics in all life forms, constructed of two nucleotide strands coiled around each other in a ladder-like arrangement with the sidepieces composed of alternating phosphate and deoxyribose units and the rungs composed of the purine and pyrimidine bases adenine, guanine, cytosine, and thymine: the genetic information of DNA is encoded in the sequence of the bases and is transcribed as the strands unwind and replicate (www.libraryadvance.org)

controversial area. For example, controversy surrounds the use of genetic tests by parents to examine the health and sex of their unborn child, the use of genetic testing by employers in vetting staff, or its use by insurance companies when making underwriting decisions concerning the acceptance or rejection of applicants or the imposition of conditions or additional premiums.

New scientific and medical advances in the field of genetics often cause public unease and present many complex social, ethical and regulatory questions (Harper and Clarke, 1990). The uncertainties surrounding the rapid advances being made in genetic testing and its subsequent use in the life insurance underwriting process is just one such development which has caused much consternation and apprehension throughout society and the insurance industry in recent years (Tyler, 2004). The world has never faced a technology with such persuasive potential to predict present and future medical health and inherent human behaviour. The diagnostic and predictive power of genetic testing has introduced a new variable for insurers to select applicants for life insurance with the possibility of using the genetic test results of an applicant to select and/or classify that individual. Traditionally, life insurance companies have sought hereditary data from applicants in the form of family history information. It has been argued that genetic tests do not raise entirely new and unique ethical and legal questions (Murray, 1997). It can equally be argued that the accuracy and reliability surrounding the results of genetic tests may not raise new issues, but genetic testing provides for greater concern regarding such ethical and legal matters. In addition, these concerns are amplified by a change in societal thinking (Giddens, 1991:133-134).

The possibility of insurers using genetic test information has raised considerable fears regarding potential risks to insurability, individual privacy and genetic discrimination (ELSI – Ethical, Legal and Social Issues in the Human Genome Program, 2003). If denied access to genetic test information, insurers are concerned that the principle foundation upon which their business rests would be sacrificed in lieu of societal peace of mind and result in the significantly detrimental consequences of adverse selection (Irish Insurance Federation - IIF, 2001). Although, these fears are based on the unknown future and incomplete knowledge rather than past experience or facts, it is this very characteristic that allows them to be equated with the undesired side effects of production processes and products commonly associated with the 'risk society' (Beck, 1992) and late modernity (Giddens, 1990). This chapter intends to examine such consequences of insurers using and not being permitted use of genetic test results in their underwriting processes in more detail. Following on from this, the risk management strategies undertaken in the Republic of Ireland shall be discussed and our development toward a risk society investigated.

2.2 Life Insurance Underwriting and Genetic Test Results

The life insurance industry in the Republic of Ireland is very competitive and fast growing, with high levels of coverage in major classes of business such as mortgage protection and pensions (Irish Insurance Federation, 2004). The number of operators in the Irish life insurance market has expanded in recent years and this trend appears likely to continue. Insurance is a cyclical business with periods of substantial losses followed by periods of robust profits. Life insurance business is currently in the profit phase of the current cycle (Irish Insurance Federation, 2006). Gross premiums written in Ireland by life insurance companies in 2003 totalled €14.52 billion comprising €7.96 billion Irish risk business and €6.55 billion foreign risk business (Greenford *et al.*, 2005). This generation of profit has allowed existing insurers to increase their capacity and write more business and it has attracted some new insurers into the market. Given the size of the industry, it can be said that life insurance business is a significant contributor to the Irish economy and one that the Government would not wish to see damaged. However, it is important to note that the Irish Government is not only faced with concerns of the economy but also the concerns of the society it is designed to protect.

On the other hand, the insurance industry is a commercial entity and is designed for profit making. The business of life insurance is based upon the scientific principle of risk-defined equity, a reasonable and long followed necessity of life insurance companies. The basic concept of risk-defined equity ensures that the premium paid by each insured life is adequate to cover the risk that an individual introduces to the life insurance fund (Brackenridge and Elder, 1998:241). This means that life insurance companies have the right to differentiate cover and premiums on the basis of the level of risk that each applicant brings to the pool. The term ‘equity’ implies that the system is one that is based upon fairness, whereby policyholders with the same or similar expected risk of loss are charged the same (Brackenridge and Elder, 1998:241). According to Pokorski (1995), an insurance scheme is equitable if it respects actuarial rules. However, as Lemmens (2000) argues, such an opinion presupposes that the use of actuarial rules itself is neutral and not questionable from the standpoint of equity.

The categories by which risks are classified are determined by a mixture of custom, practice and social acceptability, and have been known to change over time. For example, classification by smoker/non-smoker status would not have been considered acceptable 30 years ago, but it is now an almost universal category for all life insurance products (Daykin, *et al.,* 2003:8). The revolution of genetics has presented life insurers with a new factor that can be used to select and classify applicants for life insurance, thereby improving the evidence base for such decision-making. Genetic tests can reduce uncertainty surrounding certain future events, such as dying within a certain time period or suffering from a particular disease. Such a result can contribute towards the evaluation of a risk and is, therefore, usable by the life insurance industry in determining the risk status of an applicant. Insurers have been aware for some time of the genetic heritage of some diseases, hence, the inclusion of family history information as a factor accounted for in their underwriting process. Family medical history is not uniquely genetic in nature in that it may also contain elements relating to the area of residence, quality of living conditions, diet, exposure to infectious diseases, stress and other factors. However, with regards to the hereditary aspect of family history information, the results of genetic tests differ from the information derived from family history analysis only in that the results provide a more accurate and reliable source of hereditary information.

The value of genetic test results to life insurance companies is significant. Where an individual can choose whether or not to buy insurance, and how much they will buy, underwriting of all factors that affect the risk is a requirement to ensure that a fair price is paid; otherwise adverse selection may occur (MacDonald, 1999:83). Therefore, it can be said that if insurers do not take genetic test results into account in their underwriting processes, the result may be an increased susceptibility to adverse selection.[2] This occurs where applicants are not charged a premium commensurate to their risk status and arises in the absence of utmost good faith. This principle of life insurance prescribes that an applicant for insurance must reveal to the insurer all information within their knowledge

[2] There is little empirical evidence anywhere on the incidence of adverse selection, although there was some evidence of its potential impact when, for a period in the U.K., life insurance business in connection with mortgages was offered without any underwriting questions asked.

that may affect their mortality or morbidity. The principle of utmost good faith collapses if an applicant does not provide the results of a previously taken genetic test. In such a scenario, it may not be possible for the insurer to maintain profitability as a lower premium may be charged to an individual with a higher than 'standard' risk grade.

Adverse selection may also arise from and be promoted by individual applicants. There is a recognised tendency for people who believe they have a greater-than-average likelihood of loss to apply for a higher than average sum insured or continue insurance to a greater extent than lower risk lives. For insurers, there is also the risk that those who know that they have higher risk of morbidity or mortality will take advantage of this knowledge by applying for insurance which they would otherwise not take out (Fischer and Berberich, 1999). Applicants may also fail to disclose risk information due to a misunderstanding of the contract or in order to conceal or minimise risks they are subject to (Tyler, 2004). If the applicant withholds or provides misleading information, the insurance company may have to pay for investigations into any claims made and/or pay out the sum insured under fraudulent terms. The Irish Insurance Federation (IIF) estimates that insurance fraud costs insurance companies €100 million annually (IIF Press Release, 2003). It is questionable if insurance 'fraud' could be promoted by widespread use of genetic testing and an inability on the part of insurers to take genetic test results into account. Of course, taking out insurance or, larger sums of insurance on the basis of a genetic test result is not a crime and cannot, therefore, be considered fraudulent as such. However, the impact upon the insurance industry may be considerable and perhaps not much different to the affect that fraud has upon the life insurance industry today. If applicants for life insurance become informed about their genetic predispositions, and do not expose the nature of this information to the insurance company, then the symmetry of information is disrupted and this may result in negative consequences for the market equilibrium (McGleenan, 2001). The risk of adverse selection emanating from knowledge of genetic test results depends upon several factors such as developments regarding the accuracy of predictive genetic tests, the cost and availability of the tests, the prevalence and actuarial significance of the genetic diseases and the regulatory environment which persists at the time (Daykin, 2003:9).

Given our lack of knowledge regarding the future situation, a question that has not been answered and is difficult to answer in the case of the relatively small life insurance industry in the Republic of Ireland, is how much adverse selection life insurance companies will be able to take before they become bankrupt.[3] Life insurance companies implicitly cater for adverse selection through the pricing of their products. However, significant changes in the extent of adverse selection may cause unexpected changes in the insurance companies' experience (Daykin, 2003:9). Survival of the industry, should such a change occur, would require continuous profit making in the face of incalculable adverse selection, a scenario which may be difficult or impossible to achieve. Hence, life insurers are unwilling to fully relinquish their right to access applicants' results of genetic tests to protect a long established system that is considered by the insurance industry as actuarially fair (IIF, 2004:4-5).[4]

However, what is seen as actuarially fair is not always equivalent as to what is deemed socially fair by society. Indeed, a requirement that information should be actuarially relevant is not sufficient for society to consider that it is appropriate that the information should be taken into account. For example, an objection to insurance underwriting behaviour arises where particular personal characteristics that are considered to affect the risk status of an individual are also associated with historical patterns of discrimination (Daykin, 2003:8). This is why information regarding the racial background of an individual is not sought by insurers when 'setting' premiums. All forms of racial discrimination, including discrimination for insurance purposes, are legislated against by strong anti-discrimination regulations. The introduction of genetic test results as an actuarially

---

[3] Actuarial research based upon the UK insurance market suggests that multifactorial disorders are unlikely to be of much significance for life insurance. Even under quite extreme assumptions, premium increases caused by adverse selection are unlikely to be significant (Macdonald, 1997, 1999). The more serious aspect of adverse selection is any tendency to take out unusually large amounts of insurance (Macdonald, 1997; 1999). In a large life insurance market, a ban on using genetic test results for severe single-gene disorders would be unlikely to lead to significant adverse selection costs, provided excessive sums assured were controlled (Macdonald, 2001)

[4] If a restriction of genetic data is based upon a broad definition of genetic information, including genetic test results and family history information, the consequences of adverse selection may be even more severe for life insurers. If it is considered unfairly discriminatory for insurers to use genetic test results, it may be deemed unfair to use family history information also, as both factors discriminate individuals based upon their genetic heritage. Society's opinion regarding life insurers' use of family history information and the consequences upon restricted usage of this underwriting factor is not examined as part of this research as it is considered beyond the scope of the present study.

relevant factor has also given rise to policies designed to restrict insurers from accessing this information due to concerns regarding discrimination, invasion of privacy and other risks to the individual (see Section 2.3). The following section examines in more detail reasons as to why restrictions may be placed upon life insurance companies' access to genetic test results.

2.3 Consequences for Society

There are a wide variety of social and ethical concerns related to life insurers using genetic test results, ranging from fears regarding inaccurate underwriting to privacy considerations. However, perhaps the most predominant argument regarding why insurers should not be allowed to use genetic test results centres upon the issue of unfair discrimination. This is not the first time that the underwriting practices of life insurance companies have been challenged in relation to their discriminatory nature. Previously, insurers in the US and Britain asked male applicants to disclose if they were bisexual or homosexual in order to screen for HIV risk (Cumbo, 2005). Such a discriminatory practice also resulted in gay men having to take HIV tests even if they were celibate or in long-term relationships. However, gay rights campaigners called for an end to the discriminatory practice and life insurers were strongly discouraged from using this type of information to select and classify individuals for insurance. Therefore, while the practice of risk-defined equity may be defended in economic terms, it may be prohibited on ethical grounds. Genetic data may be considered as equally sensitive information by society as sexual orientation and/or sexual relationship history. If it is considered unjustifiably discriminatory for insurers to request the results of genetic tests, there may be enough public pressure applied upon the Government and the insurance industry to remove such a request from the underwriting factors that discriminate between high risks and low risks. The following paragraphs shall briefly outline several concerns related to insurers using genetic test results and the possible effects that such usage may have upon society.

2.3.1 Genetic Discrimination

As previously mentioned, the use of genetic test results by the life insurance industry has lead to the familiar cry of discrimination, albeit with a new twist. Genetic discrimination has been described as any form of differential treatment based on genetic status (Rothstein, 1994). More specifically, Gostin (1991) refers to genetic discrimination as the denial of rights, privileges or opportunities on the basis of information obtained from genetically based diagnostic and prognostic tests. Lemmens (2000) contrasts such definitions of genetic discrimination with those of Billings *et al.* (1992), Natowicz *et al.* (1992), Alper *et al.* (1994) and Yesley, (1999) who tend toward the description of genetic discrimination as discrimination against an asymptomatic individual who is found to have an abnormal genotype and, as a result, considered to be affected according to the redefinition of the disease. Given this definition, the quintessential feature of genetic discrimination is the use of genetic information about an asymptomatic person. If the disorder related to a genetic characteristic has occurred, discrimination based on the disorder may be unfair but is not customarily considered genetic discrimination (Macer, 2000:107-121). Lemmens (2000:356) considers such distinctions of genetic discrimination as flawed due to the narrow nature of their coverage. Lemmens (2000:356) argues that genetic discrimination also occurs where there is an expression of a genetic condition. Lemmens (2000:356) prefers to describe genetic discrimination as a term that can be used in a variety of 'setting's in which people are considered to be disadvantaged or at risk for being so, on the basis of factors related to genetics.[5]

Many genetic mutations are hereditary and links can be drawn between particular genetic alterations and families, whole communities and ethnic groups.[6] The prevalence of genetic disorders within communities can result in further stigmatisation of groups already affected by racial or ethnic discrimination (Lemmens, 2000:373). It is probable that no person is genetically 'perfect' and that all humans have genetic predispositions to diseases, even though at this stage, many of them are not amenable to testing. It is questionable whether or not it is fair to use genetic tests that may reveal adverse factors for some populations but not to use those that reveal risk factors for other populations. In addition, the discovery of a genetic trait today that is linked to a particular disorder may, tomorrow, be shown to be related to, or even protective from, other diseases. However,

---

[5] Lemmens (2000) argues that cases where a genetic condition is only mildly expressed and yet the individual is excluded solely on the basis of a genetic test can also be considered genetic discrimination, although the genetic condition is perceptible and the individual is mildly symptomatic.

[6] For example, a genetic mutation, most often found in people descended from Ashkenazi Jews, can double or even triple the risk of colorectal cancer. While persons of Celtic origin, such as Irish and Scottish individuals, have a genetic trait commonly associated with heamchromatosis.

awaiting the long-term work needed to calculate the actuarial implications of specific genetic variations and combinations of variations, it is difficult to see how the use of genetic test results for insurance purposes can entirely avoid unfair discrimination (O'Neill, 1999). This situation has lead some to claim that fresh tension exists between the scientific principle of insurance and the powerful appeal to remove discrimination as a result of genetic testing technologies (MacDonald, 1999:83), with some individuals warning that such a scientific principle must not have primacy over basic notions of justice (Moultrie and Thomas 1997; O'Neill 1997).

However, Husted (1998:8) argues that it is unfair to protect one risk group over all others, proclaiming that 'a risk is a risk is a risk', and it is unacceptable practice to make exemptions for some kinds of risk factors (e.g. genetic test information) and not others (e.g. cholesterol tests). If genetic test results were ignored, persons suffering 'traditional' risks (e.g. heart disease) would be charged higher premiums than applicants with as serious genetic risks. Both factors indicate a higher risk to the insurance fund, however, without knowledge of the former, insurers may not be able to calculate the correct premium to charge an individual, as is possible with the latter. Therefore, it could be said that ignoring genetic test results while taking account of results of other medical tests unfairly discriminates against persons with traditional risks. The conflict between the requirements of underwriting and the issue of discrimination may continue until it is known how much adverse selection insurance companies can handle.

Discrimination of applicants for life insurance based upon the results of genetic tests presents itself in several forms, including exclusion from coverage and the charging of a higher premium. This manifestation of discrimination raises numerous concerns pertaining to the necessity of life insurance and the requirement to be 'insurable'. Insurers may use genetic tests to select only low risk individuals, leaving many other persons excluded from coverage. This could lead to individuals having their coverage cancelled, their premium increased or, indeed, having their application for insurance rejected. Reasons for having life insurance vary. For some, life insurance may be seen as a necessity if intending to purchase a home, given that lending groups in the Republic of Ireland require life insurance as a prerequisite for a mortgage intended for the purchase of property (IFSRA, 2005:12). In the Republic of Ireland, Article 126 of the Consumer Credit Act outlines how mortgage lenders are legally obliged to ensure that an individual has adequate mortgage protection insurance or life insurance cover (Consumer Credit Act, 1995: Article 126;

IFSRA, 2005:12). Of particular relevance to this study, the article does not apply where the loan is to persons who belong to a class of individuals which would not be acceptable to an insurer, or which would only be acceptable to an insurer at a premium significantly higher than that payable by borrowers generally. However, lending institutions have the commercial right to make a decision regarding whether or not to provide a loan to an individual, and if it is not possible for an individual to obtain mortgage protection insurance, they have the right to decline such an individual (McGovern, 2005). Without adequate life insurance, the lender does not have to provide a mortgage and, for many, this would mean an inability to purchase a house. Hence, it is possible to understand why there is a perceived need for life insurance in Ireland.

In addition, given that life insurance allows an individual to provide for his or her family's financial stability in the event of death or illness, others may believe that insurance is a necessity to ensure peace of mind that loved ones are taken care of should such a disaster befall their family. Given what may be considered essential services provided by life insurers, some commentators have distinguished life insurance as a 'social good' (MacDonald, 1999:83). Lemmens (2000:390) describes how the insurance industry in the Netherlands provides a guaranteed access to life insurance for contracts under $140,000. This provision can be seen as recognition of the importance of some minimal form of financial security provided by the market. If their premium rises in accordance with the risks brought to the insurance fund, the cost of securing such social goods could increase for those diagnosed as currently having or, having the propensity for, a genetic disease. Consequently, those people on a lower income may not be able to afford life insurance and, therefore, be declined a social good. According to Lemmens (2000:391), if life insurance is a pre-condition for home ownership, it should be made available as much as possible across society, through rules which limit the insurance market. Indeed, several countries in Europe have limited life insurers' power to discriminate, suggesting that life insurance is considered as a highly valued social good.[7]

---

[7] The insurance statutes of Austria, Belgium, and Norway clearly reflect this idea. In addition, the Convention on Human Rights and Biomedicine which declares that any form of discrimination against a person on grounds of his or her genetic heritage is prohibited and that predictive genetic tests shall only be carried out for health or scientific research purposes, and requires that any person undergoing such testing must be subject to non-directive counselling has been ratified by 18 of the Council's 45 member states. These signatories include: Bulgaria, Croatia,

However, if life insurance companies are denied access to genetic test results and applicants, aware of their genetic predisposition to diseases as a result of such tests, decide to take out sums insured greater than they would have done without this knowledge, this may lead to a proportionate increase in premiums for all insured. Without prior knowledge of the risks, that is the volume and nature of the risks, being introduced to the life insurance fund, insurers would be unable to calculate cross-subsidy. This may be accelerated as genetic tests become more accurate, reliable and cover a wider range of diseases in the future. If this results in an increased number of individuals taking tests, insurers will find it increasingly difficult to quantify the increased risk as a result of being denied knowledge of applicants' genetic test results. They may try to counteract this imperfection by raising premiums, reducing the breadth of cover provided and by taking a more cautious approach to the acceptance of higher risk lives. This could certainly be counter-productive for consumers, particularly those in a lower wage bracket and least able to afford life insurance products (Tyler, 2004).

2.3.2 Genetic Testing Requirement for Life Insurance

Concerns have also been raised in relation to insurers requesting applicants to take a genetic test as a condition of consideration for life insurance (Gaivoronskaia and Solem, 2004). If such a stipulation were ever introduced, the result may be the receipt of unwanted information about health status and many applicants denied access to coverage now and in the future due to positive results for future terminal and/or incapacitating diseases. On the other hand, insurers may ask people to have genetic tests as part of preventative medicine, whereby on discovery of a genetic mutation, one may partake in medical treatment leading to a reduction in their life risk and, therefore, a reduction in their premium. However, such a request would purely be in the financial interests of the insurer, as the revelation of knowledge about future ill health may have a significant negative impact upon a person's psychological well being. With this in mind, it can be argued that individuals have the right not-to-know the results of genetic tests and, therefore, the inherent right not to have to take a genetic test as part of an application process for life insurance.

Cyprus, the Czech Republic, Denmark, Estonia, Georgia, Greece, Hungary, Lithuania, Moldova, Portugal, Romania, San Marino, Slovakia, Slovenia, Spain, and Turkey.

2.3.3 Health Issues

Due to the workings of life insurance, an individual who is at high risk of death during the term of the policy will have to pay a higher premium. This may inhibit individuals from seeking a genetic test to clarify their disposition to a particular disease, consequently denying themselves treatment that may have the power to alleviate the disease associated with their genetic make-up. Several studies have indicated that women at risk of breast cancer because of family history often refuse to undergo testing out of fear of the impact testing may have on their insurability (Jacobsen *et al.,* 1997). The end result could be a population not taking measures to reduce their susceptibility to a genetic disease due to fear of an unaffordable increase in their life insurance payments. This argument, however, can equally be applied to tests for heart disease, high blood pressure and a wide range of other conditions that are not usually seen as being genetically based in nature. However, the results of a genetic test that may determine susceptibility to one particular disease today may infer a disposition to another tomorrow. In this sense, genetic test results cannot be compared to other medical test results.

2.3.4 Inaccurate Underwriting

A life insurance employee may draw unsubstantiated conclusions from the result of a genetic test due to lack of experience and/or knowledge of genetics. As tests become more complex, increasingly specialised knowledge is required to infer accurate and useful statistics and probabilities related to the results. Consequently, a broker or an underwriter may adopt unjustified caution when using genetic information to determine an individual's life insurance contract. It is also plausible that, through confusion or ignorance, an insurer may misinterpret genetic information resulting in an individual being overcharged for a life insurance product. The case of *J.* v. *London Life Insurance* (B.C.H.R.T.D. No. 35, 1999) illustrates how such an incident may take place, whereby an insurance applicant was refused insurance due to a lack of statistical data on the part of the insurer. It could be argued that the same professional guidelines are adopted for dealings with genetic test results as other private medical data. Therefore, this issue should not be a problem for applicants to life insurance companies in the Republic of Ireland. However, information resulting from genetic testing is often premature and subject to varying interpretation (Glass *et al.*, 1996). Given that the data suggesting a genetic link between a genetic mutation and susceptibility to a disease often fluctuates according to the sample population of the underlying studies, it may be difficult for insurers to constantly update actuarial tables accordingly (Lemmens, 2000:371). In addition, given the lack of transparency associated with underwriting, it is impossible for applicants to know whether or not the correct calculation for one's risk accurately reflects the risks one brings to the life insurance fund. Therefore, it is not possible to verify the actuarial tables or the calculations of the underwriter, making speculation of improper or inappropriate calculation stronger.

2.3.5 Privacy

Another issue introduced into the debate on life insurers' right to access genetic test results, is whether or not life insurance companies may use the result of a genetic test of an individual to underwrite insurance policies for members of that individual's family or relations, thereby counteracting an individual's right to privacy. This is unlikely at present in most countries due to strong privacy laws (Data Protection Act, 1988 and 2003). There are also fears that genetic test results provided to a life insurance company may be given to parties other than those agreed to by the applicant for insurance. Given that genetic test results would more than likely be contained in a medical report, such a break in privacy is improbable. In the Republic of Ireland, the Data Protection Act 1988 and 2003 ensures that underwriters are legally bound with the same confidentiality rules as doctors when dealing with medical reports (Data Protection Act, 1988 and 2003: Section 2 (1) (c) and (d)). There are two separate provisions that apply restrictions to the disclosure of medical reports by underwriters. Section 2 (1) (c) of the Data Protection Act 1988, subsequently amended by the Data Protection Act 2003, states that data:

> "(i) shall have been obtained for one or more specified, explicit and legitimate purposes,
>
> (ii) shall not be further processed in a manner incompatible with that purpose or those purposes".

This restriction on further processing prevents medical reports being used for anything other than the purpose for which the report was drafted and prevents the report from being disclosed to third parties unless the processing is compatible with the purposes for which the report was obtained. The second restriction, contained in Section 2 (1) (d), states that "appropriate security measures shall be taken against the unauthorised access to, or unauthorised alteration, disclosure or destruction of, the data, in particular where the processing involves the transmission of data over a network and against all other unlawful forms of processing". This restriction is designed to prevent access to data by people other than those with a lawful reason. This includes restricting access to specific staff only. Therefore, it is unlikely that data provided to an insurance company should ever be leaked given such privacy laws and there are few cases whereby life insurance companies have not applied appropriate privacy measures (Sweeney, 2006).[8]

However, in line with this privacy issue is the concern that if insurers are entitled to a person's genetic information, there is a certain impingement upon a person's autonomy to make decisions over who can access their biological code to life. A person's DNA can be considered to be a blueprint of that individual's genetic makeup. The results of a genetic test may provide limited information today; however, with such rapid advances in genetic knowledge, it is quite possible that the same sample of DNA could reveal behavioural traits and genetic predispositions to a wide variety of diseases in the future. It is questionable whether or not an individual's genetic code is private property to be shared with others only at the individual's discretion.

2.3.6 Genetic Research

Finally, researchers worry that fears about 'inappropriate' uses of genetic information may deter volunteers from becoming involved in much needed research projects on genetic diseases (British Medical Association, 1998:156; United Kingdom Parliament, House of Commons, 25 Oct 2001: Hansard Vol. 373, Column 139WH). As such projects require donation of cells to be screened, individuals are worried that the information provided to such studies will be obtained and used by insurance companies in assessing their premium (British Medical Association, 1998:156; United Kingdom Parliament, 25 Oct 2001: Column 139WH). Due to strict privacy laws governing scientific research and life insurance practice in the

---

[8] See Case Study 2/99 involving the retention by an ex-employee of a life insurance company in the Republic of Ireland, of customer data. This was considered unauthorised access by the Data Protection Commissioner, http://www.dataprivacy.ie/docs/Case_Study_2/99_Life_Insurance_Company/137.htm. Sweeney, S. (3/1/06) Interview response "The Data Protection Agency is not aware of any "cases specifically involving unlawful access to medical records held by underwriters"

Republic of Ireland, it is unlikely that the above concern shall ever materialise.

With such a wide variety of societal concerns, life insurers are faced with an uphill struggle to continue their requirement to insure individuals based on their expected genetic risk status. Although, in most cases these fears can be said to be founded on theoretical speculation and extrapolation, rather than on direct evidence, the concerns can be considered no less real by virtue of the lack of indicative evidence (Daykin, *et al.*, 2003:2). If society believes that it is unfair to ask an individual to provide genetic data to an insurance company for whatever reason, this belief may indirectly act as a source of change for the Irish insurance industry. Politicians may find themselves pressed by their citizens to restrict the underwriting practices of the insurance industry with regard to their use of genetic test results. The following section shall examine how the consequences of insurers using genetic test results may be considered as risks that contribute to the 'risk society' (Beck, 1992) and the 'runaway world' (Giddens, 2000). Following this, the current policies that govern life insurers' use of genetic test results in the Republic of Ireland are examined and discussed.

2.4 Risks from Insurers Using Genetic Test Results

It can be seen from the above discussion that the consequences of insurers using or being restricted from using genetic test results affects society in a variety of ways. Advances in technology and modern social thinking are at the crux of such perceptions. Anthony Giddens' (1990; 2000) writings develop a theoretical perspective on these changes unfolding in the present day. We live today in what Giddens (2000) classes as a 'Runaway World'; a world marked by new risks and uncertainties of the sort Beck (1992) diagnoses. However, it has to be questioned if the possible corollaries discussed above can be considered as risks. To do so, it is necessary to define risk, a task that is made difficult as a result of multiple views as to the meaning of risk, as the following section explains.

2.4.1 Objective Risk versus Subjective Risk

Despite the still rapidly growing literature on the topic of risk (Short, 1984; Beck, 1992 [1986]; Krimsky and Golding, 1992; Dietz *et al.*, 1998; Renn, 1998) there is remarkably little consensus over what, in fact, is meant by 'risk'. Some risk analysts define 'risk' in terms of probabilities based on objective, verifiable experience. This traditional statistical treatment of risk calculation asserts that the general concept of risk is the chance, in quantitative terms, of a defined hazard occurring (Knights and Vurdubakis, 1993:729-764; Webster, 1983). This viewpoint combines a probabilistic measure of the occurrence of an adverse event with a measure of the consequences of that event. In other words, the probabilities and consequences of adverse events are assumed produced by physical and natural processes in ways that can be objectively quantified by risk assessment (Slovic, 1999:675). According to Jungermann and Slovic (1993), the objective, technical definitions of risk can only be used under very narrowly specified conditions, for example within insurance or the field of economic activity. Actuaries and underwriters are examples of risk analysts who view the meaning of 'risk' through an objective lens and, as a result, they offer full support to the above definition of risk.

Social scientists, on the other hand, are inclined to disagree with this approach to risk analysis, believing that there are major difficulties in attempting to view risk as a one-dimensional objective concept. They argue instead that risk is inherently subjective and, as a result, should be viewed through a subjective lens (Krimsky and Golding, 1992; Funtowicz and Ravetz, 1992; Otway, 1992; Pidgeon *et al.*, 1992; Slovic, 1992 and Wynne, 1992). In this view, risk cannot be reduced to a single objective correlate of a particular mathematical aspect of risk, such as the product of the probabilities and consequences of any event. Risk is all in the mind; it is a notion of observation and not just an object to be observed (Holzheu and Wiedemann, 1993:9). Perception of risk is inherently multidimensional and personalistic, with a particular risk or hazard meaning different things to different people, and different things in different contexts (Royal Society, 1993:7). There are a wide variety of competing views of the 'risks' related to the use of genetic test results in life insurance underwriting. Depending on one's social and cultural background, different people may see different risks when looking at the same hazard. For example, social science research has demonstrated that technical experts and lay people often differ in their conclusions about the risks and benefits of hazards (Siegrist and Cvetkovich, 2000; Fischoff, Slovic and Lichtenstein, 1982; Sjoberg, 1998).

Within the objective and the subjective views of risk lie numerous competing definitions of risk. Renn (1992) distinguishes between four objective approaches to risk and three subjective approaches to risk. The objective views of risk can be classified as an actuarial perspective of risk, a toxicology epidemiological perspective of risk, a probabilistic risk analysis approach to risk and an economic perspective where cost-benefit comparisons are made. All of these approaches are based upon statistical prediction and probability analysis. Subjective risk approaches can be categorised as a psychological view of risk, social view of risk and finally a cultural perspective of risk. Each of the seven categories can be further classified into competing views of risk (Renn, 1992). Althaus (2005:569) provides a table illustrating how various disciplines describe the concept of risk (see Table 2-1).

Table 2-1 Disciplines and risk forms applied to the unknown that determine disciplinary epistemological definitions of risk

Discipline	**How it views risk**
Logic and Mathematics	Risk as a calculable phenomenon
Science and Medicine	Risk as an objective reality
Social Sciences	
Anthropology	Risk as a cultural phenomenon
Sociology	Risk as a societal phenomenon
Economics	Risk as a decisional phenomenon, a means of securing wealth or avoiding loss
Law	Risk as a fault of conduct and as a judiciable phenomenon
Psychology	Risk as a behavioural and cognitive phenomenon
Linguistics	Risk as a concept
History and Humanities	
History	Risk as a story
The Arts (literature, music, poetry, theatre, art, etc.)	Risk as an emotional phenomenon
Religion	Risk as an act of faith
Philosophy	Risk as a problematic phenomenon

Source: Adapted from (Athaus, 2005:569)

Such a multitude and a variety of perspectives of risk can be very confusing, and has lead many to question whether or not there is, or can be, a common, true definition of risk. It would be impossible to discuss risks related to a hazard if there is no agreement on how risks can be defined.

2.4.2 'Family Resemblance' of Risk

To use Wittgenstein's (1953) terms of reference, there is a family resemblance regardless of the context or philosophical background within which the word 'risk' is defined and used. It is, therefore, necessary to demarcate the family resemblance of risk, so that risks can be defined. When searching for a precise meaning of the term risk, particular attention must be paid to the situation or context in which the term is being employed (Dolhenty, 2002). The term 'risk' implies 'chance' and uncertainty regarding the outcome of a situation or an event. Risk is therefore amenable to either a positive or a negative outcome.[9] However, one rarely talks or hears of the risk of a positive result. The notion of risk and the impact of the word provoke negative reactions among all groups of people, therefore, all meanings or definitions of the word 'risk', and/or the nature of risk, have underlying negative connotations. Risk expresses the degree of prior uncertainty with regard to possible outcomes of decisions or actions (Holzheu, 1993:241). Therefore, one can say that all approaches to define risk imply uncertainty. There is a distinction between reality and possibility. The term risk denotes the possibility that an undesirable state of reality (adverse effects) may occur as a result of natural events or human activities (Renn, 1992:57-58). The essence of risk is not that it is happening but that it might be happening (Adam, Beck and Van Loon, 2000:2). A final similarity among all definitions of risk, or a 'family resemblance' of the term 'risk', is that risk results from insufficient knowledge about future events (Holzheu, 1993:242). If a loss is expected then there cannot be a risk, as the consequence of the event is known. Taken altogether, the term 'risk' implies a possible future negative and uncertain outcome of an event or situation caused by nature or human activities where insufficient knowledge was available to prevent such an outcome. Using such a definition, the consequences related to the use of genetic test results by life insurance companies can be classified as risks.

---

[9] For example, one takes a risk in paying one euro for a lottery ticket. There is the chance of winning one million euro or the chance of losing one euro.

2.4.3 Risk Theory and Risks Perceived from Insurers Using Genetic Test Results

The risks posed by insurers using genetic test results can be associated with the risks described by Giddens (2000) and Beck (1992). According to Giddens (1990; 2000), the nature of risks we must face are changing. Risk now derives less from natural dangers or hazards than from uncertainties created by our own social development and by the development of science and technology. Technology and the awareness of capability now occupy nature's territory (Luhmann, 1993). In this new era, humanity has the ability and power, through genetic knowledge, to manipulate and control living organisms, reconstruct living species, and the information base to shape the circle of life. Although this science is extraordinary, we are consistently ambivalent about its application (Turney, 1998:1). It is widely recognised and appreciated that there may be many great benefits as a result of increasing genetic knowledge, but there are also numerous unknown and impossible to anticipate threats in biotechnology's "ability to break down old boundaries, and dissolve taken-for-granted categories" (Turney, 1998:1). Several genetic tests can predict future human health more accurately and reliably than family history information. For example, individuals predisposed to Huntington's disease, prior to genetic tests, had to wait until the first signs of the symptoms of the disease appeared, usually around age 40, before they knew if they would develop the disease or not.[10] However, with genetic tests, it is possible to tell if an unborn child is certain to get the disease. Therefore, disease that occurs during a particular stage of life can be accurately determined from youth. According to Luhmann (1993), such knowledge and technical delimitations can more easily prove destructive than constructive. As Berman (1988:15) notes, to live in this modern age and to be modern is to "find ourselves in an environment that promises us adventure, power, joy, growth, transformation of ourselves and the world" in other words immense wealth of being. However, at the same time, this wealth "threatens to destroy everything we have, everything we know" (Berman, 1988:15). The social production of wealth through technological innovation is associated with the social construction of risk (Beck, 1986:19). Genetic testing for future health problems promises many great things for some people, however, the use of genetic test results

[10] Huntington's disease is a rare, hereditary, degenerative disease of the brain marked by irregular, spasmodic, involuntary movements of the limbs or facial muscles and resulting in severe mental deterioration and eventual death.

in the life insurance industry endangers equal opportunity, psychological well-being, personal privacy, and future genetic research.

Western society is changing, in that it is becoming increasingly reflexive, with people questioning the society in which they live, the distribution of risk and demanding more access to decision-making (Beck *et al,* 1994). Rather than just accepting the benefits associated with the use of new technologies, people are questioning the distribution of benefits from genetic testing and expressing their concerns regarding the distribution of related risks and the consequences of such risks (for example, see Genetics and Insurance Committee (GAIC) annual public meetings in London). Beck (1996) argues that society's uncertainty about the future applies to all members of society, young and old, rich and poor. A risk society is developing which is more concerned about risk distribution than with wealth distribution (Beck, 1992). Fear that things can go wrong is growing rapidly parallel to the risk apportioned to decision-making (Luhmann, 1991: xii). Risk management decisions that affect the use of genetic test results by life insurance companies are examined in the following section.

2.5 Policies Governing Life Insurers' Use of Genetic Test Results

The life insurance industry in Ireland is governed by European Directives and regulated by Irish Law. Outside of official regulation, the Irish Insurance Federation (IIF), the representative body for insurance companies in Ireland, also issues codes of practice which members sign up to in respect of life insurance and pensions. A single regulatory authority governing insurers did not exist in Ireland until 1998, when the Irish Government responded to repeated calls by insurers to centralise the solvency and consumer protection supervision by creating the Irish Financial Services Regulatory Authority (IFSRA). From May 1st 2003, IFSRA overstepped the Irish Central Bank to become the single authority to ensure supervision of the banks, insurers and investment intermediaries. The Financial Regulator is a distinct component of the Central Bank and Financial Services Entity of Ireland, with clearly defined regulatory responsibilities. These cover all Irish financial institutions including those previously regulated by the Central Bank, Department of Enterprise, Trade and Employment (DETE), Office of the Director of Consumer Affairs (ODCA) and Registrar of Friendly Societies. The Financial Regulator contributes to the work of the Central Bank in discharging its responsibility in relation to the maintenance of overall financial stability. Regulation regarding solvency and authorisation are similar across the European Union however they tend to differ somewhat when it comes to rules surrounding consumer protection. For example, the laws in Ireland differ from other European countries

Looking at the Irish system of consumer protection with regard to insurance, an Insurance Ombudsman Scheme was established by the insurance industry in Ireland in 1992 to deal with individuals in their private capacity. The role of the Ombudsman is to provide information to the public and to mediate on disputes between policyholders and insurance companies. Companies that are members of the Insurance Ombudsman Scheme are bound by the decisions of the Ombudsman. The IFSRA is in the process of 'setting' up a separate ombudsman, in the form of a financial services ombudsman, however, it is unclear at present as to how the two ombudsman schemes will co-exist (Greenford *et al.*, 2006). The EU Commission has taken a more primary role with regard to recent regulation of the Irish insurance industry, with the IFSRA functioning as the authorised local regulator ensuring the enactment of EU framework documents and EU directives. Given such a prominent position, it can be stated that the European Union has a central role in shaping the current Irish insurance industry and in creating a single competitive market in the provision of insurance across borders (Greenford *et al.*, 2006).[11]

When it comes to regulating insurers' use of genetic test results in the Republic of Ireland, however, the governance of this issue is not so clear-cut. It is important to note that one cannot attribute the risks instigated by insurers using genetic test results in their underwriting process to the scientists who have and continue to discover genetic links to diseases and methods for testing an individuals' possession of disease related genes. Nor can one apply culpability to the individual with the aberrant gene. Similarly, one cannot hold insurance companies accountable for such

[11] However, the interpretation of European law has caused some consternation across authorities, in particular with regard to the "general good" principle. This principle allows national government derogation with regard to the implementation of certain legal instruments in cases where such an implementation would be contrary to the "general good" (Greenford *et al.*, 2006). It is possible to note some uncertainty surrounding the relationship between the basic concepts of freedom to provide services and the principle of the "general good" (European Commission, 2000). The application of this principle has become a legal minefield although recent procedures have ameliorated this situation somewhat.

risks, as the requirement of knowledge of all risk relevant information is a necessity of their business. However, many authorities are hesitant to threaten their electorate's standard of living, or improvements to this standard, by limiting the individual's freedom of choice (Matten, 2004:378). In addition, an authority that enforces a legal system that society considers to endorse unjustifiable discriminatory practices on the part of the insurance industry could be construed as an agent that reinforces social stigma and contributes to social injustice. Hence, on a global level, authorities have preferred to restrain insurers' use of genetic information rather than risk angering their stakeholders or body of voters by restricting their access to genetic information or by allowing any organisation to adversely or antithetically impact upon the procurement of their individual genetic data (for example, Bulgaria, Croatia, Cyprus, the Czech Republic, Denmark, Estonia, Georgia, Greece, Hungary, Lithuania, Moldova, Portugal, Romania, San Marino, Slovakia, Slovenia, Spain, and Turkey have all legislated that genetic test results cannot be used by life insurance companies; see Section 2.3.1).

This reaction follows conventions and declarations provided on an international level. For example, the World Medical Association issued in 1992 a *Declaration on the Human Genome Project,* in which it suggests that "it may be desirable, regarding genetic factors, to adopt the same tacit consensus which prohibits the use of race discrimination in employment or insurance." In 1997, UNESCO declared that "no one shall be subjected to discrimination based on genetic characteristics that is intended to infringe or has the effect of infringing human rights, fundamental freedoms and human dignity" and stressed the importance of confidentiality of genetic data, the need for informed consent and the right of individuals to decide whether they want to be informed of a genetic condition (*Universal Declaration on the Human Genome and Human Right*s, Sections 5, 6, 7 and 9). The following section examines the recommendations and regulations that have been proposed or imposed by the Council of Europe, the insurance industry and the Irish Government in relation to insurance companies' use of genetic test results. These guidelines are of particular interest to the present study as they are considered to have a direct or indirect effect on the risk management of the use of genetic test results in the Republic of Ireland.

2.5.1 Council of Europe

In April 1997 the Council of Europe adopted a Convention for the Protection of Human Rights and Dignity of the Human Being with regard to the Application of Biology and Medicine (Convention on Human Rights and Biomedicine, Oviedo IV, 1997). The Council of Europe took a strong stance on insurers using genetic test results, a factor reflected in Articles 11 and 12 of the Convention. The Council declared in Article 11, that any form of discrimination against a person on grounds of his or her genetic heritage is to be prohibited. In addition, Article 12 states that genetic testing can only be carried out for purposes of health care or research linked to health purposes, and subject to appropriate genetic counselling. The aim of human rights legislation is to forbid discrimination that may impact upon human rights, fundamental freedoms and human dignity (Lemmens *et al.,* 2004). Seventeen members of the Council of Europe have ratified this convention since its publication (see Sections 2.3.1 and 2.5). However, this convention has not yet been endorsed by the Republic of Ireland and, therefore, has no force until signed and implemented into national law.[12] Ireland may not ratify this convention for the reason that distinguishing genetic information from other grounds of disability in human rights legislation could strengthen and support genetic determinism[13], which could fuel stigmatisation and discrimination, and thus have the opposing effect (Lemmens *et al.,* 2004).

2.5.2 Irish Insurance Industry

In May 2001, the Irish Insurance Federation (IIF) in response to public and Governmental pressure issued a "Code of Practice on Genetic Testing" that declared a moratorium on the use of genetic test results by all member companies (IIF, 2001). The Irish Insurance Federation (IIF) is the trade association representing life and non-life insurance companies in Ireland and represents 54 member insurers who write over 98 per cent of all life and non-life insurance business in Ireland. The IIF indicate that the basic principle of limiting access to genetic test results is not a concern in its own right in the short term, hence the industry adoption of voluntary

---

[12] However, it is a similar document to the European Convention on Human Rights and therefore it is a significantly influential force, even in European states that have not enshrined these principles in national law

[13] Genetic determinism is the belief that carrying a specific genetic mutation has a higher detrimental and inevitable impact on people's health, well-being and behaviour than other health factors (Lemmens *et al.*, 2004)

controls in this area (IIF, 2004). This exemption from the normal duty of full disclosure of material facts was intended to apply until 31st December 2005, allowing the insurance industry a 5 year period to evaluate the actuarial relevance of genetic tests.[14] According to the Code of Practice, disclosure of the result of a genetic test was not required in new applications for life cover unless the sum insured on the new application exceeded €381,000 or the total of the sum insured on the new application and other policies, if any, taken out with any insurer between 1st April 2001 and 31st December 2005 exceeded €381,000. In the case of life cover exceeding this limit, insurers were instructed to use only approved tests by the British Genetics and Insurance Council (GAIC) in the underwriting process (IIF, 2001). The GAIC announced the reliability and relevance of only two genetic tests, both for Huntington's disease, as sufficient for insurance companies to use when assessing applications for life insurance. In the case of a proposal where genetic test results were not required to be disclosed, an applicant who had a negative result under an approved test could ask the insurer to take the result into account in the underwriting decision. Despite the establishment of the Irish Financial Services Regulatory Authority (IFSRA), the many Codes of Practice of the Irish Insurance Federation in relation to issues such as genetic testing and life insurance medical reports continued to operate as industry codes. In June, 2005, the Irish Government enacted the Disability Act, which overrides the Code of Practice on Genetic Testing. However, the moratorium instigated by the IIF shall be discussed below as this was the only form of regulatory control upon insurers' use of genetic test results in the Republic of Ireland during the time the primary research of the present study was completed.

A moratorium is designed to allow flexibility and time for the instigator to identify the various public and industry policy alternatives in order to achieve a consensus to serve the interests of affected individuals. In this case, those affected include policyholders, applicants, society, and, indeed, the insurance industry as it attempts to maintain a competitive market in the Republic of Ireland. The moratorium introduced by the insurance industry could certainly have been considered a practical approach to regulation, demonstrating self-reflective and self-critical processes within the insurance industry concerning the side effects their practice has on society (Kosseim *et al.*, 2004). The reflexive regulation encompassed by

[14] There is no regulatory entity monitoring the Irish insurance industry compliance to the "Code of Practice on Genetic Testing".

the insurance industry's Code of Practice could then encourage moral introspection and embed a sense of moral responsibility among users of genetic information, in a way that laws declared by the Government may not (Kosseim *et al.*, 2004). Incomplete knowledge, risk and uncertainty lie at the heart of such self-critical reflexivity (Benn, 2004:400). With regard to genetics and insurance risks, society and the insurance industry find themselves confronted with certain side-effects of their modern lifestyle, and as a result are making an effort to find rational solutions to these problems (Beck, 1997:11-60). This process should irrevocably change the insurance industry, one of the principal pillars of modern society, on the way to reflexive modernity, leading to the core of what Beck (1992) terms the 'risk society' (Matten, 2004:379).

However, the moratorium introduced by the IIF was not based on co-operative regulation whereby the Government and those affected were actively involved in the determination and assessment of the terms and conditions of the life insurance industry's suspension from its usual requirement of full disclosure of material facts. According to Beck's (1992) thesis, without co-operative regulation, the risks which the Code of Practice was designed to mitigate cannot be diminished. There were no Government sanctioning mechanisms to ensure that all member companies of the IIF followed the Code of Practice, a factor that the insurance industry may have abused. For example, although the Code of Practice outlined that applicants were not required to disclose the results of genetic tests that have been taken by relatives, applicants were encouraged to disclose any family history of a genetic disease. This distinction was not entirely clear from the standard wording included on insurance application forms (Society of Actuaries in Ireland, 2004). The insurance form stated that *"If this application is for life insurance only...you do not need to tell us the results of any genetic test you have had.... You must however tell us if you have a family history of a genetic disease or are having treatment for, or experiencing symptoms, of a genetic condition..."*. Without a higher authority managing the application of the Code of Practice by the IIF, there was room for insurers to gain access to genetic information from applicants without actually directly asking for genetic test results. It is therefore questionable if the Code of Practice managed the risks related to insurers using genetic test results as it was intended. In addition, there was no allowance made for the public to voice their opinion regarding the Code or to change the regulations laid down by the life insurance industry. Indeed, it could be argued that the IIF engaged in sub-political self-regulation to avoid stricter regulation by a higher governing authority, such as the Government. For example, the Code of Practice provided a narrow definition of genetic information, as in the 'results of genetic tests'. Whereas genetic information could also be described as including family history information, thereby providing a much broader definition and widening the scope of the Code of Practice. This basically means that insurers could exclude people on the basis of vague family information indicating the presence of a genetic condition in the family, rather than relying on similar, more detailed, information resulting from DNA analysis. Although, it has long been possible to obtain information on the likelihood of future disease, susceptibility to external factors, or possession of certain genes, the Code of Practice continued to allow these practices and, instead, focused exclusively on newer methods of obtaining this information. Therefore, genetic discrimination was not prevented by the IIF Code of Practice.

For the insurance industry to act as a sub-political institution without Governmental facilitation or interference may have reduced the complexity of the regulation adopted to manage the risks related to the use of genetic test results by the life insurance industry. But such oversimplification of regulatory realities could lead to erroneous decision-making that does not serve the interest of all stakeholders. As Kosseim (2004) boldly states, the insurance industry is not a legal authority in itself, so it does not, or should not, have the authority to determine what is or is not "ethical" and effectively regulate the use of genetic information in the absence of law. There is often no systematic evaluation of how well codes of conduct or practice fare after time, nor any illuminative evidence of the extent to which they are adopted, implemented and the degree to which punitive measures are enforced upon non-conformers (Kosseim, 2004). Another nail in the Code of Practice coffin was knowledge. As mentioned above, self-reflexivity is made possible by knowledge. The insurance industry possesses extensive knowledge of their processes and their risks; however, insurance companies do not have wide-ranging and all-embracing knowledge on the complicated nature and intricacies of genetic data and genetic testing. As no one has surveyed the attitudes of people in the Republic of Ireland on this issue, the Code of Practice did not acknowledge society's perceptions of the use of genetic test results by the life insurance industry. As risks are strongly dependent on the development and the spread of knowledge in society, the Code of Practice could not take into account societal perceptions of risks and, therefore, was lacking as a regulator to control or manage the risks related to the use of genetic test results by the life insurance industry.

2.5.3 Irish Government

The Irish Government has also attempted to tackle the potential risks associated with life insurance companies using the results of genetic tests when underwriting applicants. The Government published a Disability Bill in December 2001 that forbade insurers requesting, obtaining, accepting or using the results of a genetic test to underwrite an applicant and demarcated the illegality of doing so with the threat of a large fine to those who did not uphold this law. In doing so, the Bill was intended to override the previous distinction regarding discrimination made in relation to the assessment of risk for insurance purposes, where different treatment is justified by actuarial or statistical data (e.g. the Irish Equal Status Act 2002). The legislation also allowed for the 'setting' of financial limits for the health-related history of an applicants' family for life insurance, whereby, family history could be sought and used by an insurer if the sum insured was above a sum to be determined from time to time. In the case of life insurance, this sum was to be decided by the Minister for Enterprise, Trade and Employment in consultation with the Minister for Social, Community and Family Affairs. Health related insurance was to be decided by the Minister for Health and Children. Although the Bill was withdrawn in February 2002, following negative reaction from disability groups, it underwent some changes before its introduction into law in June 2005. Rather than immediately 'setting' financial limits regarding insurers access to family history information, the Minister for Justice, Equality and Law Reform has been given time to make regulations on this matter upon consultation with the Minister for Health and Children and the Data Protection Commissioner, the Minister for Social and Family Affairs, and the Irish Financial Services Regulatory Authority. The Disability Act, 2005 also requires a full review of the above provisions and regulations to be taken before 1st January 2014.

It could be argued that the Disability Act, 2005 is too broad, too restrictive, and as a result has an uncertain reach that could be potentially quite harmful. According to the IIF, the lack of a threshold above which test results must be disclosed in the Act may mean that reassurance companies would have to reconsider whether they would be prepared to continue to write policies for larger sums insured in the Irish market (IIF, 2004). Such a criticism of the Act is understandable, given that the purpose of the Disability Act, 2005 should be to ensure access to a basic level of services. Therefore, a threshold at a level that is high enough to cater for the average mortgage protection or other life insurance policy should be acceptable. In addition, the Disability Act, 2005 demonstrates how the Irish Government has chosen to adopt a 'special regime' tailored specifically to genetic information. This encourages an ''exceptionalistic'' view of genetic information by publicly differentiating genetic information from all other types of medical information requested by an insurer (Murray, 1997). It has been argued that this focused protection against genetic discrimination is unfair since so many individuals remain excluded from life insurance on other grounds (Lemmens, 2004). Therefore, the Irish Government, by disallowing insurers to take account of genetic test results, while taking into consideration results of other medical tests and risk data, unfairly causes discrimination against persons with traditional risks. This debate is furthered by the definition of 'genetic testing' provided by the Act. The debate surrounding insurers' use of genetic test results is largely related to the fairness of insurers to use predictive testing – whereby a test might indicate susceptibility to a condition which has not yet manifested itself. The definition in the Act goes considerably further than this and also prohibits access by insurers to tests/records undertaken by individuals already showing symptoms of a condition. Therefore, an 'exceptionalistic' view of genetic tests is promoted, furthering an opinion that genetic disease is entirely different to a non-genetically related disease (such as illness caused by viral infection).

Another problem influencing the effectiveness of the Disability Act, 2005 is the means of recording genetic information. Genetic information is becoming an increasingly important component of medical files (Lemmens, 2000:369). DNA tests and family histories are often accompanied in medical files by protein tests that dictate how a specific gene is affected, therefore, it may be increasingly possible for insurers to indirectly access information regarding an individual's predisposition to disease from medical records (Lemmens: 2000:369). Indeed, most medical tests are ordered in an effort to detect conditions or potential conditions that have an underlying genetic contribution. Beckwith and Alper (1998) and Alper and Beckwith (1998) demonstrate several examples of the difficulties in maintaining a distinction between genetic and non-genetic tests and conclude that a distinction between genetic and non-genetic information is artificial. Therefore, it will be interesting to note how the Disability Act, 2005 shall protect individuals against genetic discrimination and ensure genetic privacy, as the line between genetic data and non-genetic data begins to blur.

It could be argued that the Irish Government has gone too far, by intending to deny insurers their long-standing right to underwrite risk relevant information. The Act does not take into account the problem of adverse selection. Not allowing life insurers' access to all risk relevant information instigates the foundation of an uneven playing field, where the applicant knows more about his risk than the insurer does. This leads to a situation where insurers are unable to establish equitable premium rates and are instead forced to carry the costs of providing non-actuarially sound insurance. To cater for this extra burden, it is likely that all applicants will be charged higher premiums. Some companies unable to support a high number of claimants may become insolvent. While there is no evidence of adverse selection in recent years, the potential for it still remains (Lemmens, 2004). Life insurance products with relatively small markets and/or based on morbidity risk, such as critical illness insurance, have a greater exposure to damage from information asymmetry (Irish Insurance Federation, 2004:15). It is uncertain whether or not the insurance industry could maintain such a product in light of the Disability Act, 2005. Unable to use genetic test results, life insurers may have to apply stricter and more cautious underwriting procedures dealing with family history information. In addition, insurers may have to increase the prices for all of its insurance products in order to protect itself from the increased possibility of adverse selection. The Irish Government has not made any plans to insure those individuals deemed 'uninsurable' or to supply life insurance cover to those who cannot afford life insurance packages at higher rates. As discussed above, the life insurance industry is a significant contributor to the Irish economy, therefore, negative influences that affect the continued development and level of income of this industry can also be considered detrimental to the strength of the Irish economy (see Section 2.2). This will have a direct effect on Irish society which may have to forfeit Government-funded services and welfare as the country struggles to deal with a considerable loss to its income level.

The above prospective crisis illustrates the sheer rhetoric that is the Disability Act, 2005. The legislation is designed to regulate the known; however, it is flawed in that it is not adaptable to regulate the unknown. There is not enough evidence to require the legislation prescribed by the Disability Act, 2005 leading to the supposition that policy makers in Ireland have not faced the challenge of sorting through the rhetoric in order to disentangle the real problems that the legislation addresses and ought to address (Kosseim, 2004). The risks that the regulation is designed to control are uncertain and there is no indication that they are likely to occur in the near future. Given that the actual impact on the market over the years of a complete ban on access to test results should depend largely on future developments in genetic testing, the operation of the legislation should be open to review as and when warranted. There has been no public opinion survey taken as to whether or not this legislation is currently needed and, therefore, the Government cannot argue that there was pressure by the public to regulate these issues. According to Giddens (1994:186), democracy means discussion, the chance for the 'force of the better argument' to count. However, in policy decisions regarding insurers using genetic test results, the Government has not accounted for public argument and, instead, has excluded the opinion of the individuals affected by the risks emanating from the situation. Sjoberg (1996) maintains that politicians often have erroneous views of the public's beliefs and attitudes regarding risk. Without public consultation, the entities managing the risks related to the use of genetic test results in underwriting in the Republic of Ireland are incapable of making any appropriate decisions regarding the control of these risks.

According to Beck's (1992) theory on the risk society, the failure or inability of entities to enable the maintenance and effective control of risk contributes to the development of a self-critical society. Lacking technical or specialist knowledge and unable to evaluate the accuracy or reliability of information provided by numerous 'expert' sources, lay judgements of modern risks and benefits may be based on assessments of those who are responsible for managing the hazards (Siegrist and Cvetkovich, 2000:713-719).[15] Therefore, instead of laypersons making rational judgements based on complete knowledge of the 'setting', social trust is employed to select experts who are trustworthy and whose opinions can be believed as being accurate (Siegrist and Cvetkovich, 2000:713-719). It will therefore become increasingly important for Governments, and corporations, to promote credibility and trust in order to justify themselves to an increasingly sceptical public (Benn, 2004:408). This demonstrates another feature of a risk society, whereby the desire for power results in a contention for greater credibility and trust.

The Disability Act, 2005 is a classic example of symbolic politics as defined by Beck (1992). Symbolic politics is a rather deceptive method of regulation, whereby, certain rules or measures are imposed upon a body or

---

[15] A recent survey completed in the UK, demonstrated that the public have a very broad and general conception of the meaning of 'genetics'. However, a large proportion of the general population were unable to correctly respond to basic questions related to human genetic information (Human Genetics Commission, 2001).

an individual, which are designed in such a manner that they only target on a signalling effect, in this particular case, the insurance industry's need for risk-defined equity and information asymmetry (Hansjurgens, 2000:147). However, in this case the rush for legislation may negatively affect the population it is designed to protect as discussed earlier. Another challenging impact of symbolic politics is that the insurance industry, the addressee of the Disability Act, 2005, may become sceptical of this legislation and in disapproval may decide to pull out of the Irish market, furthering the negative consequences of the Disability Act, 2005. The Disability Act, 2005 serves to illustrate Beck's (1992) 'organised irresponsibility', a testament to the failure of the current political process, where no one body can be blamed. Yet the need to reduce potential risks leads to the penalising or imposition of restrictions upon a chosen, but 'blameless' entity (see Section 2.6). This further emphasises the inability of political institutions to cope alone and without full knowledge, a feature which is commonly embedded in a risk society (Matten, 2004:387). This legislation does not reduce the level of risk to which society is exposed, it merely creates other potential hazards and erodes the political power of the Government from within (Matten, 2004:391).

2.6 Risk Management in a Risk Society in Late Modernity

With such a wide variety of industrial and societal concerns, it is not surprising that entities managing the risks related to this situation are undecided with regard to the most appropriate method to manage these risks (Beck, 1992; Giddens, 1990). Such an amalgamation of legislation and ongoing debates on a global level present a situation evidently marked by much disagreement, with disputes over accountability, responsibility and how to resolve the situation (Gaivoronskaia and Solem, 2004). These discrepancies and the wide mesh of often-contradictory legislation limit the efficacy of risk management decisions made, perhaps leading to further risk situations and/or a decline in social trust and ontological security. The risk theories developed by Beck (1992; 1999) and Giddens (1990; 2000) serve as a means from which the current situation unfolding regarding insurers' use of genetic test results can be explored and analysed. Beck (1992:49) states that as a result of the size and universality of new risks and hazards introduced by modern scientific and industrial development, society is being transformed into a risk society, in which the emphasis and focus have shifted (See Figure 2-1).[16]

Figure 2-1 Beck's Risk Society

	Industrial	Risk
Motivation:	goods	danger
Result:	wealth	risks
Emotion:	need – "I am hungry"	anxiety – " I am afraid"
Reaction:	technical development	risk management

Source: Beck, 1992, Illustration taken from Wynne, S. 1999

The early underpinnings of Beck's (1992) views on society may well exist within a German and environmental context. However, they can be extended to include other first and second world industrial countries, where the consequences of recent technological innovations are being seen increasingly as potentially harmful and as having a serious negative impact upon society. As originally hypothesised by Beck (1992; 1999), the risk society is evolving beyond its beginnings and in line with industrial, technological and social developments. The risks associated with this new society remain intangible and can only be perceived psychologically, making them particularly difficult to manage as compared to more 'conventional' risks (Matten, 2004:378). A condition whereby no one body or actor can be blamed or held accountable for the risks produced, a situation which Beck (1992; 1995a; 1995b; 1999) terms 'organised

[16] Rose does not agree that we live in a risk society even though we are facing technological risks in abundance (Rose, 2000: 64). According to Rose, Beck's concept of the risk society, that it overrides the industrial society, is contingent on a strong economy and collective security. Rose explains that few European countries currently have strong economies; therefore, Europe is not part of the risk society. Also, many of the old industrial societies have not been able to hold on to modernity's promise of collective security; naming this, Rose accounts that the risk society serves to erase the still unfinished tasks of modernity (Rose, 2000:65). Perhaps as Rose (2000) states, not all European countries have strong economies, however, in this case the Irish economy can be considered strong enough to be acknowledged as an industrialised society, although the strength of its economy can be considered as a rather recent development. In addition, although collective security is still to be achieved for the large part, the features of the old and the new shall remain for some time to come, therefore it can be considered that although many European countries are slowly shaking off the features of modernity, they are still stepping precariously into late modernity and a risk society.

irresponsibility', also remains intact. These new risks are innate to the industrialised and self-advancing existence of modern society. According to Giddens (1990), society has further evolved and the risk society is unfurling into late modernity. The emphasis and theoretical standpoints of Beck's (1992) fundamental thesis persist, but the focus has shifted from environmental issues to global technological concerns, such as those relating to genetics and insurance (Giddens, 1990). The emergence of potential threats associated with the use of genetic test results by the life insurance industry followed by rushed management of the related potential hazards marks an entrance into the risk society. However, difficulties are encountered that limit the efficacy of the risk management decisions made leading to further risk situations. According to Beck, (1995b) risk decisions taken during this stage are superficial and largely cosmetic as authorities struggle to 'do something' to assuage growing public fears and demands, but with no clear indication of what that 'something' ought to be. The result is the creation of rushed policies that are neither well thought through nor suitable for managing the potentially harmful consequences of life insurers using genetic test data to underwrite applicants. Daar and Mattei (1999), when preparing for the Draft Guidelines for the World Health Organisation, warn that "hurried and premature legislation in the rapidly evolving field of human genetics can be counterproductive," but at the same time it is noted that national regulations should be developed for genetic testing and for mechanisms to monitor further developments. According to Giddens (1990), we can no longer rely on traditional risk management tactics to inform us how to shape our lives. Beck (1999) argues that 'sub-politics' or the 'shaping of society from below' will force changes in organisational structure and behaviour, while Giddens (1990) recommends that we allow reflexivity to replace the role of tradition, and allow life politics to renew or change the structures which inhibit effective risk management.

The political answer to the problems encountered by these actors may not be clear at present. However, no matter which risk management theory or strategy one adheres to, it is necessary that when regulation to manage the risks is being designed, the public, over which the legislation presides, should be consulted and involved in its construction and administration. Without the trust of the public behind the entity governing insurers' use of genetic test results, the entity's power to manage this situation decreases (Giddens, 1990). As of yet, no such consultation has taken place in Ireland and public knowledge, trust, risk perception, benefit perception and acceptance of the risks or the current management strategies is not known. It is important to understand reactions of the public to these factors in order for the risks related to this particular commercial use of genetic testing to be managed appropriately and effectively.

2.7 Summary

This chapter examined the use of genetic test results by the life insurance industry and the perceived consequences related to such usage. DNA tests may confirm suspicions, for example from family medical history, that someone has a particular genetic mutation that may indicate susceptibility to a particular disease. However, for the time being, such knowledge may not unlock any cure or even permit any worthwhile treatment. This chapter explained the potential for the results of genetic tests to be used as risk status information in life insurance underwriting by increasing the accuracy and reliability of family history information. The consequences of insurers using or being restricted from using genetic test results were examined. The dilemma surrounding insurers using genetic test results revolves mainly around the notion of unfair discrimination. However, it was noted that if insurers are not allowed to discriminate between risks, the result could initiate incalculable adverse selection. The discussion of the consequences of genetic test results leads to the argument that such negative effects can be considered as risks and the nature of these risks comply with those described as developing a risk society. The management of these risks was discussed, and it was noted that such management has arisen in the Republic of Ireland from the Council of Europe, the Irish Insurance Federation and the Irish Government. The difficulties in managing the uncertainty surrounding such risks were also noted. The development of modern Irish society into a risk society and its movement towards late modernity was discussed in the context of developing appropriate risk management strategies. The relationship between incomplete knowledge, trust, risk perception and risk acceptance in late modernity was briefly introduced. It is noted that to develop successful risk management strategies to adequately control the risks related to insurers using genetic test results, it is necessary for authorities to cultivate the trust of the public. It is deemed that trust in a particular body increases the power of that body in this scenario so that it might manage the risks related to insurers using genetic test results successfully.

The following chapter further examines the relationship between incomplete knowledge, trust, risk perception, benefit perception and risk acceptance by exploring and discussing relevant previous literature. A model is developed that illustrates how people, lacking detailed knowledge of the use of genetic test results by the life insurance industry, apply social trust in the insurance industry and the Government to formulate their assumptions of the risks and benefits related to life insurance companies using the results of genetic tests in underwriting, thereby reducing the complexity of risk management decisions. In addition, several factors that may influence this relationship are discussed.

CHAPTER THREE

FACTORS AND CONDITIONS AFFECTING RISK ACCEPTANCE

3.1 Introduction

The previous chapter explained genetic testing and the potential for the results of such tests to be used as risk status information in life insurance underwriting. The consequences of insurers using or being restricted from using genetic test results were examined and the management of these risks in the Republic of Ireland discussed. Furthermore, the influence that incomplete knowledge, social trust and risk perceptions have upon the acceptance of risk management strategies in late modernity was briefly considered. This chapter further explores this influential relationship, by investigating previous research that suggests that incomplete knowledge, social trust, risk perception and benefit perception influence risk acceptance. In addition, national culture, gender, economic status, age and life insurance ownership are examined to understand the effect that these factors may have upon risk acceptance. The interaction of all these factors shall be explored in order to understand their influential role upon society's willingness to accept or not accept life insurers using genetic test results.

In spite of two decades of studies examining the importance of trust in risk management literature, there exists little research examining the dynamics of the relationship between incomplete knowledge, social trust, risk perception, benefit perception and risk acceptance (Giddens, 1990; Earle and Cvetkovich, 1995; Siegrist, 2000). Historically, the focus of academic endeavours has centred upon the relationship between just two or three of these factors across a broad range of issues and subjects (Baird, 1986:425-435; Drottz-Sjoberg and Sjoberg, 1991:2007-2036; Flynn *et al*, 1992:417-429; Biel and Dahlstrand, 1995:25-36; Giddens, 1990; Groothius and Miller, 1997:241-257; Earle and Cvetkovich:1995; Hunt, Frewer, and Shepherd, 1999:167-180; Figueiredo and Drottz-Sjoberg, 2000:233-242;

Wright, Pearman and Yardley, 2000:681-690; Siegrist, Cvetkovich, Roth, 2000:353-362; Pisano and Woods, 2002; Siegrist *et al.*, 2003:707). It could be argued that this lack of coordination has hampered the development of new insights centred on examining the risk acceptance of individuals. While this research has yielded much valuable information on trust and risk perception, little is known of the relationship between knowledge, trust, risk perception, benefit perception and risk acceptance in relation to one specific issue and across one group of subjects.

The study of risk has a long established history in the management and psychology literature. Bernstein (1996), upon examining the historical origins of risk, asserts that pre-modern societies had different techniques for coping with external threats than modern society. For the most part, human kind in early societies trusted divinatory control (Beck, 1993:8-9). In most Western societies today, the notion that an otherworldly being predetermines future dangers has been largely rejected and superseded by the ability to socially control and avoid risks. Ironically, however, the fear and anxiety that things can go wrong is growing rapidly (Luhmann, 1991:3). As our Governments and industries have endeavoured to make life safer and healthier, it is reported that members of the public have become more, rather than less, concerned about risk, seeing themselves exposed to more serious risks than were faced by their predecessors (Slovic, 1999:675). In particular, uncertainties created by our own social development and risks from modern technological advances have become predominant concerns in the social consciousness.

According to Giddens (1990), technological development and future risk are inextricably woven together (Lomax, 2000:748). With the impossibility of preventing future social and technological advances compounding an escalating inability to manage unknown risk situations, the outlook for societal peace-of-mind could indeed be regarded as quite bleak. However, there is an intervening factor to consider. Risk perceptions and benefit perceptions may both be affected by an evaluation of trust in the authorities managing a hazardous situation (Siegrist *et al.* 2000:353-362; Siegrist *et al.*, 2003:707). According to Giddens (1990), trust usually serves to reduce or minimise the dangers to which particular types of activity are subject. This study examines if both risk perceptions and benefit perceptions are affected by judgments of trust made by those in charge of managing risks. That is to say to an extent that people perceive something to be high risk or low risk and of high benefit or low benefit because they trust or distrust the authorities or regulators of industry. If this is true, trust in those managing modern risk may alleviate anxiety caused by uncertainty, similar to the role of pre-modern trust in Divine Intervention. Therefore, if risk managers elevate social trust, societal peace-of-mind should be restored.

A number of studies have demonstrated that social trust is negatively correlated with perceived risk (Baird, 1986:425-435; Drottz-Sjoberg and Sjoberg, 1991:2007-2036; Flynn *et al*, 1992:417-429; Biel and Dahlstrand, 1995:25-36; Groothius and Miller, 1997:241-257; Hunt, Frewer, and Shepherd, 1999:167-180; Figueiredo and Drottz-Sjoberg, 2000:233-242; Wright, Pearman and Yardley, 2000:681-690; Siegrist, Cvetkovich, Roth, 2000:353-362; Pisano and Woods, 2002; Siegrist *et al.*, 2003:707). Furthermore, several theoretical and empirical studies suggest that social trust is most important where there is incomplete knowledge of the risks and benefits emanating from a new technology (Earle and Cvetkovich, 1995; Luhmann, 1989; Giddens, 1990; Siegrist and Cvetkovich, 2000; Sjoberg, 2002 and Viklund, 2003). Pushing this conjecture forward, Siegrist and Cvetkovich (2000) propose that risk perceptions and benefit perceptions are affected by judgments of trust of those responsible for managing risk. This, in turn, is influenced by the degree of knowledge an individual possesses of the associated risks and benefits.

According to this theory, if an individual has complete knowledge of a hazard and its consequences, he or she can assess the risks and benefits of that hazard based on that knowledge. However, if an individual has incomplete knowledge of a hazard, then social trust is applied to formulate assumptions regarding the associated risks and benefits. In cases of incomplete knowledge, individuals perceive something to be low risk and high benefit or high risk and low benefit because they trust or distrust the authorities or regulators of industry. In addition, Siegrist (2000:195-203) and Slovic (2000:689) hypothesise that the acceptance of risk management strategies are directly influenced by society's perceptions of the risks and benefits associated with the hazard to be controlled. This chapter proposes a model that is intended to represent the main theoretical standpoints and empirical results of previous research identifying a relationship between the above factors. It is intended that this model will be adopted to explore the ways in which risk acceptance is patterned with regard to the issue of genetic testing and life insurance and to investigate the factors and conditions that exist in order to produce the observed influential configuration.

3.2 Knowledge and Risk Perception

Social science research has demonstrated that technical experts and lay people often differ in their conclusions about the risks and benefits of hazards and in the way that conclusions about risks and benefits are made (Fischoff *et al*, 1982:240-255). An important difference is that technical experts, because of their training and experience, are more likely to have technical knowledge about hazards not available to the average individual. While experts are seen as purveying risk assessments - characterised as objective, analytic, wise, and rational – based on the *real risks,* laypersons take into account values and qualitative factors (Slovic, 1999:675). This knowledge, although less formal and more intuitive, represents a different kind of expertise that is a function of both the cognitive and the motivational system and of the conditions of the social, political and cultural environment (Jungermann and Slovic, 1993:87; Otway, 1992:220). Most people do not possess extensive and intricate knowledge of technological risks and benefits (Siegrist and Cvetkovich, 2000:714). Indeed, regarding the commercial use of genetic tests and the related results, there is little factual knowledge available on all of the risks and benefits of such employment. Genetic testing is a new technology, and the related risks and benefits associated with its various uses have to be considered as hypothetical or vague, even for those who have a good understanding of the basis of this technology and how it might be employed within the insurance industry. Therefore, the expected knowledge that laypersons or experts may have on this topic, and on the associated risks and benefits, is expected to be limited.

Lacking technical or specialist knowledge and unable to evaluate the accuracy or reliability of information provided by numerous 'expert' sources, lay judgements of modern risks and benefits may be based on assessments of those who are responsible for managing the hazards (Siegrist and Cvetkovich, 2000:713-719).[1] Therefore, instead of laypersons making rational judgements based on complete knowledge of the 'setting', social trust is employed to select experts who are trustworthy and whose opinions can be believed as being accurate (Siegrist and Cvetkovich, 2000:713-719). Because technical experts possessing expertise about

[1] A recent survey completed in the UK, demonstrated that the public have a very broad and general conception of the meaning of 'genetics'. However, a large proportion of the general population were unable to correctly respond to basic questions related to human genetic information (Human Genetics Commission, 2001).

particular hazards are better able to directly assess the risks and benefits of hazards and, because less knowledgeable lay people are more likely to assess social trust rather than direct hazard risk, Siegrist and Cvetkovich (2000) make the following assumption: social trust is significantly related to judgments of risk and benefits for hazards about which an individual has little knowledge. Establishing who to trust is most important in those situations where the individual lacks knowledge, and indeed also the interest, time, abilities, or other resources to personally make decisions and take actions. Trust minimises the complexity that people are faced with when making decisions related to risks and benefits and therefore provides heightened peace of mind (Siegrist and Cvetkovich, 2000:713-719).

Previous theoretical and empirical studies suggest that social trust is important in the absence of knowledge (Luhmann, 1989; Giddens, 1990; Beck 1992; Earle and Cvetkovich, 1995).[2] According to Beck (1992:22-23), new technological and social risks only exist initially in terms of incomplete knowledge about them and their social definition and construction. The public, incapable of making verifiable risk assessments due to incomplete knowledge, may have to surrender decisions surrounding modern risk to selected risk managers. Individuals decide which risk manager to trust and bestow decision control to that individual or entity. Giddens (1990) argues that trust can, therefore, reduce the complexity and uncertainty that prevails in modern society. However, without trust in those making decisions regarding the use of genetic test results by the life insurance industry, perceptions of risks shall not be abated by current risk management strategies.

---

[2] Giddens (1990) analyses why lay people, most of the time, trust in practices and social mechanisms about which their own technical knowledge is slight or non-existent. According to Giddens (1990:89), science maintains an image of reliable knowledge which spills over into an attitude of respect for most forms of technical specialism. However, in conjunction with this view lies an attitude of ambivalence towards science and technical knowledge. Wynne (1992:296) also recognises the basic ambivalence underlying trust. Giddens (1990) postulates that ambivalence lies at the core of all trust relations, whether it is trust in abstract systems (expert systems and symbolic tokens) or in individuals. For trust is only demanded where there is ignorance – either of the knowledge claims of technical experts or of the thoughts and intentions of intimates upon whom a person relies (Giddens, 1990:89). Ignorance is a source of scepticism or at least caution. The insurance industry's claim to specialist knowledge is often seen as a closed shop, having an insider's terminology seemingly invented to baffle the layperson. Such an industry may be observed in a hostile manner, or with an attitude of suspicion or scepticism.

This hypothesis is empirically supported in three separate studies (Siegrist and Cvetkovich, 2000; Sjoberg, 2002 and Viklund, 2003). Although, trust has often been mentioned as an important variable in determining risk perception and risk acceptance, Eiser *et al.* (2002) found evidence against the notion of trust and/or risk being factors causally related to technology acceptance. Instead, they suggest that rather than trust and risk acceptance being mediated by perceived risk, trust and perceived risk are influenced by acceptability. Sjoberg (2002) also disagrees that trust is an important dimension in risk perception and notably disputes the significance of the results from the Siegrist and Cvetkovich (2000) study. Sjoberg (2002; 2005) reports that trust in politicians in general does not significantly predict overall perception of risk. In addition, Sjoberg (1996a) criticises the use of aggregated data and the resulting mean values to investigate the relationships among knowledge and trust. According to Sjoberg (1996a), such an application creates a statistical 'illusion' as correlation based on raw data is often lower than correlations among means. However, Siegrist and Cvetkovich (2000) argue that it was unsuitable to use raw data in the case of that particular study and that the explanatory power of trust depends on how it is operationalised. For example, the measure of social trust in unspecified politicians (e.g. general trust in Government in all situations), as deduced by Sjoberg (1999a), is unlikely to yield a strong relationship between trust and risk. Sjoberg (2002) also argues that by measuring perceived risk and trust by means of attitude scale items using the same response format, the explained variance is increased. Despite such criticisms, all three studies demonstrate a significant correlation between incomplete knowledge and social trust. The results from the above studies are argued from a positivist position, whereby it is deemed possible to accurately and definitively measure trust through quantitative means and to expose rules that govern trust and its relationship with knowledge. Positivists seek to identify verifiable facts about phenomena and to devise and apply laws about their operation. Researchers following this paradigm value objectivity and replicability, hence, discussion surrounding the results from the above studies centres on acceptable statistical methods rather than the unreliability of their measurement of trust considering that society's attitudes of trust is socially constructed and, therefore, beyond what is considered positivisitically measurable. In general such technical analysis provides a narrow and limited view of trust and risk and the relationship between these factors, reducing the possibilities to numeric probabilities based on relative frequencies. However, the objectivity associated with such calculations, as performed in the above studies over-simplify the complex phenomenon that may be the relationship between trust and risk.

One of the aims of this study is to investigate whether or not knowledge of the use of genetic test results by the life insurance industry influences society's trust of authorities managing that hazard. Trust is exaFmined from the perspective of several dimensions and other factors that affect this relationship are not ignored. An effort is made to account for such factors, however, it is noted that due to the open system in which this relationship is being studied it is not possible to take account of all of the impacting factors. The following section shall examine the relationship between trust and risk perception and benefit perception followed by an investigation of the dimensions that delineate trust. Firstly, however, the correlations among risk perceptions and benefit perceptions shall be analysed.

3.3 Trust and Risk Perception; Trust and Benefit Perception

In a study conducted by Siegrist (2000:195-203), perceived benefit of genetic technology had a significant impact on perceived risk. Negative correlations among judgments of risk and judgments of benefit have been found for a number of different technologies (Siegrist, 2000:195-203; Frewer *et al*, 1998:221-235). Several explanations for the observed associations are offered. Alhakami and Slovic (1994:1085-1096) postulate that people assess hazards in terms of general attitudes (favourable or unfavourable). If people prefer consistency among their beliefs, technologies perceived as favourable will result in the devaluation of risks and the elevation of benefits related to that particular technology. For technologies perceived as unfavourable, the concluding results for perceived risks and perceived benefits are reversed, whereby, the perceived risks for the technology will be higher and the perceived benefits will be lower. This observation leads to the conclusion that it may be possible to change perceptions of risk by changing perceptions of benefit.

Gregory and Mendelsohn (1993:259-264) hypothesise that risk ratings are net measures. What is meant by this premise is that people integrate some benefits in their risk assessments. Frewer *et al.* (1998) assume that the favourableness of a technology is affected by perceptions of risks and benefits that are functionally related to each other. They also argue that it is possible to reduce perceived risks by heightening perceptions of benefits. According to Frewer *et al.* (1998), risk perceptions and benefit perceptions are inversely and causally related. This implies that the acceptability of a perceived high-risk technology can be increased by identifying and emphasising its benefits (and hence reducing the perceived risk). Frewer *et al.* (1998) conclude that people are more concerned about the extent of the benefits resulting from the development and application of a particular technology than the associated risks. That is, for as long as a risk is not so large as to be completely intolerable, public acceptance will be driven by perceived benefit (Frewer, 1999:569-574).

Reasons have been put forward to suggest that one should not automatically assume that emphasis of the benefits associated with a technology results in a decreased perception of the risks. The observed correlation between benefits and risks perceived does not necessarily mean that there is a causal relationship between the two variables. It is suggested, with the support of empirical evidence, that risk perceptions and benefit perceptions are both affected by a third evaluation concerning the introduction of a new technology. A number of studies have demonstrated that social trust is negatively correlated with perceived risk (Siegrist *et al.*, 2003:707; Pisano and Woods, 2002; Siegrist, Cvetkovich, Roth, 2000:353-362; Figueiredo and Drottz-Sjoberg, 2000:233-242; Wright, Pearman and Yardley, 2000:681-690; Hunt, Frewer, and Shepherd, 1999:167-180; Groothius and Miller, 1997:241-257; Biel and Dahlstrand, 1995:25-36; Flynn *et al*, 1992:417-429; Drottz-Sjoberg and Sjoberg, 1991:2007-2036; Baird, 1986:425-435). Of particular relevance to this research, Siegrist *et al.* (2000:353-362) examined the possibility that both risk perceptions and benefit perceptions are affected by judgments of trust of those in charge of managing biotechnology. That is to say to an extent that people perceive something to be safe or dangerous because they trust or distrust the authorities or regulators of industry. Results from Siegrist's (2000) and Siegrist and Cvetkovich's (2000) examinations of perceptions of gene technology, suggest that trust positively influences perceived benefit and negatively influences perceived risk. This theory is also evident from the findings of studies conducted by Lang and Hallman (2005), Grobe *et al.,* (1999), Jungermann *et al.* (1996), Freudenburg (1993), Flynn *et al.* (1992), Bord and O'Connor (1992) and Pijawka and Mushkatel (1991). However, not all academics agree that social trust is an important influence on perceived risk (Sjoberg, 2001:189-198; 2002:379-

400).[3] Sjoberg (2002) argues that Siegrist's (1999, 2000) positive findings are due to the use of a common methodological component. Siegrist (1999) and Siegrist and Cvetkovich (2000) measured both perceived risk and trust with Likert-type attitude scale items, unlike most other work on trust and risk. According to Sjoberg (2002), other studies employed separate ratings of risk and trust and, as a result of this, reported relatively low correlations between these two dimensions (Sjoberg, 1999; Sjoberg, 2002). Sjoberg (2001) suggests the limited importance of trust may be due to people believing that "there are clear limits to how much science and experts know, even though they can be trusted on the basis of being competent and honest (Viklund, 2003:730). While others maintain that trust should be seen as a consequence rather than a cause of risk acceptance attitudes (Eiser *et al.*, 2002).

Noting the above debate surrounding the measurement of trust, risk and benefit perception, it is important to recognise that the above researchers follow a positivist approach where causal inferences are considered valid and to have a universal application. However, arriving from a critical realist perspective, it is impossible to make collective cause-and-effect laws, as the world is generally not the same; places and cultures differ and times change, therefore, the same law cannot apply in all societies or structures and at all times. Poumadere (1995:7-15) demonstrates that even though French citizens were more trusting than Americans, they perceived risks as larger. This is an example of two different societies, where the backgrounds of the societies are completely different. The result of such differences is a lack of consistency between the application of an influential pattern deduced from studies observing social trust and risk and benefit perception in one group or country and their relevance in another group or country. It is difficult to make cause-and-effect assumptions when there are a myriad of factors and conditions influencing the cause-and-effect factors, each of which could impinge on the supposition and the outcome of the conjecture, e.g. psychological research has provided a list of factors that may influence risk perception. The main compound factors are listed in Figure 3-1. The importance of these factors might also explain why hazards judged as low-risk by technical risk assessment, are considered high-risk by the public.

---

[3] Sjoberg believes that beliefs in unknown effects of technology will add substantially to explained variance, above that accounted for by trust measures.

Figure 3-1 Risk Perception – psychological influences

RISK PERCEPTION – PSYCHOLOGICAL INFLUENCES

- familiarity with the risk source;
- voluntary acceptance of the risk;
- ability to personally control the degree of risk;
- whether the risk source is capable of causing a disaster (catastrophic potential);
- certainty of fatal impact should the risk occur (dread)
- undesired impact on future generations;
- sensory perception of the danger;
- impression of fair distribution of benefit and risk;
- impression of reversibility of the risk impact;
- congruence between benefactors and risk bearers;
- experience (collective and individual) with technology and nature;
- reliability of information sources;
- clarity of information on risk
- trust in state-operated risk control and risk management

Source: List taken from studies conducted in Austria, Germany, Britain, Netherlands and the US, Renn and Rohrmann (2000)

It is impossible to account for all of these factors and thus control them so as to determine a universal casual relationship. To explain a situation in terms of universal causal relationships, it is necessary for the researcher to have control over the behaviour of those investigated or the event investigated (Yin, 1994). However, it is impossible to control the views of a public who socially construct their own meaning in terms of their background and the constantly changing social reality surrounding them. This study recognises the differences between social realities and the constant changes in social realities. Consequently, the study is not attempting to draw a general cause-and-effect rule from its investigation of trust, risks and benefit perception. Instead, this research intends to examine if trust influences risk and benefit perception where there is incomplete knowledge of the use of genetic test results by the life insurance industry in the social reality of the members of the Irish public surveyed (see Sections 4.2.3 and 4.2.3.1 for critical realist ontology and view of causation).

3.4 Definition of Trust

To explicate individuals' trust, an understanding of trust must first be defined. Although trust is an increasingly studied topic and researchers are growing more sophisticated in their analyses, there is not yet an agreed definition of trust and, therefore, researchers continue to invent new measures (Lang and Hallman, 2005). This leads to difficulties in comparison across research projects. There exist many concepts of trust, including 'calculus-based-trust', 'knowledge-based-trust', 'identification-based-trust', 'characteristic-based-trust', 'process-based-trust' and 'institutional-based-trust' (Saparito and Chen, 2001). Cvetkovich and Lofstedt (1999: iv) note the difficulties produced by inconsistent definitions of trust, and argue that the primary and most negative effect is an overall slowing of progress in trust research. Consequently, given that such confusion has arisen from previous studies examining trust and risk, it is quite necessary to explicate trust and its attendant notions in this study.

Researchers' philosophical assumptions influence their views and opinions of the world and what can and cannot be measured. Allegiance to a particular epistemological notion can thus affect definitions of trust and associated methods for examining and exploring the concept. It is possible to study trust in accordance with the rules associated with the natural sciences or positivism and consequently provide law-like generalisations to explanations of trust. From a constructivist level, one may analyse trust from the perspective of the individual/group studied, believing that trust is socially constructed and produced through social interaction. From this view, trust remains in a constant state of instability and is un-measurable through positivistic means. However, between these two extremes there are a number of approaches to the analysis of trust. This research assumes that trust is an outcome of social construction, however, this is intervened by an ontologically real external reality. This stance allows an objective approach to be applied to the study of trust but also recognises the significance of people's socially constructed interpretations and meanings. It can be said that this is a critical realist perspective of trust (see Sections 4.2.3 and 4.2.3.1).

3.4.1 Family Resemblance of Trust

Although the term 'trust', like 'risk', means many different things to different groups of people, the word still retains its essential meaning. To use Wittgenstein's (1953) terms of reference, there is a family resemblance no matter within what context or philosophical background the word 'trust' is defined and used. Cvetkovich and Lofstedt (1994:4) distinguish the core understanding of what trust is, when they state that trust tends to imply a difference in power and control, whereby the one who trusts usually assumes a position of subordination and surrenders decision and behaviour control to the one who is trusted. Cvetkovich and Lofstedt (1994:4-5) note that the act of trusting does not necessarily produce a feeling of loss of control or power. This is because the trusting person trades behavioural and decision control for cognitive control (thinking that hazards will be controlled) and secondary control (selecting who is to be trusted). Trust involves risk, because in handing over control one resumes a secondary or inferior position. One may trust someone because of a strong belief that the trusted one will act in a certain and anticipated manner. But one can never be entirely certain that one's trust will not be broken. It can be said that trust trades a physical primary risk *(what should I do about this environmental hazard?)* for a social risk *(is the trusted one trustworthy?)* which is followed by a physical risk *(if the trusted one is not trustworthy, I may not be protected from physical harm)* (Cvetkovich and Lofstedt, 1999:4-5). Trust is an expectation about a relationship. Trust is not necessarily a characteristic of individuals. Trust implies choice, in that individuals can decide when to trust, who to trust, and why trust (Cvetkovich and Lofstedt, 1999:4-5).

If trust is considered socially constructed, it can be very difficult to study it objectively as there are numerous understandings of the meaning of trust. *" 'We trust you' may mean that we believe you can give us right answers and reliable information. It may mean that we believe that you are honest, and will tell us all that you know. Or it may mean that we trust your judgement, and rely on you for decisions, which are wise, impartial, and ethical, and in the public interest. We may trust you in one of these ways, without trusting you in the others. In this case, if a pollster asks us whether we trust you, what are we to say?"* (House of Lords, 2000, paragraph 2.29). This quotation, albeit alliterating the difficulties associated with the meaning of trust and measuring trust, demonstrates how the user's judgement of trust is dependent upon a number of factors that constitute and contribute to trust. These factors indicate reasons for trust or lack of trust in an individual or an organisation. People construct their own definition of trust and make decisions regarding trust based on a number of dimensions considered core elements of trust. Several authors have identified a number of evaluative judgements of trust that lead to a particular individual's overall view of trust and influence the level of trust they place in an individual or group (see Section 3.4.2). These dimensions shall be examined in an effort to explicate the social construction of trust and to determine the factors that directly influence trust or lack of trust so that the concept of trust may be measured.

3.4.2 Dimensions of Trust

Renn and Levine (1991) identify five behavioural and emotional attributes of trust, namely; *perceived competence* reflecting good quality science/assessment; *fairness* or the degree to which the source takes into account all relevant points of view; *consistency* which is based on the predictability of arguments and behaviour based on past experience; *faith* which reflects the perception of "good will" of the source; and *objectivity* which indicates the absence of bias in information. Kasperson *et al.* (1992:161-187) determine four dimensions that develop and maintain trust: *commitment* as in commitment of the trustee to fulfil promises; *competence* which refers to the ability of the trustee to complete a task in a competent manner; *caring* indicating the quality of being sympathetic to other people's aims and concerns; and finally, *predictability* which demarcates that people tend to place their trust in organisations that are seen as consistent. Kasperson *et al.* (1992) argue that their definition considers the multiple dimensions of trust – cognitive, affective, and behavioural. As a cognitive process, trust entails a choice based on reasoning about the available evidence and is based on a degree of cognitive familiarity with the object of trust. This knowledge is necessarily incomplete and incorporates an element of doubt or risk. The affective component involves an emotional bond among participants in a relationship, whereby violation of that bond can damage the relationship. The behavioural component reflects the actions undertaken in the belief that others will act in a similar manner (Bradbury *et al.*1999:118). Though analytically separate, the three dimensions are combined in actual human experience and indeed serve to reinforce one another (Bradbury *et al.*,1999:118).[4] Lewis and Weigert (1985:967-985) also delineate trust as a

[4] Bradbury *et al.* (1998:118) illustrate this theory by examining the behavioural display of trust of a person or an organisation. This behavioural display can increase both cognitive and affective trust in another, i.e., it assists in creating

multi-faceted concept with integrated cognitive, affective and behavioural dimensions.

Few studies have empirically identified the above dimensions of trust as aspects that the public use to determine social trust. However, Petts (1998) completed a survey on trust taking into account the various components of trust described above and found that trust was enhanced by four primary characteristics: *competence* – good quality science/ assessment, *openness and consultation*, *caring* – being sympathetic to other people's aims and concerns, and *objectivity*. Based on the concluding results of the study, Petts (1998) recommends that future studies of trust should use the measurements of a combination of these characteristics to distinguish a measurement of trust. However, a criticism of Pett's (1998) approach is that respondents were asked to use scales that are predetermined by the researcher, rather than characteristics that are selected by the respondents themselves. As a consequence, the results cannot correctly represent the public's perceptions of the dimensions of trust.

Metlay (1999) criticises the above researchers for their tendency to distinguish additional shades of meaning in the concept of trust, although it is not clear if these are indeed (empirically) discernible (Poortinga and Pidgeon, 2003). The approach distinguishing numerous dimensions of trust is questionable due to uncertainty regarding whether or not the public makes a distinction between the different theoretical features of trust. For example, the components *objectivity* and *fairness,* described by Renn and Levine (1991), may not be completely independent (Poortinga and Pidgeon, 2003). A source that is considered fair because it takes in all relevant points of view may be perceived as more objective. Similarly, people may not differentiate between Kasperson *et al.*'s (1992) aspects of trust termed commitment and caring. Perceptions of whether an institution acts in a certain way may influence perceptions of whether that institution fulfils its fiduciary obligations (Poortinga and Pidgeon, 2003).

Contrary to the above studies, Metlay (1999:101) concludes that trust is not a complex and multi-faceted concept. Rather, trust simply depends on two distinctly different components or dimensions. The first (affective component) being a tightly interconnected and intertwined set of *affective* beliefs about institutional behaviour, which Metlay (1999) terms

---

evidence of trustworthy behaviour to the other (cognitive aspect of trust), which also reinforces positive sentiments (affective aspect of trust).

"trustworthiness" is related to whether an institution is seen as having integrity and being honest, credible and fair, caring and concerned for the broad public interest. The second being how *competent* the institution appears to be, based on whether an institution is seen as being competent, open, forthcoming, reliable and consistent in actions (Metlay, 1999:101). Therefore, rather than trust being based on a large number of separate and theoretically (although not empirically) discernible components, trust is mainly a two-dimensional construct based on care (or trustworthiness) and competence (Poortinga and Pidgeon, 2003). Hovland *et al.* (1953) and Jungermann *et al.* (1996) report similar findings, with Jungermann *et al.* (1996) replacing "trustworthiness" with "honesty".

Evidence for trust as a two-dimensional construct has been mainly found using factor analysis statistical techniques (Poortinga and Pidgeon, 2003:963). Respondents did not select the characteristics that they deemed to be representative of trust. However, as trust is socially constructed, the public should predetermine all characteristics of trust. Frewer *et al.* (1996) attempted to overcome this obstacle in their investigation by personalising a questionnaire by first asking respondents to indicate their own reasons for trusting or distrusting a particular source. In this way, respondents could rate each source on the dimensions they themselves thought were important. According to Frewer *et al.* (1996) two major dimensions are important in determining trust. The first factor comprised the characteristics: truthful, good track record, trustworthy, favour, accurate, factual, public welfare, responsible, knowledgeable, and (negatively) the characteristics: distorted, proven wrong in the past, and biased. The second element included the characteristics: accountable, self-protection, and a vested interest versus sensationalism component. Poortinga *et al.* (2003) termed these two dimensions respectively as a *general trustworthiness* dimension, encompassing both *honesty* and *care*, and a *vested interest* or *accountability* dimension. French *et al.* (2002) and Poortinga *et al.* (2001) identified similar dimensions of trust using a selection of the characteristics. However, Lang and Hallman (2005) substantiate the use of a single concept to describe trust, although their study is based upon a limited number of components of trust – competence, transparency, public interest, and honesty – which may have influenced this finding. It is important to note also, that this inconsistency may also be due to cross-cultural variations.

In conclusion, there appears to be two slightly different approaches to determining the dimensions of trust. The first approach underlines trust as a complex and multi-faceted concept, while the other delineates trust as a simple two-dimensional construct. It is not the intention of this study to test which of these two approaches best describes social trust, but to select a number of characteristics that can be used to measure trust. Although there are two different approaches to explicating trust, there are certainly some aspects that overlap. These trust-related factors include competence, reliability, credibility, integrity (vested interest), care, fairness, and openness. The public identified each of these dimensions as constructing their attitude of trust toward an organisation in the survey completed by Frewer *et al.* (1996) and each of the seven factors have received theoretical backing from recent research (Poortinga and Pidgeon, 2003; Bradbury, Branch and Focht, 1999; Metlay, 1999; Petts, 1998; Jungermann, Pfister, Fischer, 1996; Kasperson, Golding and Tuler, 1992; Renn and Levine, 1991; Lewis and Weigert, 1985; Hovland, Janis, Kelley, 1953). Amalgamated, these observable and measurable dimensions can serve to provide a particular view of overall trust, meaning that it may be possible to measure social trust without applying a strict, universal definition of trust. The above seven characteristics of trust shall be applied in the present study to measure trust (see Appendix A.1). The use of various trust characteristics will also shed more light on reasons why people trust or do not trust the Irish Government and the insurance industry to manage the use of genetic test results in underwriting.

3.4.3 Trust versus Confidence

However, measuring the components of trust to gauge a specific level of trust opens a typical debate in trust studies. Delineating the distinction between trust and the related concept of confidence has proven difficult for many academics. Some argue that trust is inherently different to confidence. According to Earle and Cvetkovich (1995:61) and Siegrist *et al.* (2003:705-715), confidence is often used as a surrogate for social trust with the demarcation between confidence and social trust based on the distinction between a familiar environment (in which one may have confidence in just about anything) and a risky environment (in which one may make a judgement of trust of a person-like entity). Luhmann (1993:23) agrees, stating that trust presupposes a situation of risk. One is in a situation of confidence if one does not consider alternatives. The alternative to risk is danger, which is linked with confidence. In a situation of confidence one is exposed to dangers, but no thought is normally given to them because they are familiar and fall within one's realm of competence to manage. In a situation of trust, one must decide between alternate futures, each of which may or may not turn out to be manageable (Earle and Cvetkovich, 1995:62; Misztal, 1996; Siegrist, Earle and Gutscher, 2003:705-715). However, it can be argued that trust exists in a familiar environment as well as a risky environment.[5] In addition, there is no intrinsic connection between confidence and danger (Giddens, 1990). Danger can exist in circumstances of risk and, therefore, in circumstances of trust (Giddens, 1990). Trust is not the same as confidence; it may be that trust is a particular type of confidence. Giddens (1990) describes trust as confidence in the reliability of a person or system, regarding a given set of outcomes or events, where that confidence expresses a faith in the probity or love of another or, in the correctness of abstract principles such as technical knowledge. This definition describes trust as containing both an affective and a competence dimension.

Empirical research that supports the notion that trust and confidence are two separate constructs implies that trust is more closely affiliated with affective components of trust, with confidence closely associated with a competence dimension (Metlay, 1999:101; Siegrist, Earle and Gutscher, 2003:705-715). This indicates that the use of a 'competence' dimension to measure trust is incorrect. However, the public construct their own particular meaning of trust, regardless of theoretical announcements of obvious differences in decisions between trust and confidence. The public consider both affective and competence dimensions when making decisions regarding trust (Frewer *et al.*, 1996). The reason for this result could be that the public do not distinguish between trust and confidence or, as Giddens (1990) suggests, trust is a type of confidence and therefore the two concepts are closely affiliated in the minds of those making decisions on trust. The following section continues to pursue an understanding of trust and its associations, by exploring the indirect relationship between trust and risk acceptance, and in doing so, investigates the relationship between risk perception, benefit perception and risk acceptance.

3.5 Risk Perception, Benefit Perception and Risk Acceptance

Sheppard and Sherman (1998:422-437) put forward an argument that trust is accepting the risks associated with a given relationship. According to

---

[5] Privacy and discrimination risks related to the use of hereditary information in underwriting life insurance contracts are existent in an industry that is familiar and its use of this information familiar

Giddens (1990:35) acceptable risk is whereby danger is minimised. Therefore, risk acceptance can be understood as the acceptance of risks due to a belief that the risks are minimised or are being managed appropriately by trusted authorities. Although Sjoberg (1999:130-131) considers that risks are not necessarily accepted but, instead, are tolerated. Given this interpretation, Sjoberg (1999:130-131) defines risk acceptance as a situation whereby an individual agrees or is satisfied with current strategies designed to manage a risk. In this case, risk acceptance can be defined as allowing insurers to use genetic test results due to the perception that the risks associated with such usage are minimised. What is seen as "acceptable risk" varies in different circumstances and within different cultures, but is normally central in sustaining trust (Giddens, 1990:35).

There are few empirical studies examining the relationship between risk and benefit perception and acceptance of risks. Siegrist (2000:195-203) demonstrates that the acceptance of gene technology is directly influenced by society's perceptions of the risks and benefits associated with this technology. In a separate study, examining risks related to electro-magnetic fields (EMF), Siegrist *et al.* (2003:705-716) confirmed that social trust influenced cooperation and acceptance of risk management strategies designed to mitigate EMF risks. However, Siegrist *et al.* (2003:715) also note that the impact that social trust has upon these factors is context dependent. These results have been disputed by Sjoberg (1999:129-149), whose studies have lead to the conclusion that perceived risk does not predict acceptance of risks or a demand for risk mitigation. In contrast, Slovic (2000:689) illustrates that the regulatory policies of certain bodies have been determined by public perceptions of risk.[6]

3.6 Conclusions on Relationship Surrounding Incomplete Knowledge, Trust, Risk Perception Benefit Perception and Risk Acceptance

It is noted that there are many influences upon risk perception and it is determined that trust in those managing risk in the risk society and late modernity lies at the core of risk and benefit perception and, in doing so, indirectly affects risk acceptance. To study the influence that risk perception has on risk acceptance, it must be understood that it is not possible to control all other effects impacting upon this relationship. It is

[6] For example, the American Environmental Protection Agency.

discussed how Poumadere's (1995) and Sjoberg's (1999) studies demonstrate the effect that society has on views of trust, risk and acceptance, which, to a large extent, are ignored by the above empiricist researchers. Constant flux in society and continually changing configuration of structures and processes makes universal causal predictions impossible; therefore, there can be no universal laws that govern trust, risk and benefit perception and risk acceptance. It is discussed how trust, risk, benefit and risk acceptance are socially constructed concepts. No two societies will have the same view of these concepts, as there are inherent cultural differences between societies that ultimately influence peoples understanding and perceptions of trust, risks, benefits and acceptance of risks. It is suggested that what positivists assume as a universally given law may not apply as collectively expected.

This study accepts that there are no universal laws governing trust and risk acceptance cause-and-effect relationships, therefore, the results of this study may or may not align with Siegrist's (2000) cause-and-effect theory. However, it is intended to understand the influential nature of the relationship between trust and risk perception in order to understand the effects that trust and risk perception have upon the acceptance of life insurers using genetic test results in the Republic of Ireland. By gaining knowledge regarding these influential factors, it may be possible to extend our knowledge regarding risk management of incalculable risks. Figure 3-2 illustrates a conceptual model of the relationship between incomplete knowledge, social trust, risk perception, benefit perception and risk acceptance.

Figure 3-2 Conceptual Model of the Relationship between Incomplete Knowledge, Social Trust, Risk Perception, Benefit Perception and Risk Acceptance

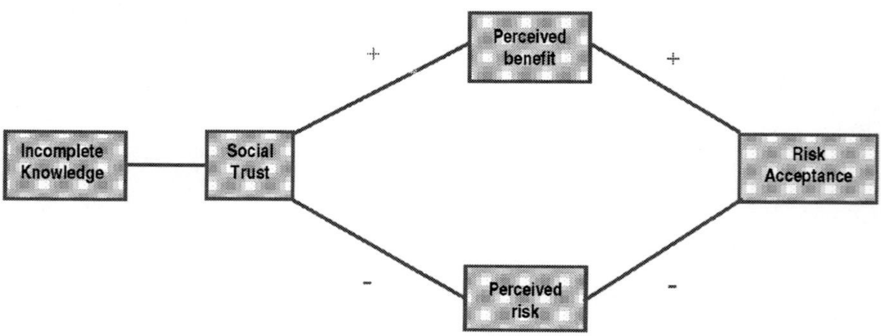

Source: Model adapted from Siegrist (Siegrist, 2000:197; Siegrist and Cvetkovich, 2000:713-719)

The model illustrates how people, lacking detailed knowledge of the use of genetic test results by the life insurance industry, apply social trust in the insurance industry and the Government to formulate their assumptions of the risks and benefits related to life insurance companies using genetic information in underwriting, thereby reducing the complexity of risk management decisions. Trust in those managing the risks related to such usage, that is trust in the insurance industry and the Government, results in a positive evaluation of the benefits related to the use of this technology in life insurance underwriting and few risks. The model demonstrates that, in this case, public risk and benefit perceptions influence society to accept the current management of the associated risks. In other words, if individuals associate the use of genetic test results by the life insurance industry with few risks and many benefits, they are more likely to be satisfied with the current management of the hazard. The opposite of this model demonstrates that relatively little trust in authorities managing these risks results in a negative appraisal of the associated benefits and an increase in awareness and concern with regard to the risks. The result depicted by the model, is a rejection or a lack of agreement with current management of the use of genetic test results in underwriting practices.

By examining this pattern in relation to society's views of insurers using genetic test results, it is possible to further understand the influences upon risk acceptance. It is acknowledged that no model, however elaborate, can mirror or replicate the uncertainties and complexities of the real relationship between trust and risk acceptance and there remains contention as to whether or not the above model accurately illustrates the relationship between these factors. For example, Eiser *et al.*, (2002) and Poortinga and Pidgeon (2005) argue that the acceptability of risk is a determinant of trust rather than the result of trust as suggested by the above illustration (see Figure 3-2). However, although the model examined in the present study may be inaccurate, it is deemed that observations regarding the relationship can, nonetheless, provide useful information regarding certain aspects of the relationship.

The following section explores other factors and conditions that exist in the study that may influence the observed influential configuration between trust and risk acceptance. The dominant culture in Ireland is examined as this may affect the responses of individuals to questions regarding trust and risk perception. In addition, the effect that gender, age and socio-economic status have upon this pattern is examined with the aid of previous literature in this area. Finally, the influence that ownership of life insurance and power has upon the pattern is explored.

3.7 Other Influential Factors and Conditions

It may be considered that every person has an individual background that shapes their knowledge and perceptions of hazards. This background may be affected by the society in which they live and the dominant culture within that society. Biographical details such as age, gender and economic status affect the society-wide distribution of resources (such as services, status, and power) and views of individuals. People may have particular perceptions of risks and benefits associated with the use of genetic test results by the life insurance industry based on their understanding and previous experience dealing with life insurance companies and the Government. Personal attributes, background, and beliefs may shape a person's view of trustworthiness of these bodies.

3.7.1 National Culture

The influence infused by the national culture of a country is a factor that may lead to a high take-up of genetic tests or a need for life insurance in a society. Cultural theory launched by Mary Douglas (1966, 1978) and Douglas and Wildavsky (1982) has had a significant part to play in discussion on risk perception and risk interpretations (Dake, 1991; Wildavsky and Dake, 1990). Cultural theory is an anthropological approach to explaining risk perception that describes forms of social solidarity that shape worldviews. Worldviews are general social, cultural, and political attitudes that appear to have an influence over peoples' judgements regarding fairness, trust and risk about complex issues (Slovic, 2000; Langford *et al.*, 2000:691-704; Siegrist, 1999; Dake, 1991). Adherence to a specific pattern of social relationships generates a distinctive way of looking at the world. This characteristic of social interaction is commonly referred to as cultural bias. According to Wildavsky and Dake (1990:42) it is possible for cultural theory to "predict and explain what kind of people will perceive which potential hazards to be how dangerous". In other words, what is perceived as dangerous, and how much risk to accept, is a function of ones cultural adherence and social learning (Douglas, 1978; Douglas & Wildavsky, 1982; Thompson, Ellis & Wildasvsky 1990).

The basis of cultural theory is Douglas' grid-group typology (Douglas, 1978; Thompson *et al.,* 1990). According to Douglas (1978), variation in social participation can be adequately accounted for by the dynamics between two dimensions: *group* and *grid*. The group variable represents the degree to which an individual is incorporated into a social unit. Where group is weak, interaction between individuals is infrequent and limited to specific undertakings, therefore there is much self-reliance and an air of competition among individuals. Strong group individuals interact frequently and there is a sense of solidarity rather than competitiveness. The grid variable represents the degree to which members of a social grouping are constrained by internally or externally imposed classifications, such as gender, race, age and profession. With high grid, social classifications severely circumscribe the social activities of individuals. With low grid, an egalitarian state of affairs exists where no one is precluded from social activities due to their membership of a particular class (Skipper, 1998). If the dimensions are placed in a two-axis system, from low to high, four outcomes occur, each representing a different kind of social environment (see Figure 3-3). The dimension a person belongs to will guide his or her interaction with the environment.

Figure 3-3 Grid/Group Topology

B FATALISTIC Life is a lottery. Risks are out of out control; safety is a matter of luck	**C HIERARCHIC** Risks are acceptable if institutions have the routines to control them
A INDIVIDUALISTIC Risks offer opportunities and should be accepted in exchange for benefits	**D EGALITARIAN** Risks should be avoided unless they are inevitable to protect the public good

Source: Adapted from Renn (1992:74), (Rayner, 1992:89) and (Skipper, 1998:407)

According to Thompson *et al.* (1990), an important and fundamental factor concerning peoples risk perceptions is their general attitudes towards nature, in that thoughts about nature and other people are interwoven with worldviews. Thompson *et al.* (1990) used people's attitudes towards ecological systems as a basis for their analysis. Oltedal *et al.* (2004) summarise the findings from this study which can be interpreted as distinguishing risk perceptions and political ideologies across the four groups. According to Oltedal *et al.* (2004), individualists support market liberalism and competition and believe that people should have the opportunity to keep their economic gains for themselves. Risks offer opportunities and should be accepted in exchange for benefits. Politically, individualists would be placed to the Right. On the other hand, egalitarians fear development that may increase the inequalities amongst people. Egalitarians are suspect of expert knowledge for fear that experts and institutions might misuse their authority. Egalitarians can be placed politically to the Left, and tend to support political action aiming to increase social equality. Unlike individualists, egalitarians see nature as fragile and vulnerable to human interventions, making egalitarians alert about new technologies that might change the state of nature. Hence egalitarians generally oppose risk that could inflict irreversible dangers on many people or future generations.

Hierarchical cultures emphasise the perseverance of the "natural order" of society and tend to fear social commotion, demonstrations, and crime. Hierarchists place great faith in expert knowledge and believe that non-adherence to such advice can only lead to severe consequences. It can be said that hierarchists accept risk as long as decisions about these are justified by government or by experts. Fatalists are often alienated individuals and take little part in social life, though they feel tied and regulated by social groups they do not belong to. Fatalists are indifferent about risk given that risks are out of their control and try not to know or worry about things they think they can't do anything about.

In addition to the four worldviews described above, one group does not fit this pattern. These people have cut all relations to their social environment and live more or less like "hermits" (Thompson *et al.*, 1990). They are withdrawn from others and oppose all the other worldviews. It is important to note that a continuum exists between the various groups, in that individuals or social organizations are not relegated to one particular category for all purposes and may display different groups characteristics when performing different operations or functions (Skipper, 1998).

Hofstede (1980) also completed a study on culture based upon the assumption that different societies or nations can be demonstrated to display specific cultural tendencies. According to Smith (1995), members of a nation are not all alike, however, they do share a dominant ideology that is built upon pre-existing kinship, religious and belief systems. Therefore, members of a nation can be considered to share a common culture or be members of a particular cultural grouping. The outcome of Hofstede's (1980) research demonstrates that Ireland's national culture is slightly different to that of Britain, therefore, it is expected that Irish society's views of trust and risk shall deviate from British. Hofstede (1980) identified and defined four cultural dimensions that differentiated national cultures based on the results of his study. The first, *power distance*, refers to the measure of interpersonal power or influence between two people or groups as perceived by the less powerful of the two (Hofstede, 1980:98). This refers to how a less powerful entity handles the inequality based on power in a single social system (Hofstede, 1980). Britain had a lower power distance score than Ireland, meaning that the Irish public could try to reduce the power that they perceive the insurers and the Government hold by insisting on a relatively higher involvement in the risk decision-making process regarding the use of genetic test information by life insurers.

The second element, *uncertainty avoidance*, describes the degree to which the members of a society feel threatened by uncertain or unknown situations, such as the risks related to the use of genetic test results in underwriting. Hofstede's (1980) results demonstrate that Britain and Ireland have the same high uncertainty avoidance scores, indicating that both societies are prepared to take risks and are less concerned about restrictions being placed upon possible risk producing events, such as the use of genetic test information in the life insurance industry. The third factor, *individualism*, refers to the degree to which members in a country define themselves by the group or organisation to which they belong, in other words, it describes the relationship between the individual and the community (Hofstede, 1980:214). Hofstede (1980) demonstrates that Ireland was relatively more concerned for the community than Britain. This may influence the need to ensure justice for the community by a willingness to prevent the risks related to the use of genetic test results in underwriting befalling members of Irish society. Finally, with regard to *masculinity*, the final dimension, Ireland was demonstrated to be slightly more masculine than Britain and hence more likely to settle disputes, such as that regarding the management of risks, by means of open debate, involving all concerned. This is not currently the case in Ireland with regard to involvement of the public in the decision-making process surrounding the use of genetic test results by the life insurance industry, which serves as evidence of temporal change.

Although Hofstede (1980) found only minor differences between British and Irish cultures, there still remain some distinctions, each of which may influence societies trust, risk and benefit perception and risk acceptance of the use of genetic test information in life insurance underwriting. Based on Hofstede's (1980) work, it could be considered a mistake for the Irish Government and the Irish insurance industry to make decisions on the use of genetic test results by the life insurance industry based on British views and ignoring Irish opinion. The risk management policies that were in effect in the Republic of Ireland during the course of the primary research of the present study reflected the response of the British Government to insurers using genetic test results. The main difference between both countries' management of these risks is that the British Government organised Peoples Panels and public conferences in order to debate the best methods of dealing with the risks emanating from this usage of genetic tests. The Disability Act, 2005 in Ireland takes a different approach to managing the risks related to insurers using genetic test results than Britain. However, public debate was not pursued in the development of this legislation. Therefore, this Act cannot be considered to reflect the attitudes of the Irish public toward genetic testing. To conclude, following the work of Douglas (1966, 1978), Dake and Wildavsky (1982, 1990), Dake (1991) and Hofstede (1980), national culture may be a condition that influences societal perceptions of trust, risk and risk management and, if public attitudes are reflected in future legislation regarding this situation, national culture may have a role to play in the content of these policies.

Wildavsky and Dake (1990) and Dake (1991) have empirically verified cultural theory as determining risk perceptions, although they did not measure fatalism as part of their studies. Rippl (2002) and Rundmo (1995) tested all four worldviews and found the Fatalist dimension to be important in predicting risk perceptions. However, the majority of recent studies have failed to confirm the findings of Wildavsky and Dake (1990). Marris, *et al.,* 1998, Sjoberg, 1997 and Palmer, 1996 each reported that cultural theory explains only a minor part of the variance in how people perceive risks. No further empirical work has been completed on the basis of Hofstede's (1980) study therefore it is impossible to directly compare the findings of Hofstede's (1980) study with any other.

There are several reasons as to why recent research has indicated that cultural theory may not determine risk perceptions in the manner described above. Society is in a constant state of flux and, given that Douglas's (1978) and Douglas and Wildavsky's (1982) studies were completed decades ago, it is plausible that society may have changed quite dramatically since then. Individuals can exhibit different cultural biases depending on the context and the time (Thompson, 1990). Therefore, cultural theory may be less appropriate or efficient in determining risk perceptions in modern society. For example, Hofstede's (1980) study, which examined the work related values of employees of a specific international organisation between 1967 and 1973, may not reflect social culture in the Republic of Ireland during the current time frame. In comparison to the present day, Ireland was an agriculturally based and economically poor country when Hofstede's (1980) research was conducted.

Another aspect of cultural theory is the nature of the continuum that exists between high and low group and high and low grid. When providing respondents with a questionnaire designed to determine the cultural group to which they belong, it could be questioned which of the cultures determines their answers. If someone is hierarchical at home but individualistic at work, how will this be captured by a questionnaire that presupposes that people adhere to only one culture? In addition, the development of more extensive communication between different cultures may also contribute to less diversity between cultures and perhaps more diversity within the same culture. There may exist individual differences within cultural groups and individual differences among these, making it very difficult to group individuals or, indeed nations, as attempted by Hofstede (1980), into one cultural box. Hofstede's (1980) work is limited to political boundaries however societies cannot always be defined as such. Earlier in this chapter (see Sections 3.3 and 3.6) it was discussed how there may be a relationship between trust and risk perception. Indeed, it has been demonstrated that worldviews have a significant influence on judgements of trust in institutions responsible for or, regulating, gene technology and on the assessment of risks and benefits associated with different technologies (Siegrist, 1999; Peters and Slovic, 1996). However, the level of trust in a society may also affect the degree of hierarchy one finds in a culture. For example, if there is dissatisfaction with a government in a particular society, the level of hierarchy may decrease within that society. Given that Governments change frequently in a democratic society, it is likely that trust in the Government may have also fluctuated causing the cultural grouping of a particular society to change throughout the years. Overall, Oltedal *et al.* (2004) suggest that the concept of a joint cultural theory is naïve and the creation of four groups is wrong and very optimistic.

Given the limitations and criticisms of previous work on cultural theory, it would appear foolish to assume that one can test for a national culture that distinguishes its perceptions of risk and trust attitudes from other countries. However, the cultural categories defined by Douglas (1978), Dake (1991), Douglas and Wildavsky (1982), Wildavsky and Dake (1990) and Hofstede (1980) stand as a significant contribution to cultural research, and it cannot be denied that culture appears to influence risk perceptions to some extent. Therefore, the present study shall assume that culture influences individuals' perceptions of the risks related to insurers using genetic test results. However, it is beyond the scope of the present study to define what this culture is and how it might affect risk perceptions and trust attitudes.

3.7.2 Gender and Economic Status

According to Slovic (2000), sex is significantly related to trust and risk judgements and consequently influences acceptance of risk management strategies applied by various bodies. A number of studies have demonstrated that men tend to judge risks as smaller than women, characterised by higher trust in institutions and authorities (Satterfield *et al.,* 2004; Slovic, 2000; Siegrist, 2000; Gustafson, 1998; Davidson and Freudenburg, 1997; Barke *et al.* 1997; Slovic *et al.,* 1997; Gutteling and Weigman, 1993). Women and persons who are less advantaged economically, tend to report a perception of higher risk, express greater concern about environmental hazards and express a greater distrust in science and Governmental agencies, than do men. Women also tend to assess all risk objects to be more threatening and more unacceptable than men (Wester-Herber and Warg, 2002:70; Finucane *et al.*, 2000; Flynn *et al.*, 1994). In addition, Dosman *et al.* (2001) found gender to be a strong predictor of risk perceptions related to health and food safety issues (Dosman *et al.* 2001:307-318). In a further study, completed by Siegrist *et al.* (2005), it was found that gender is a significant predictor for technological hazards but not for non-technological hazards, with females perceiving more risks than males. It is important to note that not all studies find gender and economic status to be strong predictors of risk perception. Brenot *et al.* (1998:729-740) found that gender and social class were weakly related to *risk perceptions*, while findings from other studies further indicate that gender is not an important explanatory factor of risk perception (Pardo *et al.*, 2002; Cutter *et al.*, 1992).

The explanations as to why women and economically disadvantaged persons have a higher risk perception than men and economically advantaged individuals has tended toward a focus on issues such as differences in technical knowledge, the influence of social and biological 'roles', and power. Finucane *et al.* (2000:19-172) speculate that the above disparity emerges because of power differentials: those with more power (men) and greatest socio-economic advantage are less likely to see the world as dangerous as others. Slovic (2000) suggests that economically advantaged males view less risk in society because they create, manage, control, and benefit more from many of the modern technologies and activities. If this is the case, it may be that such individuals feel that they have more influence over the use of such technologies and decision-making processes regarding the management of risks and, hence, their trust in those managing the risks will be higher. A study completed by Bord and O'Connor (1997:832) suggests "women consistently exhibit much stronger perceptions of vulnerability to illness and physical debilitation". This finding may influence the results in the present study given the health related possibilities of genetic testing. Siegrist (2000) calls for more studies to be completed on the influence of gender on risk perception and the acceptance of technologies, in order to increase an understanding of the causes of such or any differences (Siegrist, 2000:202). The present study shall analyse if gender and economic status are factors that influence knowledge, trust, risk perceptions, benefit perceptions and risk acceptance with regard to the use of genetic test results by the life insurance industry. An increased comprehension of the influential nature of the above factors could facilitate a better understanding of how power may influence risk acceptance.

3.7.3 Age

It is questionable whether age plays a role in influencing an individual's knowledge, social trust, risk perceptions, benefit perception, and risk acceptance. In a study completed by Siegrist *et al.* (2005), it is reported that age is positively correlated with risk perception. In a study on food related health risks, it was also found that age is a significant predictor of risk perceptions (Dosman *et al.*, 2001:307). In addition, age is found to influence the extent to which individuals are willing to allocate present resources to alleviate a future problem (Nielsen *et al.* 2003:39). Specifically, people above the age of 50 appear more reluctant to pay for an intervention against a future potential health threat. Age may also be related to an interest in purchasing life insurance, a factor that will be explored in the following section. Given the above research, this study shall question whether or not age is a factor that may influence risk acceptance, and in doing so knowledge, trust, risk perceptions and benefit perceptions. If this is the case, it may be that age is a factor that affects the influential pattern of trust and risk acceptance.

3.7.4 Irish Economy and Life Insurance Ownership

The economic state of the country in which the participants of the study inhabit may be considered a condition that could affect life insurance ownership and risk acceptance of insurers using genetic test results. The Republic of Ireland is a small, modern, trade-dependent economy with real growth averaging a relatively high 6.1 per cent over the period 1994-2002 (Haughton, 2005). Although growth has slowed slightly since 2002, it still remains far higher than any of Ireland's EU counterparts. Agriculture, once the most profitable sector, is now overshadowed by industry and services. Although exports are essential for Ireland's growth, the economy has also gained from an increase in consumer spending, construction activity, and business investment. This growth in consumer spending is reflected by a rise in premium income for the life insurance industry in recent years (Irish Insurance Federation Factfile, 2005). Over the past decade, the Government has implemented a series of national economic programs designed to restrain inflation, lower Government spending, augment labour force skills and encourage foreign investment. Campaigns encouraging consumers to 'shop around' for all types of products, including those distributed by the life insurance industry, are expected to cause increased awareness of the differences in prices for relatively similar products (IFSRA, 2004a). This price difference consciousness may amplify competition between life insurers in order to attract a price and value attentive customer base. This may or may not affect the use of genetic test results by the life insurance industry. It is possible that insurers wishing to attract customers in the short term may avoid demanding genetic test results, in order to appear as a favourable choice for those who have or, intend to obtain, a test for their genetic susceptibility to disease.[7] However, it is quite likely that if this does occur, it will only be a policy enacted for the short term, as developments in genetic testing improve the accuracy of results and the range of diseases detectable.

The population growth rate in the Republic of Ireland is a steady 1.16 per cent per annum (Central Statistics Office (CSO), www.cso.ie), therefore, there will be a continual need for life insurance products to financially protect families. Approximately 67.5 per cent (www.cso.ie) of the population are within the age group that is most likely to be interested in or who have purchased life insurance (aged 15-64 years). The average life expectancy for males is 74.1 years of age and 79.8 years of age for females (Haughton, 2005). This could be very simply interpreted as there being a greater need for life insurance covering ill health during working years rather than death. With increasing life expectancy, it is possible that there will be a rise in demand for life insurance products providing financial protection in times of sickness. This could certainly affect the use of genetic test results by the life insurance industry, as people become more concerned about the diagnosis and/or prognosis of genetically inherited diseases that may lead to a present or future claim. If this results in an increase in numbers taking genetic tests, there will be a greater need for life insurers to be aware of the results of these tests.

It is plausible that individuals who own life insurance products or, who are interested in purchasing a life insurance product(s) in the future, may respond differently to questions regarding knowledge, social trust, risk perceptions, benefit perceptions and risk acceptance than those individuals who do not have life insurance and do not wish to ever purchase any life insurance products in the future. This may be due to an increased interest in, or a perceived need for, life insurance products and hence, concern regarding the effects that genetic testing may have upon the price or the likelihood of being able to obtain such products. This theory shall be tested and, if differences are observed, further analysis shall be pursued in order to better understand the reasoning behind such differences.

---

[7] Examples of insurance companies who do not use genetic test results to determine life insurance premiums in the UK -Britannic Assurance (asks for information but doesn't use it for life insurance), Equitable Life, Hiscox, Ideal Benefit, London Life Ltd., National Mutual Life, Pinnacle, Standard Life, Virgin

3.7.5 Power and Decision-Making

As discussed in Sections 2.5 and 2.6, disputes between experts in areas where the risks are not fully known provoke major difficulties for decision-making and policy implementation. Traditionally, the neutral, disinterested and objective expert has been promoted as the rational and authoritative arbiter of disputes over scientific or technical issues. However, this ideal whereby laypersons interpret objective facts and decisions based upon experts' advice has been eroded. The issues surrounding insurers using genetic test results demonstrate the increasingly obvious limitations of expert knowledge in resolving issues of public controversy. In this case, it can be noted that experts disagree as to whether or not insurers should be allowed to use genetic test results, and where this is the case it is quite likely that one or all of the so-called experts are not infallible. Indeed, the objectivity of experts can also be questioned, whereby their purportedly "disinterested" advice may be influenced by professional, economic, or political considerations (Martin and Richards, 1995). Given such confusion among experts, the question must arise as to who should be making the decisions with regard to policy formulation regarding this matter.

Politics is the process by which groups make decisions. In the Republic of Ireland, as in most democratic countries, it is the Government which has the final authority or power to determine and to judge public risk situations and to enforce laws based upon decisions made regarding such evaluations. However, according to Bachrach and Baratz (1970) this formal political arena can be influenced by behind the scenes agenda-'setting' by elite groups who could be either politicans, industrialists or special interest groups. Each of these groups may have their own hidden agenda that the public may not be aware of, but this agenda may filter through to the ultimate decision made by the Government. It can be said that the Government possesses legitimate power to manage the risks related to insurers using genetic test results, with this power based upon its relative position and duties to protect society. At the same time, one could refer to the insurance industry and special interest groups as possessing expert power derived from skills and expertise that can be considered very specific and limited to the particular area in which they are trained or qualified (French and Raven, 1959). In this case, the Government has bestowed authority regarding the management of genetic test results by life insurance companies to the IFSRA, given this agency's charter to manage insurance related issues (see Section 2.5.3). The IFSRA was given the legitimate power to make decisions regarding this matter, while the insurance industry and other interest groups had the expert power to advise the IFSRA on their decision.

Section 2.5 outlined the manner in which the insurance industry's use of genetic test results was managed during the time period within which this study was completed. The insurance industry relied on its own Code of Practice on the use of genetic test results during the primary research stage. After the Irish Government published a Disability Bill in 2001, the Irish Insurance Federation remained the only interest group within Ireland to submit an analysis of the Disability Bill to the Irish Government (IIF Submission on the Disability Bill 2004 to the Joint Committee on Justice, Equality, Defence and Women's Rights). On saying that, however, the Irish Government did appear to take on board the advice of 'outside' bodies such as the World Medical Association, UNESCO and the Council of Europe in that they introduced the Disability Act in 2005 which withdrew the rights of insurers to access applicants' genetic test results.

This presents a rather interesting scenario. It can be observed that the advice of European institutions appears to take precedent over the wishes of industry. According to the declarations from these European bodies, the reasoning behind denying insurers access to genetic test results is to protect the citizens of Europe from genetic discrimination. Therefore, on the outside it appears that the well being of society lies at the backbone of Governmental legislation, indicating some power on the part of citizens in determining risk related regulation. Looking at this from another angle it could be said that society appears to have the ultimate power in influencing Governmental decisions as opposed to industry. However, this conclusion could be considered rather simplistic, given previous reports on private industry influencing Governmental decisions despite citizen requests to the contrary (see Hogan, 2001). It is also possible for the state to influence its citizens, in that if the Government wished to allow insurers to use genetic test results, the positive financial reasons as to why insurers should be permitted such usage could be emphasised in media commentaries. Media reporting is demonstrated to influence risk perception (Kone and Mullet, 1994). It will be interesting to note in the present study, who respondents believe to have power in influencing decisions regarding insurers' use of genetic test results.

3.8 Summary

This chapter examined several factors and conditions that may generate risk acceptance of insurers using genetic test results. Previous literature indicates a relationship between incomplete knowledge, trust, risk perception, benefit perception and risk acceptance, in that each of the factors either directly or indirectly generate risk acceptance. The connection between the above factors may be seen as a relatively recent occurrence, given that trust in institutions may be a uniquely modern phenomenon. In modernity, the importance of personal interactions is declining as a result of this form of exchange becoming less practical. Without personal communication, it becomes necessary for people to develop trust in institutions that determine many of the conditions of modern life (Giddens, 1990; Lang and Hallman, 2005). Trust can be seen as moving from an unexamined and habitual confidence in science and technology to a trust that is given by autonomous, reflexive individuals to governments and organisations that are seen as having earned their trust (Giddens, 1990; Beck, 1992; Lang and Hallman, 2005). According to Giddens (1990), trust in modernity is fundamental to the emergence and prominence of organisations and institutions in daily life. In addition, trust aids society by enabling individuals to tolerate increasing uncertainty due to progressive technological complexity (Luhmann, 1979; Barber, 1983). The theorised nature of the relationship between knowledge, trust, risk perception, benefit perception and risk acceptance appears to indicate that for the incalculable risks related to insurers using genetic test results to be accepted by society, trust needs to be bestowed upon those managing the risks related to such usage. Although trust does not necessarily eliminate the risks related to insurers using genetic test results, it appears to be an important device in mitigating the perceived risk of unpredictable and harmful consequences resulting from such usage (Lang and Hallman, 2005).

Moreover, national culture, gender, economic status, age and life insurance ownership and power in decision-making were revealed as factors and conditions that may each affect the biographical background of the individuals that make up a society and hence may influence the factors that generate risk acceptance. Previous research conducted on several of these factors indicates that they can directly influence trust in risk managers and risk perceptions, therefore, feeding into a trust – risk acceptance pattern. The reasoning for these factors to act as generative mechanisms in risk related decision-making lies deep within issues of power and perceived need. The result of such an interaction of factors and their underlying dynamics may influence individuals to accept, or to not accept, life insurers using genetic test results.

The intention of the following chapter is to describe the methodology used to explore the trust-risk acceptance pattern and the factors and conditions that underlie the influential nature of this relationship. The philosophical approach of the researcher shall be explained and its influence upon the study and its associated methodology shall be illustrated. The instruments used to gauge society's perceptions of risk and trust shall also be described and the limitations of the choices made explained.

CHAPTER FOUR

RESEARCH METHODOLOGY

4.1 Introduction

Chapters 2 and 3 explained the background and the social world surrounding the objectives of the present study. The extant literature pertaining to risk and trust was presented, and the influence that the nature of genetic test results and the life insurance system have upon the risk management of this situation was analysed. The discussion highlighted the need for a greater understanding of the influence that trust has on risk perception, and the need to consider their effect upon acceptance of life insurers' use of genetic test results. To fulfil the aims of the research, it is necessary to decide upon the research approach. The intention of this chapter is to justify and explain the research design of the present study. In doing so, a complete description of the data collection process shall be provided (Fowler, 1988:141). An introduction to methodology was provided in Section 1.4 and the present chapter aims to build on that introduction.

This chapter begins by evaluating the theoretical considerations and methodological approaches influencing the choice of research design. Among the theoretical decisions are those relating to the researcher's own philosophical assumptions. Carson *et al.* (2001) outline the purpose of considering a philosophy of research as a way of underpinning the choices and decisions involved in staking a research position, thus providing a clearer purpose to specific research projects within the wider context. Saunders, *et al.* (2003) illustrate this concept through the example of an onion, whereby the outer layer, representing the research philosophy, binds all other layers together and dictates the choices made in each layer (see Appendix B.1). Decisions made in each further segment after that, in turn influence choices made in the following layers, until the core, representing the data collection method(s), is reached.

In order to choose an appropriate data collection method, the philosophical approach of the researcher must firstly be considered in light of the research issues. Before committing to research methodology, Guba and Lincoln (1994:105-117) recommend that consideration be given to the question of paradigm. "Questions of method are secondary to questions of paradigm, which we define as the basic belief system or worldview that guides the investigator, not only in choices of method but in ontologically and epistemologically fundamental ways" (Guba and Lincoln, 1994:106). Paradigm is comprised of three elements: epistemology, ontology, and methodology. It is considered that each of these elements is important to this research; therefore, an examination of the ontological and epistemological issues that affect the current study shall be discussed. This discussion is followed by an analysis of the research paradigm adopted in the present study and the methodological techniques applied. In addition, the development and administration of the research instruments shall be described and reviewed.

4.2 Research Philosophy

When examining philosophical assumptions, ontology describes the reality which the researcher holds to exist, whilst epistemology is concerned with how knowledge can be held about the world or, indeed, reality, and affects the process through which we develop what we take to be warranted knowledge of the world (Perry *et al.*, 1996:547; Guba and Lincoln, 1994; Johnson and Duberly, 2000:7). In simple terms, epistemology is the philosophy of knowledge or how we come to know (Trochim, 2000). To discover information about reality, the researcher must adopt a methodological technique. There is, therefore, an implied relationship between the researcher, reality, knowledge about reality, and the methodology. Epistemology is intimately related to ontology and methodology; as ontology involves the philosophy of reality, epistemology addresses how we come to know that reality, while methodology identifies the particular practices used to attain knowledge of it. Perry *et al.* (1997) and Guba and Lincoln (1994) indicate how a researcher's basic belief system, in other words, their adherence to a particular paradigm, can influence the methodology chosen (see Table 4-1). A research paradigm has been defined as "a loose collection of logically held together assumptions, concepts, and propositions that orientates thinking and research" (Bogdan and Biklan, 1982:30). A theoretical paradigm is the underlying basis that is used to construct a scientific investigation. To understand how a researcher's paradigm can influence the methodology chosen, the following sections shall examine the main paradigms or worldviews that guide the social science investigator. However, it is important to note that except for positivism, the paradigms discussed are still in their formative stages, in that there remains disagreement among the proponents regarding their definitions, meanings and their associated implications for research.[1]

Table 4-1 Paradigm Ontology, Epistemology and Methodology

PARADIGM			
	Positivism	**Realism**	**Constructivism**
Ontology	Naïve realism: reality is real and apprehensible	Critical realism: reality is "real" but only imperfectly and probabilistically apprehensible and so triangulation from many sources is required	Critical relativism: multiple local and specific "constructed" realities[2]
Epistemology	Objectivist: findings true	Modified objectivist: findings probably true	Subjectivist: created findings

[1] An example of the manner in which the definitions, meanings and implications of paradigms other than positivism cannot be considered collective are the divergent understandings of constructivism (see Section 4.2.2)

[2] It is important to note that this is not a collective agreement as to the ontological position of constructivism (see Section 4.2.2)

Methodology	Experiments/surveys: verification of hypotheses: chiefly quantitative methods	Case studies / convergent interviewing: triangulation, interpretation of research issues by qualitative and quantitative methods such as structural equation modelling	Hermeneutical / dialectical: researcher is a "passionate participant" within the world being investigated

Based on Perry, Alizadeh and Riege, (1997:547) and Guba and Lincoln (1994)

4.2.1 Positivist Paradigm

A central issue in this extensive debate is the question of whether the social world can and should be studied according to the same principles, procedures, and ethos as the natural sciences (Bryman, 2001:11). Adoption of a similar philosophical stance as that of the natural sciences reflects the epistemological stance of positivism. Knowledge is considered to be discovered and verified through direct observations or measurements of phenomena, in that facts are established by taking apart a phenomenon to examine its component parts. Smith (1998) provides a useful insight into positivist thinking within social sciences, describing positivist approaches to the social sciences as assuming that things can be studied as hard facts and the relationship between these facts can be established as scientific laws. The world and the universe are considered to be deterministic, in that they operate by laws of cause-and-effect that are discernable through scientific method. According to Smith (1998), for positivists, such laws have the status of truth, and social objects can be studied in much the same way as natural objects. Positivist followers believe that all real knowledge should be derived from human observation of objective reality. The senses are used to accumulate data that are objective, discernible and measurable; anything other should be rejected as transcendental (Comte, 1853). Thus, science is largely a mechanistic or mechanical affair in positivism.

Deductive reasoning is used to postulate theories that can be tested. Research findings may indicate that a theory does not fit facts determined from previous studies well; hence, the theory must be revised to better predict reality. Positivists believe in empiricism, the idea that observation and measurement are at the core of the scientific endeavour. The key approach of the scientific method is the experiment, the attempt to discern natural laws through direct manipulation and observation (Trochim, 2000).

Kolakowski (1972) asserts that words, generalisations and abstractions are merely linguistic phenomena and do not give new insight into the world. The positivists' antipathy to metaphysics within scientific enquiry is well illustrated by David Hume who states that if we take in our hand any volume "of divinity or school metaphysics, for instance; let us ask, does it contain any abstract reasoning concerning quality or number? No. Does it contain any experimental reasoning concerning matter of fact and existence? No. Commit it then to the flames: for it can contain nothing but sophistry and illusion" (Hume, 1748/ 1984).

The importance of induction and verification, and the establishment of laws, are stressed by logical positivists and in this respect differ from the earlier tradition of positivism proposed by Comte (1853). Most interpretations still acknowledge positivism's conviction toward an observable social reality, with the study of such resulting in context free law-like generalisations, similar to those produced by the natural and physical sciences (Saunders, *et al.*, 2003:83, Guba and Lincoln, 1994:109). This objective reality exists independent of human behaviour and is, therefore, not a creation of the human mind. Positivists assume that the researcher is independent of, and neither affects nor is affected by, the subject of the research and, thereby, introduces no bias to the research as such (Remenyi *et al.*, 1998:33). This belief that reality is stable and can be observed and described from an objective viewpoint without interfering with the phenomena being studied, directs one towards the ontological position associated with a positivistic stance, that is objectivism (Levin, 1988). Objectivism affirms the notion that social phenomena and their meanings have an existence that is independent of social actors or their influence (Bryman, 2001:17). According to this paradigm, it is possible to converge on the 'true' state of affairs using objective research methods (Guba and Lincoln, 1994:109). By adopting prescribed procedures, it is possible to prevent values and bias from influencing outcomes. Replicable findings are considered as true; in other words, if findings are replicated it is understood that the findings mirror the truth about reality and can be considered as natural laws (Guba and Lincoln, 1994:110).

The general elements of positivist philosophy have a number of implications for social research based on this approach. Easterby-Smith *et al.* (1997) and Hughes (1994) discuss several of these implications as follows. With regard to the methodology, propositions in the form of questions and/or hypotheses are subjected to empirical testing in order to verify them. All research should be quantitative; indeed, from a positivist view only research which is quantitative can act as the basis for valid generalisations and laws (Easterby-Smith *et al.,* 1997; Hughes, 1994). This in turn implies that concepts need to be operationalised in a way that enables facts to be measured quantitatively. According to the positivist paradigm, problems are better understood if they are reduced to the simplest possible elements (Easterby-Smith *et al.,* 1997; Hughes, 1994). Positivist proponents assume that a state of independence exists in positivistic studies, in that the role of the researcher is independent of the subject under examination. That is, researchers view the world through a "one-way mirror" (Healy and Perry, 2000). With regards to value freedom, the choice of what to study, and how to study it, should be determined by objective criteria rather than by human beliefs and interests (Easterby-Smith *et al.,* 1997; Hughes, 1994). The aim of research should be to identify causal explanations and fundamental laws that explain human behaviour (Bond, 1989; Easterby-Smith *et al.,* 1997; Hughes, 1994). Causality is identified in positivist research as a series of regularities that occur when examining certain aspects of a phenomenon. If there appears to be continuous replications of these findings, the results of such studies are then considered to indicate universal causality regarding that particular phenomenon.

There are a number of disagreements surrounding this philosophical position. A major criticism of the positivist approach is that it does not provide the means to examine human beings and their behaviours in an in-depth way. Ayer (1969) questions the use of positivist and empirical approaches to the study of human behaviour, and suggests that it may be something about the 'nature of man' that makes impossible the establishment of laws and ability. Humans are not 'objects', and are subject to many influences on behaviour, feelings, perceptions, and attitudes that positivists reject as irrelevant and belonging to the realm of metaphysics. That is not to say that a positivist approach is not useful, but given such restrictions of view, it can be considered as providing limited data, in that it appears to only provide a superficial view of the phenomenon it investigates. Positivism's view of causality is also sceptical, in that it is assumed possible to make universal predictions by identifying sets of empirical regularities within a local environment. In physics, it is possible to formulate laws relating to the expansion of metal when heated. From such laws, the amount of expansion that will occur in particular circumstances can be predicted. However, with regards to the study of society, given the amount of intervening subjective forces that may affect the nature of the event under study, universal causality based upon a series of local and objective regularities cannot be assumed.

Another criticism of the positivist approach is the assumption of the ability to separate the researcher form the object of research. However, there is evidence to suggest that this cannot be the case, regardless of the quantitative techniques adopted. Guba and Lincoln (1985) quote the Heisenberg uncertainty principle[3] and the Bohr complementarity principle[4] as evidence that it is impossible for the inquirer not to influence the phenomena under investigation. Human experience includes a process of interpretation, in addition to sensory reception and apprehension of the external, material world. In connection with this criticism, the strive for value-free or value-neutral knowledge to arrive at 'true' knowledge is also suspect. It is considered that having feelings or preconceptions regarding a particular phenomenon or state of affairs might distort a researcher's beliefs and findings about what is the case (Sayer, 2000:58). It is not always desirable to be value-free, although it is advisable to interrogate one's research to ensure that wishful thinking is not engaged and to make sure that it is capable of acknowledging possible circumstances of which one may disapprove. Indeed, values may even have a positive effect by directing a researcher to issues and aspects of situations that others miss, for example, in the case of feminism and gender. Sometimes attempts at value-freedom may result in blindness to values that are present in a particular social phenomenon, and hence blur the 'truth' that is so sought

[3] The Heisenberg uncertainty principle is frequently, but incorrectly, confused with the "observer effect", as it relates precision in measurements related to changes in velocity and position of certain particles relative to the perspective the observer takes on them. In the natural sciences, the observer effect refers to changes that the act of observing has on the phenomenon being observed. For example: observing an electron will change its path because the observing light or radiation contains enough energy to disturb it. It is likely that Guba and Lincoln (1985) mistakenly used the Heisenberg uncertainty principle to illustrate the observer effect in this case.

[4] According to the Bohr Complementarity principle there is no quantum world. There is only an abstract physical description. It is wrong to think that the task of physics is to find out how nature *is*. Physics concerns what we can *say* about nature (Pertersen, 1963)

after by positivists (Sayer, 2000:59-59). Finally, it could be argued that positivist theories resulting from quantitative research are themselves theories and, hence, positivist research is value-laden rather than value-free (Guba and Lincoln, 1994:105).

Given such criticisms, positivism may be seen as highly questionable as a philosophical basis for research. However, the philosophical underpinnings of positivism continue to pervade most theory and research in the social sciences. Within the history of risk analysis, positivist approaches derived from science have tended to dominate much research in this field (for example Starr, 1969; Okrent, 1975; Lowrance, 1976; Rowe, 1977; Inhaber, 1979; Wilson and Crouch, 1982; Cohen, 1983; Whipple, 1992) (Rosa, 1998:19). According to Morgan (1995), as a result of positivist quantitative methods being the strongest aspect of risk analysis, there are data gaps that exist in the results of research findings that other methods could fulfil. With regard to the present study, the researcher does not take a positivistic approach. Indeed, it could be said that positivism is inappropriate for the study of trust, perceptions and risk acceptance, since it ignores the 'human ability to reflect on problem situations and to act on these' in an interdependent way (Robson, 1993:60). Positivism does not allow for the study of phenomena discerned by the higher senses; it is, therefore, not possible to study evaluations of trust, or perceptions of risk, or risk acceptance from an entirely positivistic stance. In addition, given that there are various understandings of trust (see Section 3.4), it is not possible to articulate one definition of trust which everybody understands and adheres to in order to study trust in the authorities governing insurers' use of genetic test results, a requirement of research from a positivistic stance. The positivist orientation is, thus, reductionistic, in that it seeks to reduce risk to a pure scientific reality where the only world is a real world and that is where risk resides (Shrader-Frechette, 1991; Rosa, 1998:20). Moreover, an understanding of risk as a neutral product of science, devoid of values, bias, ethics, or sociological shaping is questionable, given the influence that these values have been determined as having upon risk perceptions (see Section 2.4.1).

In addition to the above reasons, positivism was not the accepted philosophy in this study as its necessity for value-neutrality or to be value-free was not considered possible. It is argued above how value-neutrality may be justiable as an ideal in pure science, despite the fact that it is difficult to find it in practice (as pointed out by Kuhn (1970 [1962]; Rosa, 1998:20). According to Rosa (1998:20), risk analysis is seldom conducted with pure science as its singular goal; indeed, risk analysis's principal *raison d'être* is to inform policy and other public choices. Risk analysis can be considered as the bridge between theory (what is) and public philosophy (what ought to be). Thus, to claim it is value-neutral is to make the unfounded claim that it concerns itself with facts and theory alone (Rosa, 1998:20). Given the objective of this research in informing policy regarding the risks perceived and accepted by society regarding insurers' use of genetic test results it is impossible to pursue the positivist assumption of value-neutrality.

Finally, positivism insists on the development of universal law-like generalisations from research; however, it is impossible to make such rules that define the social world, given the state of constant flux which is the social reality surrounding the phenomena under study. According to Popper (1972:1-31), it is impossible to be certain that an explanatory statement about one set of conditions will work just as well elsewhere. Given the assumptions of positivism, it can be argued that a positivistic paradigm cannot be applied to social research.

4.2.2 Constructivist Paradigm

"Epistemological alternatives to positivism define themselves through reference to, criticism of and, controversy with, what they take to be positivist assumptions" (Johnson & Duberly, 2000:12). Constructivism, also known as 'social constructivism' and 'social constructionism', is at the opposite side of the scales to positivism, in that constructivists adopt a subjective view of social reality and are consequently critical of the application of the scientific model to the study of the social world. Ontologically, constructivist assumptions of reality differ from positivists, in that reality is considered to be subjective and mentally constructed by individuals. Since each individual experiences reality from their own point of view, each person experiences a different reality. As such, the phenomenon of "multiple realities" exists. Verifying this perspective are Guba and Lincoln (1994:111), who acknowledge that the ontology of constructivism is based upon a belief that realities are apprehendable in the form of multiple, intangible mental constructions that are socially and experientially based, local and specific in nature, and dependent on the individual persons or groups holding the constructions. This perspective implies that social phenomena are produced through social interaction and are consequently in a constant state of flux (Bryman, 2001:18). Constructions are alterable, as are their associated 'realities' in that they may change as their constructors become more informed and sophisticated (Guba and Lincoln, 1994:111).

Constructivism views knowledge as constructed, in that it is established through the meanings attached to the phenomena studied, and it does not necessarily reflect any external transcendent realities. Knowledge is viewed as contingent on conventions, human perceptions and social experience. While positivists separate themselves from the world they study, constructivists acknowledge that they have to participate in real-world life to some extent so as to better understand and express its emergent properties and features (Healy and Perry, 2000). Indeed, the constructivist paradigm is supportive of the position that the knower and the known are co-created during the inquiry (Al Zeera, 2001). Researchers interact with the subjects of study to obtain data and such inquiries change both researcher and subject. Given that a researcher is a unique individual with their own view of reality, all research is essentially biased by the researcher's individual perceptions. Constructivists are opposed to positivist methods that attempt to aggregate across individuals. Given that each individual is unique their particular perspective of reality is also unique. Therefore a completely common view of reality is unlikely. Constructivists operate under different epistemological assumptions from positivist researchers. For instance, many constructivists believe that the best way to understand any phenomenon is to view it in its context. They see all quantification as limited in nature, looking only at one small portion of a reality that cannot be split or unitised without losing the importance of the whole phenomenon. For many constructivist researchers, the best way to understand what is going on is to become immersed in it and to move into the culture or organisation being studied and experience what it is like to be a part of it. Rather than approaching measurement with the idea of constructing a fixed instrument or set of questions, constructivist researchers choose to use qualitative analyses that allow questions to emerge and change as one becomes familiar with the study content (Krauss, 1995:759-760).

Constructivists are of the view that knowledge is very much context and time dependent; hence a universal law based upon quantitative data is not considered viable (Coll and Chapman, 2000; Cousins, 2002). It is considered that constructions are not more or less 'true' in any absolute sense, but simply more or less informed and/or sophisticated. Constructions are alterable, as are their associated 'realities' (Guba and Lincoln, 1994:108). Guba and Lincoln (1994: 111) define a constructivist epistemology as one in which the investigator and the object of investigation are assumed to be interactively linked so that the 'findings' are literally created as the investigation proceeds and, in doing so, the line between ontology and epistemology is blurred somewhat. Schwandt (1994:118-137) agrees with this view, stating that constructivism means that human beings do not find or discover knowledge so much as construct or make it.

The goal of constructivist research is to offer meaning or an interpretation that challenges, provokes, or encourages further questions rather than one that provides definitive explanations. Constructivists assume that there are many possible interpretations of the same data, all of which are potentially meaningful. In an effort to construct meaning, people impose order on the world perceived, where meaning lies in cognition and not in elements external to the individual (Lythcott and Duschl, 1990). Information impinging on an individual's cognitive system is screened, translated, altered, and perhaps rejected by the knowledge that already exists in that system; the resulting knowledge is idiosyncratic and is purposefully constructed (Lythcott and Duschl, 1990).

There are many varieties of constructivism with differing areas of commonality and divergence (Neimeyer and Raskin, 2001). However, there are two basic conflicting streams of thought regarding constructivism. The core argument of the first alternative viewpoint is that of ontological realism, whereby it is considered that the world, although unknowable, exists apart from our thinking about it (Matthews, 1992). This view of constructivist-thinking shares an (ontological) 'real world' assumption with positivism (Becker and Niehaves, 2006). Weber (2004) argues that although realist constructivism assumes that a real world exists beyond the realms of human cognition, unlike positivism, this definition of constructivism assumes that it is impossible to achieve objective knowledge about this real world. From this perspective of constructivism, it is understood that knowledge is always determined by the subject and, thus, no such thing as objective knowledge exists (see Table 4-2).

The second and more traditional view of constructivism, radical constructivism, drifts further away from positivistic notions of ontological realism, whereby reality is considered as being independent of the observer. Instead, it follows the hermeneutic and phenomenological tradition where such observer-independence of reality is not usually accepted (Moran 2000; Varey *et al.* 2002) (see Table 4-2). According to this view, the object of research is considered to exist only within the human mind, rather than having qualities that exist independently of the mind. Such radical constructivists are often termed as idealists and, according to Ernst von Glasersfeld (1989:104), radical constructivism is perhaps the best known and the most widely discussed form of constructivist thinking. He succinctly compares the two views of constructivist-thinking, whereby the realist constructivist "believes his constructs to be a replica or reflection of independently existing structures, while the (radical) constructivist remains aware of the experiencer's role as originator of all structures…for the (radical) constructivist, there are no structures other than those which the knower constitutes by his very own activity of coordination of experiential particles" (Von Glasersfeld, 1989:104). However, given their divergence of ontological thought, both perspectives of constructivism share the opinion that objective cognition of reality is impossible (see Table 4-2).

Table 4-2 Epistemological Differences Between Two Polar Views of Constructivism and Positivism

		Epistemological Position	
		Objective Cognition is impossible (Constructivism)	**Objective cognition is possible (epistemological realism)**
Ontological Position	**A real world exists (ontological realism)**	Realist Constructivism	Positivism
	No real world exists (ontological idealism)	Radical Constructivism	

Adapted from Niehaves and Stahl (2006)

According to Niehaves and Stahl (2006), realist constructivism and positivism do not lead to greatly differing paradigmatic views on social phenomena; however, their means of analysing such phenomena differ. In this sense, constructivism and positivism do not inevitably lead to contradictory theories of social phenomena and are, thus, not incommensurable in the sense of Kuhn (1962).[5] Such an interpretation would explain the possibility of successfully combining positivist and constructivist approaches in the form of critical realism (Trauth and Jessup 2000). In the second view of constructivism, radical constructivism, it is difficult to reconcile the two paradigms if it is assumed that the fundamental difference between positivism and constructivism is the ontological question of reality, or our knowledge thereof (Darke *et al.*, 1998; Miranda and Saunders, 2003). Given the understanding of reality as a social construction, the second and most popular view of constructivism ensures that the epistemology and methodology of constructivism and positivism differ vastly. Comparing the two polar views of constructivism, it can be said that both have methodological implications that differ from each other in accordance to whether or not reality is considered independent of the observer. This difference extends to the role of the researcher, who, in the first view, is an external observer to a certain extent, while in the second view the researcher is considered involved in the situation whereby reality and the individual who observes it are considered inseparable (Introna, 1997; Trauth, 2001). Indeed, the understanding of the meaning of research results also differs between the two paradigms. The criteria of truth and, thus, of good research, are fundamentally different (Niehaves and Stahl, 2006). Constructions are not more or less 'true', but more or less sophisticated and/or informed (Guba and Lincoln, 1994:111).

A common constructivist approach to risk theory is that of cultural theory, which posits that risks are the dangers that society define as troublesome (Douglas and Wildavsky, 1982; Krimsky and Golding, 1992:xv) (see Section 3.7.1 for review of cultural theory). The identification of risk is considered to be an entirely social process, whereby risks do not exist in objective reality, but in the collective consciousness of cultures. Risk is

[5] Kuhn (1962) argues that rival paradigms are incommensurable, in that it is not possible to understand one paradigm through the conceptual framework and terminology of another rival paradigm. For many critics, this thesis appears to entail that theory choice is fundamentally irrational. If rival theories cannot be directly compared, then one cannot make a rational choice as to which one is better.

thus perceived from a radical, social constructivist point of view as a cultural phenomenon, not a physical one. Our perceptions of risk, and our choices of which risks to be concerned about, are equivalent to risk itself (Hilgartner, 1992). Individuals and researchers view the world through lenses mediated by experiences in the world, values, biases and social forces and, in doing so, man is continually negotiating the world's meaning. The social construction of risk is described by Wynne (1992:295) as an understanding that the "definitions of risk, and knowledge, and responses to information and uncertainty are based ultimately on the attempted maintenance of familiar social identities . . . Physical risks thus have to be recognised as embedded within and shaped by social relations and the continual negotiation of our social identities'. This fresh look upon risk perceptions provides risk analysis research with new insights into the role of social forces in individuals' perceptions of risk. Indeed, the present study acknowledges the constructivist account that knowledge is socially constructed and that the findings of research are context and time dependent. However, constructivist approaches, including cultural theory, are open to much debate (see Section 3.7.1 for several criticisms of the cultural approach to risk perception).

There does not appear to be numerous internal methodological flaws in constructivism; however, the power of social constructivist methods at the micro-level of sociological analyses to explain the dynamics of risk acceptance should be questioned. Constructivist methods need to be supplemented with macro-level analyses, or analyses that involve reference to non-social factors. The radical constructivist paradigm does not acknowledge the 'real' economic and technological dimensions of business, such as those that encompass the life insurance industry (Hunt, 1991). According to Popper (1963), these abstract things born of peoples minds exist within an external reality, in that they are not affected by the perceptions of any one individual (Magee, 1985: 61). This separate reality is not socially constructed, but to study it requires the researchers' observation and interpretation. However, followers of radical constructivism do not believe in a separate reality and, therefore, do not acknowledge this duality. Possible external structures of conflict in a society are disregarded, and hence, possible sources of social change are ignored to a large extent.

Proponents of realist constructivism believe that an external reality exists; however, an understanding of this reality is only available through the understandings and meanings of individuals. The present study accepts that the world (for example, economy and society) can only be known in terms of available descriptions or discourses. However, the view that one cannot judge between discourses, and decide that some accounts are better than others is not agreed with. The constructivist difficulty in deciding upon conditions of falsifiability is considered a major flaw in the paradigm by Popper (1963), who argues that irrefutability is not a virtue of a theory, hence constructivist studies are incapable of producing theory. A common defence of judgemental relativism invokes the theory-laden character of observation – as if this meant that theories could not be falsified by observations, or that observation is necessarily *purely* theory-determined. It is hard to see why observations would be necessary if our concepts already specified everything about what could be observed (Sayer, 2000).

As noted by Rayner (1992:98), the claim that all views of the world are equally valid, is one of the criticisms of cultural theory. Indeed, Giddens (1978:251) extends this observation to the constructivist paradigm in general. Constructivism is correct in stressing the inventive, humane, culturally and historically dependent aspects of creating the *theoretical* objects of science — but none of this bears, of itself, upon truth. According to Von Glasersfeld (1989), the constructivist assumption of correspondence as a mark of truth should be rejected because we cannot see reality, we only have our sensations to reflect upon, and so we are never able to judge correspondence between our ideas and the world. And given that constructions are alterable, their associated 'realities' are subject to change as constructors become more informed and sophisticated (Guba and Lincoln, 1994:111). Therefore, according to all streams of the constructivist view, the criterion for progress is that, over time, everyone formulates more informed and sophisticated constructions and become more aware of the content and meaning of competing constructions (Guba and Lincoln, 1998:113). This feature of constructivism leads (Winner, 1991) to argue that the findings from constructivist research do not take evaluative stances or invoke moral or political principles; indeed, Winner (1991) believes constructivism to disdain all evaluative stances.

Given the nature of the present study, in that its objectives include exploring risk perceptions and risk acceptance of insurers using genetic test results in an effort to inform political thinking regarding this matter, constructivist thinking does not appear to be appropriate. Indeed, why would anyone listen to the findings of a survey and take action upon these findings when they are not considered to be representative of a common or shared reality. The constant state of flux in which society finds itself means that constant conjunctions of events do not occur in the social world, hence the inability to formulate generalisations. Instead, the knowledge and beliefs gained from constructivist research are considered equally valid, in that there are no correct or incorrect theories, thus making it impossible to make any form of generalisation from the study of society. This understanding of the constructivist paradigm is not entirely suitable for this study, as it is considered that without the ability to formulate or justify generalisations, one cannot further develop knowledge of the factors that influence society's risk acceptance of insurers using genetic test results. The narrowness of scope and consequently the limited social and political relevance of constructivist studies, are reasons why the present study does not follow a constructivist paradigm.

4.2.3 Critical Realism

Critical realism, as a philosophical paradigm, has elements of positivism and constructivism (Healy and Perry, 2000). Critical realism provides the potential for a broad basis for the conduct of social research (Carter *et al.*, 2004; Danermark *et al.*, 2003; Miles, *et al.*, 1994; Robson, 2002). It has been claimed that its importance lies in its capacity to provide a sound foundation for research, given the decay of positivism as a viable model, and the concomitant rise of scepticism in the form of postmodernism (Miller and Brewer, 2003:258). This study acknowledges that there is a 'real' world to discover; however, it is deemed that this reality is imperfectly and probabilistically apprehensible (Guba and Lincoln, 1994; Merriam, 1988). According to Sayer (2000), the defining feature of realism is the belief that there is a world existing independently of our knowledge of it. While positivism concerns a single, concrete reality, and radical constructivism believes in multiple realities, realism concerns multiple perceptions about a single, mind-independent reality (Healy and Perry, 2000). The world is considered to exist regardless of what people think about it. People can make fallible claims about the existence of objects or relationships in the world; however, the world continues to exist despite the evident fallibility of our knowledge (Sayer, 2000:2).

The realist paradigm assumes that there is something out there which can be understood if the conditions that activate structures can be identified, so creating an event (referred to as a generative mechanism) (Smith, 1998:298). The aim of critical realist research is the understanding and explanation of these mechanisms, by investigating the structures of objects and how they produce events in different conditions (Houston, 2001: 850). Critical realism aims to discover observable and non-observable structures and mechanisms independent of the events they generate (Outhwaite, 1983; Tsoukas, 1989). In other words, researchers working from a realist perspective observe the empirical domain to discover, by a "mixture of theoretical reasoning and experimentation" (Outhwaite, 1983:332), knowledge of the real world, by naming and describing the generative mechanisms that operate in the world and result in the events that may be observed (Krauss, 2005). This inherent complexity that exists within the world of the social scientist, thus posits a reality that may be considered "real but fallible" (Wollin, 1995:80). Indeed, social systems are open and changeable, in that they evolve because people have the capacity to learn and change their behaviour. If the world was a product or construction of our knowledge, then one could say that our knowledge was infallible. It would be considered impossible to be mistaken about anything (Sayer, 2000:2). Critical realism is a "fallibilist philosophy", whereby the world can only be known under particular descriptions (Sayer, 2000:2). Despite such constructivist criticisms, critical realism recognises the significance of peoples' socially constructed interpretations and meanings. Indeed, critical realism acknowledges this subjective reality within the context of seeking to understand broader social forces, structures or processes that influence, and perhaps constrain, the nature of people's views and behaviours (Sayer, 2000:3). The concept of reality embodied within realism thus extends beyond the self or consciousness, but which is not wholly discoverable or knowable (Krauss, 2005:761).[6]

So far, it may appear that the discussion of critical realism closely correlates with the ontological view of realist constructivism, whereby it is considered that an ontological real external reality exists outside of the observer (see Section 4.2.2). However, followers of the realist constructivist paradigm believe that knowledge is always determined by the subject and, thus, no such thing as objective knowledge exists. Therefore, it is impossible to examine the world from an objective standpoint. Critical realism is based on a belief that an objective approach can be applied to the study of social phenomena. In some way, it is argued, an external reality must impinge as a discipline upon the construction of theory (Holmwood, 2001:949). Critical realism distinguishes

[6] For example, although explanations of gravity change throughout time and are culture dependent, there is an inter-subjective recognition of the constraints of gravity, suggesting that this physical feature of the world is sending compellingly similar signals to all percipient observers. Moreover, it implies the coming of these signals from somewhere outside our own phenomenological context of interpretation (Rosa, 1998:25).

between the world and our experience of it; in other words, the intransitive and transitive dimensions of knowledge (Sayer, 2000:11).[7] When theories about the world change (transitive dimension), this does not mean that what they are about changes too (Sayer, 2000:11).[8] From the critical realist view, social phenomena are considered an outcome of social construction intervened by an ontologically real external reality (Sayer, 2000:11). Therefore, given that the social world can be said to be socially constructed and includes knowledge itself, it cannot be said that the social world exists completely independently of some knowledge. However, a change within our sphere of thinking does not necessarily produce a significant change in the phenomena studied (Sayer, 2000:10-11). This distinction between the transitive and intransitive dimensions of science implies that the world should not be conflated with our experience of it (Bhaskar, 1975). The world does not correspond to the range of our senses and it is not identical to what we experience (Sayer, 2000:11).[9] This understanding of reality and our knowledge of reality pervades all aspects of the critical realist paradigm. As the present study follows a critical realist paradigm, the ontological view of this belief influences the entire research process and dictates, to a certain extent, how the findings of the study can be interpreted.

4.2.3.1 Critical Realist View of Causation:

The intention of realist research is to seek to identify both necessity and possibility in the world, to explore what things must go together and what could happen given the nature of the objects (Sayer, 200:11). Contrary to realist contructivism, critical realists make evaluative judgements that can be explained using a stratified view of reality. Critical realism acknowledges a difference between the real, the actual, and the empirical (Bhaskar, 1975). "The real is whatever exists, be it natural or social, regardless of whether it is an empirical object for us, and whether we happen to have an adequate understanding of its nature" (Sayer, 2000:11). According to Sayer (2000:11), the real is the realm of objects, structures

---

[7] The transitive dimension relates to our theories about the world (for example, that the world is flat or round), while the intransitive dimension is the world which the theories are about (Sayer, 2000:11)

[8] For example, there is no reason to believe that a round earth theory was accompanied by a change in the shape of the earth (Sayer, 2000:10)

[9] Individuals' constructions do not mirror reality and empirical observations are not always consistent with reality, in that the world is not identical to what is experienced by its inhabitants.

and powers. Casual powers indicate a capacity to behave in a particular way, while causal liabilities are specific susceptibilities to certain kinds of change. The *actual* refers to what happens if and when those powers are activated, concentrating on what they do and what eventuates when they do. Thus, realists argue that it is necessary to distinguish between the 'real' and the 'actual' since underlying causal mechanisms can be understood as 'real' even where they are not 'actualised' (Holmwood, 2001:750; Mouzelis, 2000:743; Giddens, 1984:17). The *empirical* is defined as the domain of experience. It is possible to observe the real and the actual and doing so may allow us to be more confident about what we think may exist, but the existence of the real and of the actual is not dependent upon the results of our observations. Contrary to constructivist thought, rather than relying on observability for making claims about what exists, proponents of the critical realist paradigm are conscious of, and accept, a causal criterion also (Sayer, 2000:12). Although realism shares with constructivist social science the view that social phenomena are concept dependent and have to be understood, unlike constructivism, it argues that this does not rule out casual explanation (Sawyer, 2000:17). This is because material change in society has to be explained too, and reasons can also be causes, in that they prompt us to do things, think differently, etc. (Sayer, 2000:18).

According to Sayer (2000:12-13), critical realism proposes that the world is characterised by emergence, that is, situations in which the conjunction of two or more features or aspects gives rise to new phenomena, even though the latter are necessary for their existence. In the social world, people's roles and identities are often internally related, so that what one person or institution is (or can do), depends on their relation to others (Sayer, 2000:13). Thus, to be an insurance company can only be explained at the level of their relation to insurance applicants and insurance policy holders and vice-versa. The present study explores how the introduction of genetic test results as a factor that may be underwritten by life insurance companies has changed the insurance system. Genetic testing as used by life insurance companies, may affect individuals with insurance policies or those intending to take out insurance polices. As a consequence of how they are affected, the insurance industry may be influenced to transform their underwriting system to exclude this factor from the information they request from such individuals. The powers which both elements of the equation can draw upon depend partly on their relations to one another, and to relevant parts of the context, such as the legal environment. Social systems commonly involve "dependencies or combinations (which) causally affect the elements or aspects, and the form and structure of the elements causally influence each other and so also the whole" (Lawson, 1997:64).

This world can change or evolve as a consequence of emerging objects and structures. For example, the use of genetic test results in underwriting applicants for insurance can be considered a new or emergent object within the insurance system, and the insurance system, as it is currently understood, may evolve and change as a consequence of society's perceptions and acceptance of this information use. It has been argued that the social world is changing more rapidly than the natural world, because the social world incorporates the perception, reflection and action of human beings and social collectives engaged in social experience, learning and innovation (Wad, 2001). According to the critical realist view, human agency, that is the capacity for human beings to make choices and to impose those choices on the world, is made possible by social structures that themselves require the reproduction of certain actions or pre-conditions. For example, in order for society's acceptance of the risks related to insurers using genetic test results to have an effect upon insurance company's use of this information, involvement of social structures, such as the Government, are necessary as a means to apply political pressure on the insurance industry. Although individuals, as part of a society, require the Government to ensure that their reasoning is heard and acted upon, the structure, and even the existence, of the Government is dependent upon society (in that the existence of a democratic Government is dependent upon voters and agreement among a population to have this form of Governmental rule). Further, the individuals that inhabit these social structures are capable of consciously reflecting upon, and changing, the actions that produce them—a practice that is in part facilitated by social scientific research. However, human agents do not produce social structures out of nothing, instead agents reproduce or transform structures (Danermark *et al.,* 2002:181).

There are approximately three theories on the relationship between social structures and agents (Danermark *et al.,* 2002:179). The 'social fact paradigm' outlines that all influence moves from structures to agents, whereby social structures have a real existence comparable to that of material things. From this view, it is considered that structures are the object of study in science, and even the actions of individuals can be traced back to their positions in different structures. A second paradigm is the 'agency paradigm', where the individuals are agents acting for their own objectives and goals in the light of the meaning they convey to their social environment and the interpretation they make of the phenomena of their context. Individuals act based upon their own objectives and goals, and these actions taken together are what make up social structures. It can be concluded that both paradigms have different starting points, and in doing so conflate the importance of either structure or agency in social scientific study. A third theory is proposed by Giddens (1984), is known as 'stucturation theory', and proposes that structures do not exist separately from individuals, in that they are always the medium as well as the outcome of social action. A duality exists between structure and agent. This implies that agent and structure constitute one another and cannot be separated. Analysis of this duality involves 'methodological bracketing', whereby in the analysis of social structures and institutions, agency and its properties must be disregarded; likewise, when analysing agency, social structures must be ignored (Danermark et al., 2002:180).

Archer (1995) believes the structuration theory suffers from central conflation, whereby structure and agent can only exist by virtue of each other. According to Archer (1995), agency and structure are not two elements of the same process; instead, they are two different phenomena. Indeed, social structure exists prior to the present agent in that structures, such as society or the insurance system, are not created out of nothing, they are reproduced or transformed. While social structures cannot be reduced to individuals, the former are a prerequisite of any human action – structures enable actions and constrain what actions are possible. Bhaskar (1993:55) demonstrated a realist transformational model of the connection between social structure and agency which is consistent with each of the three theories (see Figure 4-1 below).

Figure 4-1 The transformational model of the connection between social structure and agency (Bhaskar, 1993:55)

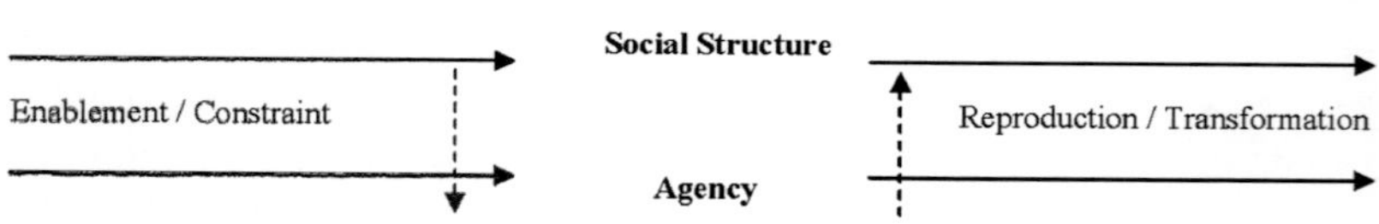

According to the realist transformational model, social structure and agency are regarded as two separate phenomena with different powers and properties (Danermark *et al.,* 1993:181). However, although structure and agency are separate strata, one is essential for how the other is or will be moulded. Hence, when conducting scientific inquiry, it is necessary to move away from the view that only one side should be examined at a time; instead, critical realists should study the interplay between structure and agency. Archer (1995) further developed Bhaskar's (1993) model by introducing a time dimension to the analysis, taking into account that emergence is a process whereby the interplay between structure and agency takes place over time. The properties and powers within a particular stratum precede another stratum because the latter emerges from the other, but as soon as emergence has taken place, the powers within these strata have a relative autonomy with respect to one another (Archer, 1995). Archer (1995) terms this procedure 'analytic dualism'.

The present study adopts a critical realist view of the dualism between human agency and structure, and this view is applied within the present study. The study looks at how human developments within the natural sciences, in the form of genetic testing for disease, have been introduced to the system of insurance, an element of the world which is classified in the present study as belonging to the stratum of structure. The insurance system has, therefore, been transformed by this new data as a result of the interaction between human agents and the insurance system. The present study focuses on the reaction of agents to this new circumstance in which they find themselves. Perceptions and acceptance of the risks related to insurers using genetic test results are explored in an effort to understand whether individuals feel constrained by this transformed structure or; indeed, view it as a positive change. In doing so, it may be possible to obtain a closer view of whether human agents will push for further transformation of the insurance system, whereby genetic test results are not considered in insurance premiums, or insurance companies allowed to process this information. This study, therefore, addresses human agents' reactions to change in the insurance system.

The intention of realist research is to identify necessity and possibility in the world, whereby one explores what things must go together and what could happen given the nature of the objects (Sayer, 200:11). The present study focuses on the transformation of the insurance system with the introduction of the use of genetic test results in the underwriting process, and the effect that this change has upon human agents. Reactions to insurers using genetic test results are explored in terms of knowledge, social trust evaluations, risk perceptions, benefit perceptions and risk acceptance. Previous research has declared that these items are connected, in that they go together and result in risk acceptance. It is noted, however, that this line of assumed causation is contingent on the context of the study, whereby the context is formed by the external effects of other mechanisms within an open system of mechanisms, events and experiences, i.e. the domain of the real world. Such external effects are thought to include culture, the political system, the insurance system, etc. Given the diverse range of effects, the findings regarding risk acceptance may indicate to some extent future changes in the insurance system as a consequence of human actions based upon risk acceptance attitudes and the objects affecting risk acceptance.

Internal relations fall outside the ontological grids of positivism, which systematically misrepresents society by presenting such phenomena as reducible to independent individuals or atoms (Sayer, 2000:13). Given the nature of chance encounters and peoples' sensitivity to context and time, social phenomena do not tend to have the consistency or durability of the objects studied by the natural sciences (Sayer, 2000:13). Indeed, the same casual power can produce different outcomes[10] or sometimes different causal mechanisms can produce the same results[11] (Sayer, 2000:15). In an open system where there are multiple mechanisms, the effects of any causal mechanism may not occur because they are cancelled out, or otherwise distorted, by the operation of another mechanism (Holmwood, 2001:949). Therefore, the regularity of phenomena is rarely a constant in the open system of the social world[12], and it cannot be expected that social science's descriptions shall remain stable or unproblematic across time and space (Sayer, 2000:13). According to critical realists, positivism is based upon a mistaken ontology, one which is too 'flat' and fails to recognise the ontological 'depth' – or stratification – of the world as being made up of many causal mechanisms and their associated hierarchies (Bhaskar 1979, 1986; Sayer 1984; Archer *et al.*, 1998). Unlike positivistic universal casual predictions regarding the social world, critical realists attempt to make generalisations regarding the social world in the form of a description of local mechanisms (or tendencies) (Danermark *et al.*, 2002:54). The world is deemed to have ontological depth, whereby events

---

[10] This is dependent on how the conditions for closure are broken, for example, economic competition can prompt firms to restructure and innovate or to close.

[11] For example, one might lose a job for a variety of reasons

[12] If regularity of phenomena can be observed within the social world, this is usually as a result of a conscious effort to maintain continuities through change, rather than as a result of doing nothing (Sayer, 2000:13)

arise from the workings of mechanisms which derive from the structures of objects, and they take place within geo-historical contexts.

Critical realism rejects the positivist view that causation involves regularities among sequences of events. Instead, a realist view of causality focuses on the structures of objects.[13] Objects are, or are part of, structures, whereby 'structure' is defined as a set of internally related elements whose causal powers, when combined, are emergent from those of their constituents (Sayer, 2000:14). Thus, acceptance by individuals who make up a society of the risks related to insurers underwriting genetic test results may influence the life insurance system to use or to discard genetic test results in their underwriting practices. Whether these powers are exercised depends on other factors and conditions. This study explores whether a sample of the Irish populations' risk and/or benefit perceptions of insurers using genetic test results, trust in the insurance industry and the Irish Government, and knowledge of insurers using genetic test results influence risk acceptance. Indeed, when causal powers are activated (as when risks are accepted), the results depend on a further set of conditions, for example, the political environment, the legal environment, etc.

For critical realists, causation is not understood on the model of regular successions of events, and, hence, explanation does not depend on finding them or searching for putative social laws (Sayer, 2000:14). Instead, Sayer (2000) views regularities as merely suggesting where one might look for candidates for causal mechanisms. According to Sayer (2000: 14), what causes something to happen has nothing to do with the number of times it has been observed to happen, instead, explanation depends on identifying causal mechanisms and how they work, and discovering if they have been activated and under what conditions. To explain why a certain mechanism exists involves discovering the nature of the structure or object which possesses that mechanism or power (Sayer, 2000:14). Therefore, to

---

[13] In critical realism this is a question of the nature of the object, which determines what a certain object can and cannot do (Danermark *et al.* 2002:205). "To ask what has caused something is to ask what made it 'happen', what 'produces', 'generates', 'creates' or 'determines', or, more weakly, 'enables' or 'leads' to it (Sayer, 1992:104). From a realist perspective, it is not a matter of a relation between two events, that are separated and demarcated from each other. "Causes are about objects or relations and their nature. It is a matter of what causal powers there are in a certain object or relation. In more general terms it is a matter of how objects work, or a matter of their mechanisms" (Danermark, 2002:54; Sayer, 1992:103)

understand if, or why, a causal mechanism exists between risk perceptions and risk acceptance, it is necessary to identify the underlying factors that exist as part of the structure of risk acceptance and risk perceptions

According to Sayer (2000:17), critical realism goes beyond looking for associations, whereby it is explored if associations found could be otherwise. What Sayer (2000) terms 'associational' thinking is a form of empiricism which fails to recognise the distinction between the real and the actual (Holmwood, 2001:950). From Sayer's (2000) view, associational thinking mistakenly regards the non-occurrence of 'necessary effects' as indication of the falsity of a theoretical claim. Risk perceptions might regularly be associated with risk acceptance, but this does not mean that they have to co-exist as mutual preconditions. According to Sayer (2000), associational thinking is also mistakenly inclined to regard found conjunctions of effects or events (again, such as risk acceptance, risk perceptions and trust) as 'necessary', when they may be no more than the contingent associations of different structures whose 'real' tendencies are quite other than might be inferred from empirical conjunctions. The task of the critical realist is to use abstract thinking to identify these 'real' structures, by abstracting from the contingent conjunctions which mask their operation (Holmwood, 2001:950).

The present study diverges slightly from what can be considered as Sayer's (2000) disregard to some extent of associations. A conjunction of effects is considered to be one of the indications of how a researcher's understandings might be reconstructed. Following the distinction between the 'real' and the 'actual', it should be acceptable that sometimes, the observed non-actualisation of effects might be an indication of the misrecognition of underlying causal mechanisms, rather than a derivation from countervailing tendencies generated by the operation of other relevant mechanisms (Holmwood, 2000:950-951). The non-occurrence of 'necessary effects' *might* indicate that the claimed mechanism is false. Thus, according to Holmwood (2001:951), if a causal mechanism is misidentified, its tendencies are not *real* and that error of identification explains the non-occurrence of anticipated effects. If misrecognition is a possibility, the task is then to transform understandings to correctly identify the appropriate causal mechanism. It is argued that recognition of non-associations, where causal associations were previously thought true, makes possible sociology as a *critical* undertaking, or what constitutes realist social science as critical realism (Archer *et al.,* 2000:470). Therefore, although the study diverges somewhat from Sayer's (2000) approach to association, the study still reflects critical realist thinking regarding this matter.

Within this stream of thought, it is still possible for research studies with this understanding to move beyond conjunctions to reconstruct theoretical understandings, and to provide an account of explanatory mechanisms (Holmwood, 2001:951). According to Holmwood (2001:951), any reconstruction is likely to re-define and re-specify the identification of effects, because the prior understandings of mechanisms (of objects and relations) through which their initial identification as divergent effects were made (and which made us aware of our problem of understanding), have been transformed. This can be considered as what is meant by the theory-dependence of observations in that where theory is reconstructed, so too will observations be reconstructed. Observations that do not support pre-existing explanatory claims will be re-defined in the process of being fitted to new explanatory schemes (Holmwood, 2001:951). Given this understanding of the importance of exploring associations, the present study shall explore the association between knowledge, trust, risk perceptions, benefit perceptions and risk acceptance. Observations indicating either association or non-associations between these factors, and indeed any findings relating to the effects of other factors and conditions upon those given priority in the study, shall allow for the reconstruction of our knowledge regarding trust, risk perception and risk acceptance.

According to Sayer (2000:16), many of the controversies in social science relate to the problems that social scientists encounter when exploring mechanisms within an open system, as there are many interacting structures and mechanisms. This can be seen by the wide ranging opinion of the factors that influence risk acceptance and risk perception (see Sections 3.2 to 3.7). This problem of identifying causal responsibility in the complex open system inherent in the observations of the present study are dealt with by asking a series of characteristically realist questions designed by Sayer (2000:17):

1. *What does the existence of this object/practice presuppose? What are its preconditions?*
   An example of this question asked in the present study refers to risk acceptance, whereby the factors and conditions underlying the existence of the acceptance of insurers using genetic test results are explored within the confines of a sample population of Irish society. Secondary and primary research is used to respond to this question.

2. *Can or could object A exist without object B? Answering this question helps to sort out the conditions of existence of social phenomena.*
   For example, this study questions whether acceptance of the risks related to insurers using genetic test results can exist without being influenced by risk perceptions of this situation.
3. *What is it about this object which enables it to do certain things?*
   An example of this question asked in the present study refers to risk acceptance, whereby it is questioned how risk perceptions can transform the insurance system.

In answering these questions, much depends on how the objects are conceptualised, for example, what is meant by risk, and what is included in this understanding of risk (Sayer, 2000:17). Sections 2.4, 3.4 and 3.5 clarified the understanding adopted in the present study of knowledge, trust, risk and risk acceptance, and further analysis of this understanding is undertaken in the primary research. Section 2.4 demonstrated how at one extreme risk is understood as an objective property of an event or activity, and is measured as the probability of well-defined, adverse events (Kates and Kasperson, 1983). This view of risk is widely accepted to be positivistic (Rosa, 1998:27). At the other extreme is the strong constructivist view that risk is all in the mind, in that it is a notion of observation, and not just an object to be observed (Holzheu and Wiedemann, 1993:9). As extremes, both views provide a poor description of reality. This study provides a definition of risk, taking into account both the objective and subjective views of risk by determining the family resemblance among the competing views (see Section 2.4.2). The term 'risk' is considered to imply a possible future negative and uncertain outcome of an event or situation caused by nature or human activities, where insufficient knowledge was available to prevent such an outcome. According to Rosa (1998:28), this definition of risk expresses an ontology of risk, an ontological realism that specifies which states of the world are to be conceptualised as risk.[14] The definition also acknowledges risk as an

[14] "A linguistic conundrum for the constructivist position on risk, but not for the ontological realism position, emerges with the ordinary language assertion that someone or something is at risk, or that certain situations are risky or hazardous. These make perfect sense if risk is defined as an ontological state of the world. A person is at risk because they are subject to an uncertain outcome in the world where the outcome contains something of stake to that individual. On the other hand, if risk is entirely a social construction, a collective agreement about a perception, what is the meaning of these ordinary language assertions? If someone

objective state of the world, whereby risk arising from nature or human activity exists independent of our perceptions and of our knowledge claims, subjective judgments about what is at risk, and how likely a risk will be realised.

This differs from the constructivist view of risk whereby risk is seen as arising from our perceptions of it (Douglas and Wildavsky, 1982; Freudenburg, 1989; Rayner, 1992; Wynne, 1992; Slovic, 1992; Shrader-Frechette, 1991). From this view, it is impossible to distinguish risk from risk perception, a position that is reflected by absolute relativists. However, some risks are undeniably real (for example, the risk of death) and exist outside of our experience and perceptions of them. Such risks can be presented as probabilities; however, until they become a certainty, they remain as "purely perceived, theoretical, or estimated" (Rosa, 1998:32). From this point of view, risk identification can only occur through risk perception; hence, risk is at once both real and perceived (Rosa, 1998:33). Risk and risk perception can be considered as representative of both domains of the world, whereby actual risk exists external to the human mind, while risk perception is an internal and subjective understanding of the state of risk. Risk represents a state of the world where there is a conjunction between uncertainty of outcome and human concern about the outcome (Rosa, 1998:40). The influence that the research paradigm followed by the present study is paramount, not only to the thinking behind the objects studied, but it also pervades the methodological approach used to explore these objects and their relationship.

Reasons were provided as to why positivist and constructivist paradigms were not considered suitable for guiding the present study (see Sections 4.2.1 and 4.2.2). By applying critical realism in the present study, it was deemed possible to overcome the weaknesses and flaws associated with the positivist and constructivist paradigms. In short, by applying critical realism in the present study, an alternative is offered to both hopes of a law-finding science of society associated with a positivist philosophy, and the constructivist reductions of social science to the interpretation of meaning (Sayer, 2000:3). Critical realism is deemed suitable for the present study as it allows for more than the empirical positivist approach

---

is at risk, does that mean that the person is simply in a position to violate the collective agreement? Within a constructivist orientation it is difficult to know exactly the meaning of the assertions but easy to see the strained awkwardness that orientation produces" (Rosa, 1998:28).

to the objects of research that is too limiting and consists of a flawed conception of objectivity in science. By bringing realistic conceptions of causality (mechanisms) and conditions for their triggering (context) back into view, the reality of observations can be fore-grounded and statistical theory can be used as an aid to scientific investigation, rather than the most frequently employed explanation of observational differences (Connelly, 2004).

From the critical realist view, social phenomena are considered an outcome of social construction intervened by an ontologically real external reality; hence, it is necessary to account for the subjective nature of peoples accounts to provide the best picture of reality. The constructivist paradigm follows that knowledge is always determined by the subject and, thus, no such thing as objective knowledge exists, making it impossible to examine the world from an objective standpoint. However, an external reality must impinge as a discipline upon the construction of theory, a concept that is not accepted by constructivists (Holmwood, 2001:949). As a consequence constructivists have difficulty in deciding upon conditions of falsifiability; hence, constructivist studies are incapable of producing theory, which is an aim of the present study. A critical realist analysis identifies the generative mechanisms at work that both necessitate and constrain acceptance of the risks related to insurers using genetic test results. This study intends to explore whether several generative mechanisms in particular, trust, risk perceptions and benefit perceptions, could ultimately lead to risk acceptance and the transformation of the insurance system. It is the intention of the present study to allow the findings from the research to inform and build upon risk theory.

4.2.4 Dimensions of Social Reality

This research explores the complex reality surrounding society's knowledge, trust, risk and benefit perceptions and acceptance of the risks related to insurers using genetic test results. To fully understand the complex social reality that influences public views, the study has teased out the different, although interconnected and interdependent, dimensions that shape reality. By nature, all research is partial in relation to the real world (Danermark *et al.*, 2002:169). Therefore, the study focuses on only some of the elements of the research layers that are of particular relevance to this study. Layder (1993) outlines five layers of social reality that intertwine to produce a multifaceted or stratified model of society. These five dimensions are context, 'setting', situated activity, and self, plus the overriding dimension of history. Figure 4-2 displays Layder's research map in the form of a web, representing the entangled nature of all five dimensions (Greenford, 2001).

Figure 4-2 Layder's Research Map in the form of a web

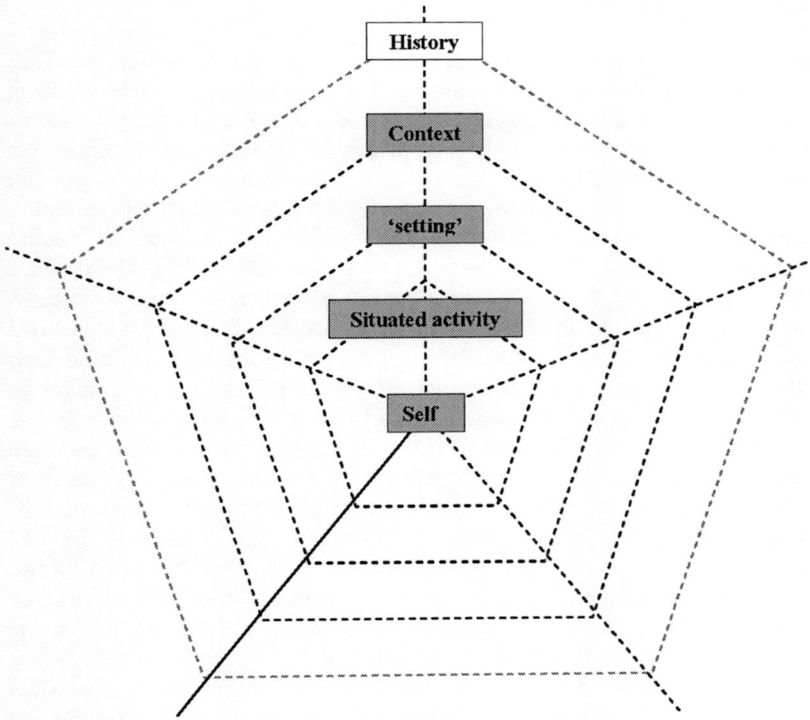

For the purposes of this study, an attempt is made to disentangle four of the elements that constitute Layder's (1993) interpretation of social reality, and separately examine them, and their effect on the research questions *('history' shall not be examined as it is beyond the scope of this study).* The four elements indicate and build upon the objectives of this research clearly and precisely. Layder (1993) also notes the impact of power, which encircles and influences all of the aforementioned elements. This realist map endorses the interwoven nature of the macro and micro aspects of society and, despite the difficulties in differentiating and separating the dimensions, the map is used as a resource tool from which several aspects of society can be studied. By applying Layder's (1993) guide to examining social reality, all the dimensions that impact on public opinion are considered. With respect to this research, the map will be used as a basis in building an understanding on if and how knowledge, trust, and risk intertwine and the nature of the web in which knowledge, trust, risk, and risk acceptance are interwoven. In doing so, the study is both extensive and intensive by design.

At the centre of the web of social reality lies self, a dimension in which individuals make decisions regarding their acceptance of insurers using genetic test results. Decisions are made according to their concerns surrounding the risks related to such usage and the benefits from insurers accepting genetic test results. These perceptions and views are not "reality", but merely a window on to reality from which a picture of reality can be triangulated with other perceptions and the social reality surrounding the self (Perry, Alizadeh and Riege, 1997:1952). Individuals perceive risk and make decisions within a particular situated activity. 'situated activity' directs attention to the nature of social involvements and the dynamics of social interaction encompassing the self. The situated activity, in this research, occurs where individuals and representatives of life insurance companies interact to process an application for life insurance. The way in which individuals perceive insurers and life insurance products, and the perceptions of life insurance companies of applicants, affect the relationship and interaction between the two parties. This demonstrates the fluidity of the self and 'situated activity' dimensions, which, in turn, affect and are influenced by the 'setting'. ''setting'' represents the immediate social environment of an individual/ group in this case, the self and 'situated activity' are surrounded by the insurance and risk management 'setting', a dimension of social reality where the developments in genetic testing and the nature and requirements of the insurance industry exist. The context lies at the outer edges of the web, and directs attention to macro social forms such as the national and dominant culture. The dominant culture influences the attitudes and evaluations of individuals, and shapes their perceptions of risk and risk acceptance decisions. These perceptions feed into the nature of the dominant culture. This illustrates the influential nature that each of the dimensions of social reality have upon each other, and the weaving of the dimensions to create a picture of the social reality surrounding insurers' use of genetic test results in the Republic of Ireland. The previous two chapters broadly examined the 'setting', 'situated activity' and context surrounding this particular study. The remaining 'self' dimension is explored using the methodology prescribed in the subsequent sections. The influence of 'setting', 'situated activity' and context are also noted throughout the primary research process. The following paragraphs explain the research approach applied to obtain information regarding the social reality surrounding risk acceptance.

4.3 Research Approach

Shih (1998) lists four areas for consideration when deciding on a research approach: the philosophical paradigm and goal of the research, the nature of the phenomenon of interest, the level and nature of the research questions, and practical considerations related to the research environment and the efficient use of resources. However, linking philosophical stances directly with research approaches has been criticised as potentially misleading and as having no practical value (Bryman, 2001:20). Instead, it is thought that the nature of the research question should dictate the research approach adopted. However, there is evidence to suggest that this form of distinction is within continued use (Saunders *et al.*, 2003:85) and according to Easterby-Smith *et al.* (2002), failure to think through philosophical issues relating to the relationship between data and theory can seriously affect the quality and value of the research output. Exploration of philosophy can help the researcher to refine and specify the research methods to be used in a study, that is, to clarify the overall research strategy to be used. This would include the type of evidence gathered and its origin, the way in which such evidence is interpreted, and how it helps to answer the research questions posed (Easterby-Smith *et al.*, 2002). Through the knowledge of research philosophy and its assistance in evaluating different methodologies and methods, it is possible that the researcher could avoid inappropriate use and unnecessary work by identifying the limitations of particular approaches at an early stage. Furthermore, an analysis and understanding of research methods and the limitations of certain approaches may enable the creation of designs that may be outside of the researcher's previous experience. Indeed, Cresswell (1994) goes so far as to state that the ontological and epistemological position of the researcher *has* to be considered when deciding between quantitative and qualitative methodologies.

There are advantages of considering one's ontological and epistemological position when making a choice regarding research strategy. Therefore, it seems appropriate to allow critical examination of the grounds for the researcher's fundamental beliefs to further guide the choice of the research approach. In doing so, philosophy plays a useful role as "underlabourer" to the research - the term under-labouring is taken from Locke (1894:14) as "clearing the ground a little... removing some of the rubbish that lies in the way of knowledge". The confidence provided by understanding different philosophical positions and deciding on one's own epistemological and ontological beliefs provides the researcher with the power to argue for different research approaches and allows one to confidently choose one's own sphere of activity (Dobson, 2002). The "bothering" with philosophy can also provide the potential for emancipation from domination by one's social or academic grouping. By understanding and following one's own epistemological and ontological beliefs, it may be that the researcher could deviate from the research approaches applied by their academic peers and thus produce research that provides a 'new' view surrounding a particular phenomena.

Both Walsham (1995 and 1993), and Garcia and Quek (1997) argue that a more coherent (i.e., consistent, rational, and logical) research process can be achieved through philosophical reflection. Indeed, according to Archer (1995:16-17) expains, "…the nature of what exists cannot be unrelated to how it is studied…the social ontology endorsed does play a powerful regulatory role vis-à-vis the explanatory methodology for the basic reason that it conceptualises social reality in certain terms, thus identifying what there is to be explained and also ruling out explanations in terms of entities or properties which are deemed non-existent". There is, therefore, a strong argument that ontology and methodology should be closely linked. As Archer (1995:17) explains "what social reality is held to be, also is that which we seek to explain". Given the above advantages and what some have intimated as a necessity in making decisions regarding the research approach, the present study links the underlying philosophical stance of the research with the approach and methods adopted.

There are two contrasting approaches concerning the design of research and the relationship between theory and research, the first being the deductive approach and the second being the inductive approach. The following section analyses both approaches to research and their associated philosophical background, in order to decide upon the research approach best suited to the present study.

4.3.1 Deductive Approach

The deductive approach entails the development of a theory and hypothesis, and the subsequent testing of that hypothesis via a chosen research strategy (Saunders *et al.*, 2003:86). The deductive approach is commonly associated with a positivist ontology, epistemology and methodology. There are a few general characteristics that should be highlighted in relation to the deductive approach to research. Firstly, there is a desire to examine and explain influential relationships between variables and generalise about such regularities. Secondly, the deductive approach tends toward quantification in the collection and analysis of data, preferring a 'scientific', objective and impersonal standpoint (Saunders *et al.*, 2003:86). Followers of the inductive approach criticise the deductive approach for its rigid methodology and highly structured research design, resulting in a lack of appreciation for possible alternative theoretical explanations (Saunders *et al.*, 2003:88).

4.3.2 Inductive Approach

The inductive approach, on the other hand, emphasises the generation of theories through the collection of data, and prefers a more qualitative slant to data collection and analysis. Critics of this approach claim that it is essentially descriptive and thus fails to really conclude anything, as it is impossible to make the infinite number of observations required to prove a universal statement true in all cases. In other words, inductive reasoning only supports a conclusion to some degree of probability, a problem which logicians are quick to point out (Johansson, 2002). The inductive approach is closely associated with a constructivist paradigm.

4.3.3 Amalgamation of Deductive and Inductive Approaches

To summarise, the deductive approach, with regard to philosophical background, is more inclined toward positivistic epistemologies and objectivist ontologies, while the inductive approach is orientated toward more constructivist philosophies. However, it is possible to amalgamate the two research approaches. According to Layder (1993:16), the deductive and inductive approaches to research are inherently limited by the theory of knowledge that underpins them and their explanatory power can be enhanced by "modifying and extending their terms of reference". A combination of the two approaches forms a research approach similar to that recommended by Layder's (1993) realist approach to research, where theory-testing and theory-building approaches are integrated.

Compared to positivism and constructivism, critical realism endorses or, is compatible with, a relatively wide range of research methods. It implies that the particular choices should depend on the nature of the object of study and what one wants to learn about it (Sayer, 2000:19). According to Sayer (1992), it is firstly necessary to ensure careful abstraction and conceptualisation of the various components or influences within the study and consider how they combine or interact. The objects of study in this research were defined and their influence discussed in Chapters Two and Three (see Sections 2.4, 3.4, 3.5 and 4.2.3).[15] Critical realism recognises the use of multiple research methods across intensive and extensive boundaries to increase our understanding of the nature of objects and to decrease the possibility of misattribution of causality (Sayer, 2000:20). The differences between extensive and intensive research designs are illustrated in Table 4-3.

Table 4-3 The Differences Between Intensive and Extensive Research (adapted from Sayer, 2000:21)

	Intensive	**Extensive**
Research Question	How does a process work in a particular case or small number of cases? What produces a certain change? What did the agents actually do?	What are the regularities, common patterns, distinguishing features of a population? How widely are certain characteristics or processes distributed or represented?
Relations	Substantial relations of connection	Formal relations of similarity

[15] This study has also attempted to describe and explain the objects of study – knowledge, trust, risk perceptions, benefit perceptions and risk acceptance using previous research on these objects and through qualitative interviews described in Chapters Six and discussed in Chapter Seven.

Type of groups studied	Causal groups	Taxonomic groups
Type of account produced	Causal explanation of the production of certain objects or events, though not necessarily representative ones	Descriptive 'representative' generalisations, demonstrate associations
Typical methods	Study of individual agents in their causal contexts, interactive interviews, ethnography, qualitative analysis	Large scale survey of population or representative sample, formal questionnaires, standardised interviews. Statistical analysis
Limitations	Actual concrete patterns and contingent relations are unlikely to be 'representative', 'average', or generalisable. Necessary relations discovered will exist wherever their relations are present, for example, causal powers of objects are generalisable to other contexts as they are necessary features of these objects.	Although representative of a whole population, they are unlikely to be generalisable to other populations at different times and places. Problem of ecological fallacy in making inferences about individuals.
Appropriate tests	Corroboration	Replication

It can be said that intensive and extensive research methods have complimentary strengths and weaknesses. Intensive research is strong on causal explanation and interpreting meanings in context. However, intensive research tends to be very time consuming, so one can normally only deal with a small number of cases. Intensive research is commonly associated with inductive thinking. Extensive research, which arrives from a positivistic background, indicates the extent or quantitative dimensions of certain properties and relations and demonstrates the existence of associations between objects (Sayer, 2000:22). Extensive research is commonly associated with deductive reasoning.

Layder (1993) states that empirical and theoretical research if arriving from a critical realist perspective, should involve both deductive and inductive thinking and extensive and intensive research approaches. In general, Layder (1993) is sympathetic to constructivism, and even to some of the ideas in grounded theory, although he is critical of "pure" grounded theory. Healy and Perry (2000) agree that within a critical realist framework, both intensive qualitative and extensive quantitative methodologies are appropriate for researching the underlying mechanisms that drive actions and events. Methods such as case studies and unstructured or semi-structured in-depth interviews are deemed as acceptable and appropriate within this paradigm, as are statistical analyses, such as those derived from structural equation modelling and other techniques (Bisman, 2002; Perry *et al.,* 1997). With critical realism, the seeming dichotomy between quantitative and qualitative is therefore replaced by an approach, whereby it is considered appropriate to amalgamate both forms of research methods given their suitability in accordance with the research topic of interest and level of existing knowledge pertaining to it (Krauss, 2005:762). It is the intention of the following section to break down this dichotomy further, in order to assess which research strategy(s) should be used to achieve the objectives of the present study.

4.4 Research Strategies

The research strategy is a general plan depicting how the research questions will be answered, leading to a means of achieving the research objectives. Assessment of the relevance and the value of the different research methodologies that can be employed to investigate the questions posed within a thesis are influenced by the researchers' underlying epistemological commitments. (Johnson and Duberley, 2000: 3). According to Layder (1993), theory-testing (deductive, extensive and positivistic) research involves formal techniques or strategies of collecting and analysing data, such as a social survey or an experimental strategy. Easterby-Smith *et al.* (2002:42) summarise theory-testing methods as fast and economical, however, they tend to be inflexible and lack the ability to understand process or the significance that people attach to actions. Finally, as they tend to focus on the past or the present, it is difficult for a policy-maker to infer what changes or actions are required in the future. Theory-building (inductive, intensive and constructivist in nature) research involves the use of qualitative strategies of conducting research. These strategies include ethnography, action research and case studies and look at how change processes over time. These constructivist methods tend to enable an understanding of peoples' meanings and can adjust to new issues and ideas as they emerge (Layder, 1996:3; Easterby-Smith *et al.*, 2002:42). On the other hand, analysis of the data can be difficult to interpret and depends, quite highly, upon the tacit knowledge of the researcher (Easterby-Smith *et al.*, 2002:42). Realism advocates the use of both theory-testing and theory-building methodologies as presented in Figure 4-3.

Figure 4-3 A representative range of methodologies and their related paradigms

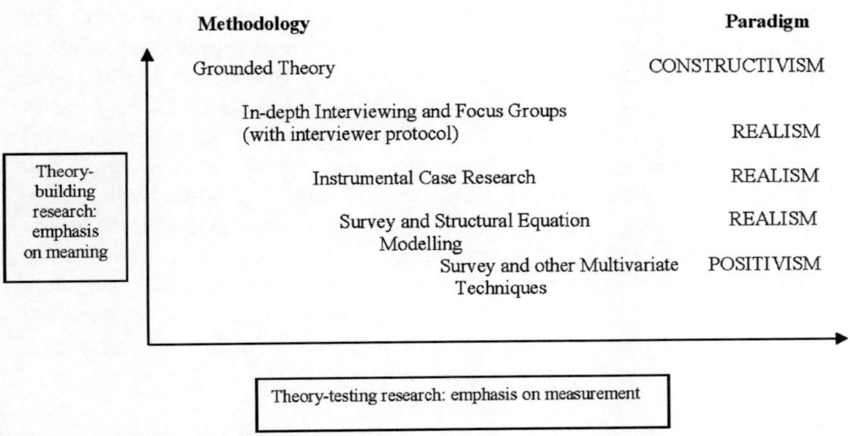

Source: Healy and Perry (2000)

Figure 4-3 demonstrates how realism does not insist on the use of any particular method of empirical research (Robson, 2002:43). From the above analyses of inductive and deductive, extensive and intensive, theory-testing and theory-building methods, it appears that there is no right or wrong choice of methodology. For this critical realist study; indeed, each approach has its own particular advantages and disadvantages (Leedy, 1989:88). Therefore, rather than universally advocating any single methodological approach for all evaluation situations, the selection of method(s) is driven by the kind of research questions asked, from whom and under what circumstances (Patton, 1990:492; Robson, 2002:223).

The present study intends to explore a theory regarding the factors that influence risk acceptance, and to identify the conditions that impact upon this relationship. To do so, a theory that suggests that incomplete knowledge, trust, risk perception and benefit perception are factors that affect risk acceptance is tested and built upon and a final theory of risk acceptance is presented. Therefore, it can be said that the study involves both theory-testing and theory-building. However, it is important to note that it is not the intention of the present study to test the overall theoretical findings of this study. Instead the data will be presented in a form that may be tested by future research. It can, therefore, be said that the overall aim of the research is theory-development rather than theory-testing.

4.4.1 Theory-testing and Theory-building

To develop a theory regarding the factors and conditions that affect risk acceptance, it is necessary to test the application of a previous theory to the present study in relation to life insurers using genetic test results and to then develop this theory in the context of the study. The first requirement is clearly theory-testing whilst the second is theory-building. Yin (1994) states that quantitative methods are more appropriate for answering *what* and *how many* (theory-testing) questions, whilst qualitative methods are useful for finding information on *how* and *why* (theory-building) things occur (Robson, 1993:6; Layder, 1996:3). According to Kohler (1994), quantitative measurement is measurement with numbers, while qualitative measurement can be described as measurement with words. Combining methodologies in the study of the same phenomena enables the researcher to focus on a research question from a variety of angles.

Multi-method research is considered increasingly valuable with regards to validity and reliability, when compared to a study adopting a single method approach. A combination of methods forms alternate views and attacks research questions from different methodological perspectives, thereby, minimising systemic bias. According to Denzin (1989), the rational for across-method triangulation is that the flaws of one method are often the strengths of another. By combining methods, researchers can achieve the best of each method while overcoming the unique deficiencies related to each method. However, Easterby-Smith *et al.* (2002:41-42) warn that mixing methods to obtain a richer picture may lead to contradictions and confusions, particularly if it is hard to reconcile discrepant sources of data that suggest different conclusions. Different methods show up different aspects of reality. Hence, the findings from each of the methods used in the present study may not perfectly correspond with each other. Indeed, sometimes the reality under investigation may be considerably more complex than a particular data collection technique is capable of identifying or demonstrating (Easterby-

Smith *et al.*, 2002:41-42). Quantitative analysis is rather 'surface-based', in that it does not allow the researcher to explore beyond the minimal single phrase responses of participants to the survey. Therefore, analysis of these responses may lead to conclusions that cannot possibly explain the underlying reasons for the responses. Qualitative analysis, on the other hand, allows the researcher to dig underneath the simple responses of the survey in order to explore the factors and conditions that lie behind such simple responses. Given the different levels of reality explored by each method, what may appear as inconsistent findings may merely be suggestive that the methods used are investigating separate dimensions of the reality and therefore revealing different elements of interest within the study.

According to Yin (1994), the decision regarding choice of strategy should depend on whether the research requires control over behavioural events and whether it focuses on contemporary events. In this case, it is not possible for the perceptions or views of the public to be controlled by the researcher; therefore, an experimental approach cannot be used. The objectives of this research require the researcher to explore and describe society's risk perceptions and trust. Robson (2002) concludes that where this is the case, surveys and case studies should be used. This research asks the question whether associations occur between incomplete knowledge, social trust, risk perceptions, benefit perceptions and risk acceptance. In doing so, this study tests a model developed by Siegrist (2000), which suggests an influential link between these factors can be applied in relation to the use of genetic test results by the life insurance industry in the Republic of Ireland. This objective orientates around the question "what is happening in society's world of risk perception?" and the answer will be largely descriptive in nature. A survey is a useful strategy for obtaining a quantifiable description of society's views. Social trust and risk perception are largely measured through self-assessments in interviews, focus groups and surveys in the form of questionnaires. Previous research testing similar objectives to the first main objective of the study tended to use the questionnaire approach. The research also asks the question "how are results from the statistical analysis of the present study produced or generated?" and "why do individuals respond to trust and risk questions in the manner recorded?". According to Yin (1994:6), when a "how" or "why" question is being asked about a contemporary set of events over which the investigator has little control, the use of a case study is particularly appropriate. Case studies allow for description, understanding and explanation of the area of study. For these reasons, a survey, followed by a case study, were used to gain the information required for this research.

Thos research thus accommodates the deductive and inductive; the extensive and intensive; and theory-testing and theory-building approaches to research. Therefore, it is possible to recognise and accommodate the central features of critical realism by attempting to preserve a 'scientific' attitude towards social analysis as well as acknowledging the importance of people's meanings (Layder, 1993:16). The aim of the study is to convey the 'textured' or interwoven nature of the different dimensions of social reality as depicted by a realist philosophy, giving the study the ability to capture real-world complexity (Perry *et al.,* 1999). The deductive and extensive approach was used to elucidate associations between knowledge, trust, risk perception, benefit perception, and risk acceptance. Statistical analyses were used to obtain a general feel for the area of research given little pre-existing knowledge of the relationship between trust, risk perception and risk acceptance in Ireland with regard to insurers using genetic test results. From this perspective, it was noted that the non-occurrence of 'necessary effects' *might* be indicative that the claimed mechanism between trust, risk perceptions and risk acceptance is false or; indeed, this lack of association *may* be indicative of a separate prevailing mechanism influencing the actual events.[16] Statistical analyses also provide descriptive generalisation that impart an understanding to some level of trust evaluations, risk perceptions, benefit perceptions and risk acceptance across society.

The inductive, intensive approach was employed to develop an explanation of the results from the deductive study, and to further investigate the factors and conditions that lead to the acceptance or rejection of insurers using genetic test results. The goal of a qualitative investigation is to understand the complex world of human experience and behaviour from the point-of-view of those involved in the situation of interest. It was considered that the subjective (emotional) nature of an individual's responses was better captured by qualitative research. Interviewees' emotions can be externalised and observed by the researcher, providing an opportunity to explore the phenomenon from a subjective position. Semi-structured interviews present many possibilities for interpreting and explaining risk acceptance and risk perceptions from

[16] Qualitative research shall explore whether the non-occurrence of effects may indicate the reasoning of the latter or the former.

the different perspectives of the interviewees; a possibility not achievable using traditional positivistic methods. This research strategy is based on interpretation and understanding. It produces social scientific accounts of risk acceptance of insurers using genetic tests results by drawing on the concepts and meanings used by individuals. The individual's background, beliefs and experiences regarding insurance, insurance companies and genetic testing must be grasped in order to provide an understanding of their acceptance of risks related to insurers using genetic test results. By capturing the subjective nature of peoples' responses it is possible to construct more integrated knowledge about different dimensions of the acceptance of risks from insurers using genetic test results.

Qualitative methods are also required to deal with specific issues in the present study. One of the features of this study is that it is researching an area where knowledge is low due to the relative newness of genetic tests being used by life insurance companies and; indeed, the lack of exploration of this situation by the media. To add to this issue, it should be noted that the consequences of insurers using genetic test results lie sometime in the future and uncertainty also underlies these consequences, in that the future is open to interpretation. Therefore, there is an element of time or temporality to be found within the relationship between knowledge of insurers using genetic test results and knowledge of the consequences of such usage. Given the nature of this issue, it is expected that individuals may bring past experience to their understanding of these issues (Sellars, 1908:600). Therefore judgements of the consequences of insurers using genetic test results may not be based upon present knowledge, but upon past experience and interpretation of those incidents or encounters in light of insurers using genetic test results. It is not possible for the quantitative survey analysis to gain an understanding of these experiences and how they influence the 'knowledge' of individuals. However, qualitative methods are perfectly designed for availing and allowing for an interpretation of this aspect of the present study. Hence, the present study relies quite heavily upon qualitative methods in explaining knowledge, trust evaluations, risk perceptions, benefit perceptions and risk acceptance.

4.5 Validity

It is important to consider the quality of the research being conducted and whether the research can be considered valid. Research validity refers to the degree in which the research carried out measures what it is designed to measure (Oppenheim, 1979). If the measure corresponds accurately to the concept it is designed to measure, the measure can be considered valid. Yin (1994:33) outlines four principle tests of validity as external validity, construct validity, internal validity and reliability. Fulfilment of these criteria ensures that factors external to those that the research is analysing do not influence the results from the study. However, according to Healy and Perry (2000:121) such an analysis of validity is positivist in nature. Constructivists also identify the following critical criteria as withholding the quality of research – (those being) credibility, consistency or dependability, and applicability or transferability (Guba and Lincoln, 1985). Given that a judgement regarding the quality of the research should be made in accordance with the paradigm followed by the research, the above two sets of criteria cannot be applied to the present study, as they do not reflect the realist paradigm on which the research is based. Healy and Perry (2000) note the validity criteria for realism as ontological appropriateness, contingent validity, multiple perceptions of participants and of peer researchers, methodological trustworthiness, analytic generalisation and construct validity. The following section shall examine each of these criteria and their impact upon the present study.

4.5.1 Ontological Appropriateness

According to Healy and Perry (2000:125), the first step in assessing this study is to ask if the research is dealing with complex social phenomena outside people's minds involving reflective people. This reality consists of the independent creations of people's minds, such as art, science, language, ethics and institutions (Healy and Perry, 2000:121). The present study examines public acceptance of insurers using genetic test results and investigates the factors and conditions that influence this relationship. In doing so, the influence that science (genetic testing) has upon the institution of the life insurance industry is examined, and the impact that ethics and a further institution, the Irish Government, has upon this situation is explored. The use of language is also accounted for, in that interpretations of the factors that may influence risk acceptance are determined from the point of view of those individuals participating in the study. In addition, the research acknowledges that people make choices regarding risk acceptance and are influenced by phenomena, such as ethics and institutions, and in turn people are affected by these creations. Hence, the 'context', 'situated activity', 'setting' and 'self' are investigated and described in order to account for these complexities and the reality of an open system. This research can, therefore, be deemed as ontologically appropriate.

4.5.2 Contingent Validity

According to Lincoln and Guba (1985:290), in positivist research internal validity is assessed by the extent to which changes in a dependent variable can be attributed to controlled variation in an independent variable. However, this is dependent upon a closed system, whereby it is possible to control the conditions required for a causal mechanism to be created. From a realist position, this is not possible, as social research is conducted within an open system, where, unlike positivists' assumptions, actors do not make decisions in a mechanical manner which can then be manipulated to develop direct cause-and-effect paths (Healy and Perry, 2000). Causal impacts are not fixed but are contingent upon their environment. Realism research discovers knowledge of the real world by naming and describing broad, generative mechanisms that operate in the world (Bhaskar 1979; Perry *et al.*, 1999). According to Pawson and Tilley (1997:152), the intention of realism research is to develop a family of answers that cover several contingent contexts and different reflective participants, albeit imperfectly. Thus, the second quality criterion for realism research is contingent validity, that is, validity about generative mechanisms and the contexts that make them contingent (Healy and Perry, 2000). This study intends to understand if and why there is a relationship between risk perception and risk acceptance, rather than just describing an influential pattern. To obtain information on this relationship, a variety of information sources are obtained through theoretical exploration and by surveying society's knowledge, trust, risk perceptions, benefit perceptions and risk acceptance of life insurers using genetic test results (Patton, 1990; Yin, 1994). Descriptions of the cases to be studied are given and the 'context', 'setting' and 'situated activity' of the study discussed. It can, therefore, be upheld that the study should satisfy the criterion of contingent validity.

4.5.3 Multiple Perceptions of Participants and of Peer Researchers

Rather than being supposedly value-free, as in positivist research, or value-laden as in constructivist research (Lincoln & Guba, 1985), realism is instead value-cognisant; conscious of the values of human systems and of researchers. Within this study's attempts to be value-aware, it relies upon multiple perceptions of a single reality in order to imperfectly and probabilistically apprehend the real world (Healy and Perry, 2000:123). It is recognised that perceptions have a certain plasticity (Churchland, 1979) and that there are differences between reality and people's perceptions of reality. According to Dobson (2002), the critical realist agrees that our knowledge of reality is a result of social conditioning and, thus, cannot be understood independently of the social actors involved in the knowledge derivation process. However, it is possible to take issue with the belief that reality itself is a product of this knowledge derivation process. This study acknowledges that real objects are subject to value-laden observation. However, reality and the value-laden observation of reality operate in two different dimensions, one intransitive and relatively enduring; the other transitive and changing.

Hence, although this study is value-aware, and is conscious of the value-laden nature of individuals' views, it is believed that these views are not entirely representative of reality. This view is in accordance with the assumptions underlying the critical realist paradigm which requires the researcher to acknowledge that the perceptions of an individual do not reveal reality; rather a participant's perceptions reveal a particular view of reality and this view, along with other participants' views, provide a clearer picture of reality (Healy and Perry, 2000:123). Hence, to have a more holistic view of the 'real' situation surrounding the risks related to insurers using genetic test results, multiple perceptions of this situation are obtained and analysed. These multiple perceptions involve triangulation of several data sources and of several peer researchers' interpretations of those triangulations (Healy and Perry, 2000). This study examines previous researchers' work regarding the risk perception and risk acceptance of individuals (see chapters 3 and 4), surveys the perceptions of 180 individuals and conducts in-depth analysis of four cases regarding their perceptions and acceptance of risk.

4.5.4 Methodological Trustworthiness

Methodological trustworthiness refers to the degree to which the study can be audited by developing a case study database and the use of quotations in the written report (Healy and Perry, 2000). The study clearly outlines the case selection and data analysis, and provides quotations in the appendices (see Appendix D.1). In addition, it is possible for individuals to request a copy of the transcripts (prior approval from the interview participants has been sought). In doing so, it is attempted to ensure that the study satisfies the criterion of methodological trustworthiness by ensuring the ability of future research to audit the study. It is also possible for the survey to be audited by individuals as there is a survey response database.

4.5.5 Analytic Generalisation

The fifth criterion is 'analytic generalisation' (Yin, 1994) that is theory-building (Healy and Perry, 2000). Healy and Perry (2000) note that realism researchers build and confirm a theory before its 'generalisability' is tested, which is the primary concern of positivism. The study intends to explore a theory regarding the relationship between risk perception and risk acceptance and to investigate the factors and conditions that underlie this relationship. To do so, a theory that suggests that incomplete knowledge, trust, risk and benefit perception are factors that affect risk acceptance is tested and built upon and a final theory of risk acceptance is presented. It is not intended in the present study to test the theoretical findings from the study, although the data is presented in a form that may be tested in future research. Therefore, the aim of the research is theory development rather than theory-testing.

4.5.6 Construct Validity

Construct validity refers to how well the information about the constructs in the theory being built is measured in the research (Healy and Perry, 2000). Construct validity of the survey and the case study process is ensured through the use of multiple sources of data (Yin, 1984). Triangulation is a feature of the present study, and aids the fulfilment of the requirement of validity. In addition, to ensure construct validity of the survey and interviews, prior theory is used from risk management and trust theory in order to define the constructs of the study. Through the use of face-to-face questionnaires and semi-structured interviews, it is possible to obtain multiple perceptions about the single reality surrounding acceptance of insurers' use of genetic test results. In addition, a 'chain of evidence' is maintained, permitting an external observer to follow the origin of the research questions, through the survey findings and to the ultimate case study conclusions (Yin, 1994:98). The questionnaire and interview schedules are, therefore, included in the Appendices (see Appendix A.1 and Appendix D1). All interviews are recorded by tape and transcribed in order to maintain evidence of the raw data and to ensure methodological transparency and reliability. The following section shall examine, in more detail, the methods used to obtain the data required to complete the present study.

4.6 The Survey

As mentioned above, a survey is considered suitable to detect an influential link between incomplete knowledge, social trust, risk perception, benefit perception and risk acceptance. A survey can be defined as research that "…entails the collection of data on a number of units and usually at a single juncture in time, with a view to collecting systematically a body of quantifiable data in respect of a number of variables which are then examined to discern patterns of association" (Bryman, 1989:104). Questionnaires are a common form of obtaining data for this type of research and involve the development of questions that can be scored, followed by statistical analysis of these scores.

4.6.1 Survey Approval

Approval was sought and received for the final draft of the questionnaire from the University of Limerick Research Ethics Committee. The Ethics Committee agreed that the study design, the related investigation techniques and the questions involved did not promote any risks to participants' well-being or safety (August, 2004).

4.6.2 The Questionnaire

Surveys can be classified according to the method of communication used, that is, whether the surveys are conducted via mail, telephone or personal (face-to-face) (Tom and Hawkins, 1976). There is no right or wrong choice of survey approach; indeed, each approach entails its own particular advantages and disadvantages. With regard to the mailed questionnaire, despite the advantages related to cost savings and the distribution of the sample, there may be a very low response rate as people may not wish to give the time needed to respond to the questions, feeling no obligation to respond due to the lack of presence of an interviewer. The design of the mailed questionnaire must be simpler than those used in face-to-face interview and telephone interview, thereby, limiting the complexity of the questions asked. Unlike face-to-face interview and telephone interview, the mailed questionnaire provides no means for control of question order or for developing a rapport. The quality of the response is poorer in mailed questionnaires than personal interviews.

Unlike the mailed questionnaire, in all interview situations, there is the opportunity for greater care in communicating questions and in eliciting information. Information obtained by mail and telephone questionnaire is limited to the written and verbal responses of subjects to prearranged questions. Unlike face-to-face interviews, telephone questionnaires may result in respondents' unwillingness to disclose personal information or their personal opinion over the phone, but if allowed to talk in the comfort and security of their own home in a face-to-face situation, individuals may be more forthcoming with information. In addition, in a face-to-face interview situation, the interviewer has the advantage of observing the behaviour of the person interviewed, allowing perceptions of the interviewees' attitude toward certain questions to be noted. By completing the questionnaire via face-to-face survey, it is easier to establish a rapport with the respondent and, therefore, it is possible to attract a higher number of respondents than a mailed questionnaire or a telephone interview. For these reasons, it was decided to conduct a self-completed questionnaire through face-to-face interview in order to obtain the views of the general Irish public in relation to knowledge, social trust, risk perception, benefit perception and acceptance of risk management strategies.

The survey was divided into six sections, with each section further sub-divided into sets or groups of questions. Each section was designed to obtain information regarding either knowledge, social trust, risk perceptions, benefit perceptions, risk acceptance or socio-demographic information (for full copy of questionnaire see Appendix A.1, for breakdown of main sections of questionnaire see Appendix B2).

4.6.2.1 Section One – Knowledge

The first section was related to the degree of knowledge individuals believe themselves to have about the use of genetic test results by the life insurance industry. Three distinct questions were asked and were differentiated based upon the specificity of the question. For example, the first question asked individuals to indicate how much they know about genetic testing, while the second question went one step further, asking how much the respondent knows about the use of genetic test results by life insurance companies in their assessment of whether or not they will insure an applicant (see Appendix B.2). Three specific contexts of knowledge were questioned to ensure that a full range of the respondents' knowledge of genetic testing and its use by life insurance companies was accounted for, from the very general to the more specific. By providing three alternative assumptions of knowledge, the present study differs somewhat from the Siegrist and Cvetkovich (2000) study, where only one assessment of knowledge was documented. Siegrist and Cvetkovich (2000) questioned respondents' knowledge regarding 25 items making it unsuitable to ask more specific knowledge questions as was possible in the present study. Given that this study explores knowledge regarding only one item, it does not have the same wording constraints as the Siegrist and Cvetkovich (2000) study. By asking more than one question regarding a respondent's knowledge, a fuller picture is provided regarding the level of knowledge of an individual.

Other differences between the knowledge questions in the present study and the Siegrist and Cvetkovich (2000) study relate to the type of question asked. Siegrist and Cvetkovich (2000) questioned knowledge regarding the risks and benefits associated with numerous items, unlike the present study which examines more general knowledge relating to the use of genetic test results by the life insurance industry, and not just knowledge of risks and benefits. Reasoning for this difference is related to the necessity of questioning knowledge of risks and benefits, for it can be assumed that lacking both specific and general knowledge about the use of genetic test results by the life insurance industry, most people would not directly assess the risks and benefits associated with the use of this technology. Hence, there was little requirement to question respondents' knowledge of the risks and benefits relating to insurers using genetic test results as part of the survey. Furthermore, the knowledge assessment scales also differ between the two studies. Scales for the Siegrist and Cvetkovich (2000) study are based on a scale ranging from 1 (know almost nothing) to 7 (know a lot). Respondents of the present study were asked to specify their knowledge to each question based upon a five-point knowledge scale, where the endpoints are labelled "no knowledge" (1) and "perfect knowledge" (5). As a five point Likert scale was used for the other sections measuring trust, risk perceptions, benefit perceptions and risk acceptance, it was decided to pursue this uniformity with regard to knowledge assessment to enable ease of response and analysis. The wording also differed, as it was believed that "know almost nothing" and "know a lot" were not specific enough, leaving too much room for individual interpretation. "No knowledge" and "perfect knowledge" were considered less ambiguous and more precise.

To understand if the three items designed to measure knowledge were indeed measuring the condition that they were designed to measure,

Pearson's coefficient was used to analyse the correlation between the three items measuring knowledge. It was predicted that at a higher level of knowledge of genetic testing, knowledge of insurers' use of genetic test results should be higher, as should knowledge regarding current policies guiding life insurers' use of genetic test results. A positive correlation between knowledge of current policies and insurers' use of genetic test results with knowledge of genetic testing was, therefore, predicted. Hence, the tests for these variables were one-tailed, given that before the data was collected, a special kind of relationship was predicted.

Table 4-4 Knowledge Correlations

	Knowledge 1	**Knowledge 2**
Knowledge 1		
Knowledge 2	.372(**)	
Knowledge 3	.304(**)	.549(**)

(n = 180)
** Correlation is significant at the 0.01 level (1-tailed).

The SPSS output above provides a matrix of the correlation coefficients for the three variables, N = 180 (see Table 4-4). Knowledge item 1 is positively related to knowledge item 2, with a Pearson Correlation Coefficient of $r = 0.372$ (see Appendix B.2 for list of items). The output also demonstrates that knowledge item 1 is positively related to knowledge item 3, although this relation is weak, $r = 0.304$, $p<0.01$. In addition, knowledge item 2 is strongly related to knowledge item 3 ($r = 0.549$, $p<0.01$). This illustrates that as knowledge of the use of genetic test results by the life insurance industry increases; knowledge of policies governing life insurance companies' use of genetic test results also increases. Knowledge of the use of genetic test results by life insurance companies accounts for 30.14 per cent of the variability in knowledge of the policies governing life insurance companies' use of genetic test results (R^2). The relatively weak relationship between knowledge of genetic testing and the other two variables suggests that this particular variable may not be measuring the same dimension of knowledge as the other two variables (accounts for just 13.8 per cent of the variability of knowledge of the use of genetic test results by life insurance companies, and just 9.2 per cent of the variability of knowledge of the policies governing the use of genetic test results by life insurance companies).

To obtain a pure measure of the relationship between the three knowledge items, it was necessary to perform a partial correlation where knowledge of genetic testing was held constant. The partial correlation between knowledge items 2 and 3 is $r = 0.493$, $df = 177$, $p \leq 0.001$, which is just slightly less than the correlation when the effect of knowledge of genetic testing was not controlled for ($r = 0.549$, $p<0.01$). In terms of variance, the value of R^2 is 0.243, which means that knowledge of the use of genetic test results by life insurance companies still accounts for 24 per cent of the variation in knowledge of the policies governing the use of genetic test results by life insurance companies.

When controlling for knowledge item 2, the partial correlation between knowledge item 1 and knowledge item 3 is as follows: $r = 0.128$, $df = 177$, $p \leq 0.087$, which is noticeably less than the correlation when the effect of knowledge of life insurance companies' use of genetic test results is not controlled for ($r = 0.304$). In terms of variance, the value of R^2 is 0.016, which means that knowledge of genetic testing only accounts for 1.6 per cent of the variation in knowledge of the use of genetic test results by life insurance companies. As such a truer measure of the role of knowledge of genetic testing has been measured.

When controlling for knowledge item 3, the partial correlation between knowledge item 1 and knowledge item 2 is $r = 0.258$, $df = 177$, $p \leq 0.001$, which is considerably less than the correlation when the effect of knowledge of the policies governing life insurance companies' use of genetic test results is not controlled for (0.372). In terms of variance, this accounts for 6.7 per cent of the variation in knowledge of the use of genetic test results by life insurance companies. As such, a truer measure of the role of the knowledge of genetic testing has been measured.

The above results suggest that the first item to gauge knowledge does not explain much of the variation in the other two items. Perhaps this advocates the non-use of the first item when attempting to understand this sample population's knowledge of the use of genetic test results by the life insurance industry. It may be that it was too general a question and not as specific for the risk measures which this study is designed to investigate. However, before making any such decision, principal components analysis was completed on the three variables to see if more than one dimension of knowledge was being measured.

Principal components analysis can be used to ensure against the inclusion of unrelated items in a scale, thereby, increasing reliability of the survey. One single Principal Components Analysis with varimax rotation was conducted on all three knowledge items. A correlation matrix was used to eliminate problems associated with variables with the same measurement but with very different variances. When the significance values from the correlation matrix were examined, there were no variables where the majority of values were greater then 0.05. There were also no correlation coefficients above 0.9. Therefore, multicollinearity was not a problem for this data, meaning that there was no need to consider eliminating any variables at this stage. The Kaiser-Meyer-Olkin test reveals a value of 0.62, which identifies that factor analysis is appropriate for this data. Kaiser (1974) recommends regarding values greater than 0.5 as acceptable (values below this reflect a need to rethink which variables to include). The KMO values for individual variables were all above 0.5, which suggested that there was no need to exclude any variables from the analysis. For these tests, Barlett's test was highly significant ($p<0.001$), indicating that there are some relationships between the variables and, therefore, factor analysis is appropriate.

The initial three items were described by just one main component, which accounted for 60.84 per cent of the variation. This component loaded highly on knowledge of the use of genetic test results by life insurance companies (item 2) and knowledge of the policies governing that usage (item 3) (0.843 and 0.810 respectively), while knowledge of genetic testing (item 1) loaded considerably lower on 0.677. When the number of factors to be extracted was increased to two, taking Joliffe's criterion of retaining factors with Eigenvalues above 0.7, it was possible to complete a rotated component matrix (Field, 2004: 437). It was obvious that the two items measuring knowledge 2 and 3 were one component, while knowledge item 1 loaded highly as a second component (see Table 4-5). This test further suggested that the three items measuring knowledge should be considered separately, and only the two items measuring knowledge of the use of genetic test results by life insurance companies and knowledge of the policies governing that usage should be used as the factor to measure knowledge of the use of genetic test results by the life insurance industry in this sample.

Table 4-5 Rotated Component Matrix(a) Knowledge

	Component	
	1	2
Knowledge item 3	.894	
Knowledge item 2	.828	
Knowledge item 1		.978

Extraction Method: Principal Component Analysis.
Rotation Method: Varimax with Kaiser Normalization.
a Rotation converged in 3 iterations.

4.6.2.2 Section Two – Risk Perception and Benefit Perception

The second section was designed to explore risk and benefit perceptions of life insurers using genetic test results to decide if they will insure an applicant. The questions were not directly related to any particular study, however, they were designed to measure risk and benefit perceptions similarly to the Siegrist (2000) and Siegrist and Cvetkovich (2000) studies. Unlike these studies, where 'riskiness' of 25 items to society were explored, the present study concentrates on just one item and can, therefore, afford to question risk perception in a more direct and specific manner (see Appendix B.2). The present study also differs from the Siegrist (2000) and Siegrist and Cvetkovich (2000) studies, in that two sets of statements were constructed to measure risk and benefit perceptions, with each set measuring risk or benefit perceptions firstly to Irish society and secondly to the respondent and their family (see Appendix B.2). When measuring risk perception it is important to take into account the framing of the risk in relation to perceived personal or general risk. According to Sjoberg (2002), people do not make the same estimate when they rate personal risk (to themselves), family risk and general risk (to people in general). The fact that people most often claim to be less subjected to risk than others is referred to as risk denial or a phenomenon termed unrealistic optimism (Sjoberg, 2000:2). Therefore, perception of risk in the present study was constructed in such a manner as to include both personal and general risk and benefit targets, ensuring a more accurate measurement of such perceptions. Finally, rather than questioning the riskiness of certain technologies, agreement with statements regarding the risks and benefits related to life insurance companies using genetic test results was examined. A Likert five-point scale was used, measuring agreement from strongly disagree (1) to strongly agree (5). The notation of a number from this scale after the risk perception/benefit perception statement indicates level of agreement. It can be considered that strong disagreement with each of the statements is representative of no risk/benefit at all, while strong agreement is indicative of very high risk/benefit perceived as a result of life insurers using genetic test results to decide if they will insure an applicant. This differs from the Siegrist and Cvetkovich (2000) study in that assessment of risk and benefit perceptions were based upon a seven point scale ranging from "not at all risky/beneficial" (1) to "very risky/beneficial"(7). Reasoning behind this difference is that in order to ensure ease of analysis, each set of factors used the same Likert scale. This allows responses to be directly comparable.

It was predicted that at high levels of risk perception for Irish society there should be high levels of risk perception for the family. And where there are low levels of risk perception for society, there should be low perceived risk observed for family. Therefore, correlation tests for these variables were one-tailed, because prior to data collection a special kind of relationship was predicted. Looking at the correlation between perceived risk for family and society, Pearson's r reveals a strong correlation between the two measures of risk perception, $r = 0.82$, $p<0.001$. Perceived risk for society accounts for 67.2 per cent of the variability in perceived risk for family. Due to such a strong correlation between these two items, it can be said that they are measuring the same condition.

Similar to risk perception, it was predicted that at high levels of benefit perception for Irish society there should be high levels of benefit perception for the family. And conversely, where there are low levels of benefit perception for society, it was expected that there should be low perceived benefit for family. Similar to the tests for risk perception, the test for these variables were one-tailed given the predicted relationship. With regard to the correlation between perceived risk for family and perceived risk for society, Pearson's r reveals a strong correlation between these two measures of benefit perception ($r = 0.793$, $p<0.001$). Perceived benefit for society accounts for 62.9 per cent of the variability in perceived benefit for family. Due to such a strong correlation between these two items, it can be said that they are measuring the same condition.

4.6.2.3 Sections Three and Four – Trust in the Government and Trust in the Insurance Industry

The third and fourth sections of the questionnaire were designed to measure trust in the Irish Government and trust in the insurance industry to manage life insurance companies' use of genetic test results to decide if they will insure an applicant (see Appendix B.2). Eleven statements measured respondents' trust in the two entities, each statement measuring a particular dimension considered representative of trust. The statements were selected from a review of previous work on trust (see Section 3.4.2), in particular the research conducted by Poortinga and Pidgeon (2003). Answers were given on a five point scale from strongly disagree (1) to strongly agree (5), similar to the Poortinga and Pidgeon (2003) study, whereby five responses ranged from totally disagree (1) to totally agree (5).

4.6.2.3.1 Trust in the Government

To examine whether the 11 statements on trust, designed to measure the underlying dimensions of trust, all measure trust, one single Principal Components Analysis with varimax rotation was conducted for the case of the Government's management of insurers' use of genetic test results. A correlation matrix was used to eliminate problems associated with variables with the same measurement but with very different variances. When the significance values from the correlation matrix were examined, there were no variables for which the majority of values are greater then 0.05 and there were no correlation coefficients above 0.9. Therefore, there it was considered that there should be no issues with regard to the singularity of data and multicollinearity was not considered a problem for this data. This meant that there was no need to be concerned about eliminating any variables at this stage. The Kaiser-Meyer-Olkin test revealed a value of 0.77; which identifies that patterns of correlations are relatively compact, therefore factor analysis was expected to yield distinct and reliable factors. Given these finding, factor analysis was considered appropriate for analysis of this data. The KMO values for individual variables were all above 0.5, once again suggesting that there is no need to exclude any variables from the analysis. For these tests, Barlett's test was highly significant ($p<0.001$) and, therefore, it was determined that factor analysis is appropriate for this set of data. Principal Components Analysis revealed four factors with Eigenvalues above 1 (see Table 4-6).

Table 4-6 Rotated Component Matrix(a) – Trust in the Government – Four Components

	Component			
	1	2	3	4
Government – competent	.883			
Government – skilled	.849			
Government - fair decision-making	.609	.427		
Government - does good job	.596			
Government - listens to public concerns		.896		
Government - listens to ordinary people		.875		
Government - acts in public interest	.461	.498		
Government - changes policies without good reason			.843	
Government - distorts facts			.816	
Government - provides information				.758
Government - influenced by life insurance companies				-.714

Extraction Method: Principal Component Analysis.
Rotation Method: Varimax with Kaiser Normalization.
a Rotation converged in 6 iterations.

Items in Factor 1 represent those items designed to measure *Fairness, Competence* and one item out of three, measuring *Care*. This factor accounts for 23.93 per cent of variability after rotation. The second factor accounts for 20.73 per cent of variability and includes items representing *Fairness* and all of the items related to *Care.* The third factor, with 13.91 per cent of variability, relates to *Credibility* and *Reliability* and the fourth factor accounts for 11.94 per cent of variability include items designed to measure *Openness* and *Integrity*.

It was decided to reduce the number of factors presented in the above study to two, due to the obvious overlap, the small percentage of variability represented by the first two factors and the last two factors, and in order to compare the dimensions of trust obtained through the present study with previous studies. The scree plot associated with the PCA illustrates that the first two factors have Eigenvalues well above 1 (4.071 and 1.478 respectively), however, the third and fourth factors rest barely above the Eigenvalue 1 (1.172 and 1.037 respectively) (see Figure 4-4). Although Kaiser recommends retaining all factors with Eigenvalues above 1, the last two factors represent much lower variation and provide less meaning to the analysis.

Figure 4-4 Scree Plot – Trust in the Government

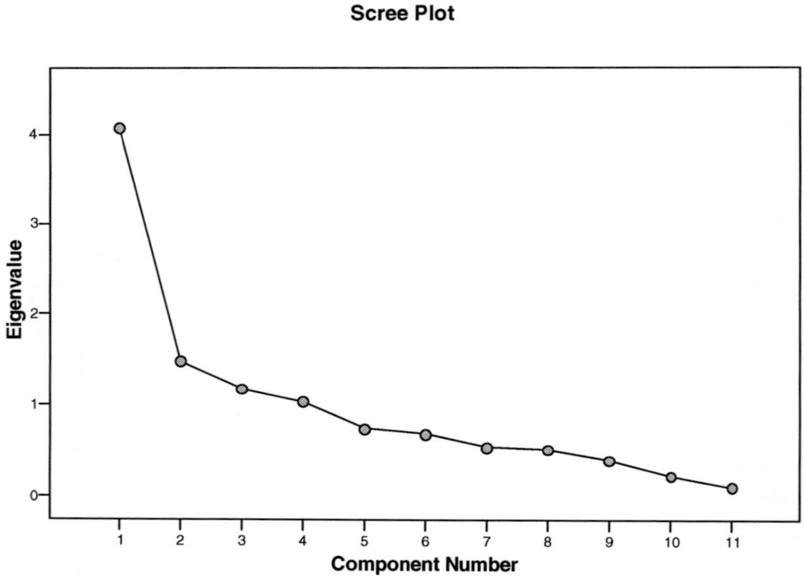

The total variance explained by the two-factor model is 50.45 per cent, this represents 20.1 per cent less than the four-factor model presented above. The first component, which accounts for 33.84 per cent of the variance, comprises the items designed to measure *Care, Fairness* and *Competence*. This could be interpreted as a general trust component. The second component accounted for 16.6 per cent of variance of the original items, is concerned with *Credibility, Reliability and Integrity*. The latter component found here reflects the sceptical view on how the Government manages the risks related to insurance companies accessing genetic test results, as illustrated by Poortinga and Pidgeon (2003). However, one item, designed to measure *Openness,* is excluded from the analysis (see Table 4-7).

Table 4-7 Rotated Component Matrix(a) – Trust in the Government – Two Components

	Component	
	1	**2**
Government – competent	.758	
Government – skilled	.753	
Government - fair decision-making	.741	
Government - listens to ordinary people	.737	
Government - listens to public concerns	.713	
Government - acts in public interest	.701	
Government - does good job	.639	
Government – changes policies without good reason		.806
Government – distorts facts		.756
Government - influenced by life insurance companies		.472
Government - provides information		

Extraction Method: Principal Component Analysis.
Rotation Method: Varimax with Kaiser Normalization.
a Rotation converged in 3 iterations.

Upon completion of a three-factor analysis, the same result is found (see Table 4-8). It is perhaps plausible to believe that this item (Openness) does not measure trust in the Government with regard to the management of insurers' use of genetic test results and, in fact, measures something entirely different.

Table 4-8 Rotated Component Matrix(a) – Trust in the Government – Three Components

	Component		
	1	2	3
Government – competent	.859		
Government – skilled	.830		
Government - does good job	.630		
Government - fair decision-making	.611	.439	
Government - listens to public concerns		.902	
Government - listens to ordinary people		.881	
Government - acts in public interest	.506	.526	
Government - provides information			
Government - distorts facts			.765
Government - changes policies without good reason			.755
Government - influenced by life insurance companies			.582

Extraction Method: Principal Component Analysis.
Rotation Method: Varimax with Kaiser Normalization.
a Rotation converged in 6 iterations.

The above analysis demonstrates that it is likely that Openness is not measuring the same condition as the remaining items. Therefore, a PCA on trust in the insurance industry was conducted to decide whether or not this item should be kept in future analyses of trust in this section of the study.

4.6.2.3.2 Trust in the Insurance Industry

To examine whether the 11 statements on trust evaluate the underlying dimensions of trust, one single Principal Components Analysis with varimax rotation was conducted for the case of the insurance industry management of insurers using genetic test results. When the significance values from the correlation matrix were examined, there were no variables for which the majority of values were greater then 0.05 and there were no correlation coefficients above 0.9. Therefore, no problems were anticipated with regard to singularity of data, and multicollinearity was not expected to be a problem for these data. This meant that there was no need to consider eliminating any variables at this stage. The Kaiser-Meyer-Olkin test revealed a value of 0.815, signifying that patterns of correlations were relatively compact, hence, factor analysis was expected to yield distinct and reliable factors. Therefore, it was believed that factor analysis should be appropriate for this data. The KMO values for individual variables were above 0.5, which once again suggested that there was no need to exclude any variables from the analysis. For these tests, Barlett's test was highly significant ($p<0.001$) and, therefore, it was determined that factor analysis is appropriate for this set of data.

The initial 11 items were described by three main components which accounted for 66.29 per cent of the variation. The first component, which accounted for 42.81 per cent of the variance before extraction and 26.6 per cent after extraction, comprised the items designed to measure *Care, Fairness* and to a lesser extent *Competence* (see Table 4-9). This could be interpreted as a general trust component. That is, it represents a general evaluation of insurance industry's management of the use of genetic test results by life insurance companies. The second component accounted for 20.52 per cent of variance of the original items, and was concerned with, *Credibility, Reliability* and *Integrity.* This, once again, reflects a sceptical view on how the insurance industry manages the risks related to insurance companies accessing genetic test results and is, therefore, labelled as a 'scepticism' component (Poortinga & Pidgeon, 2003). The third factor accounted for almost the same percentage of variance as the second factor, 19.2 per cent, and represents *Competence.* Item 11, designed to measure *Openness,* was not considered as part of any of the above three factors (see Appendix B.2 for list of Items). As in the above PCA on trust in the Government, this result suggests that item 11 is not considered as part of any components of trust.

Table 4-9 Rotated Component Matrix(a) – Trust in the Insurance Industry – Three Components

	Component		
	1	2	3
Insurance industry - listens to ordinary people	.924		
Insurance industry - listens to public concerns	.910		
Insurance industry - acts in public interest	.625		
Insurance industry - fair decision-making	.565		
Insurance industry - does good job	.460		.440
Insurance industry - distorts facts		.871	
Insurance industry - changes policies without good reason		.784	
Insurance industry - influenced by life insurance companies		.763	
Insurance industry – skilled			.876
Insurance industry – competent			.825
Insurance industry - provides information			

Extraction Method: Principal Component Analysis. Rotation Method: Varimax with Kaiser Normalization.
a Rotation converged in 5 iterations.

It was decided to explore what the table of factor loadings would look like when the number of factors are reduced to two, given that the third factor accounts for an Eignenvalue of just 1.028, barely above 1 (see Figure 4-5). Once again, although Kaiser recommends examining all factors with Eigenvalues above 1, the third factor represents much lower variation and provides less meaning to the analysis and, therefore, it was considered that a decision should be made with regard to whether or not this factor should be discarded from future analyses.

Figure 4-5 Scree Plot – Trust in the Insurance Industry

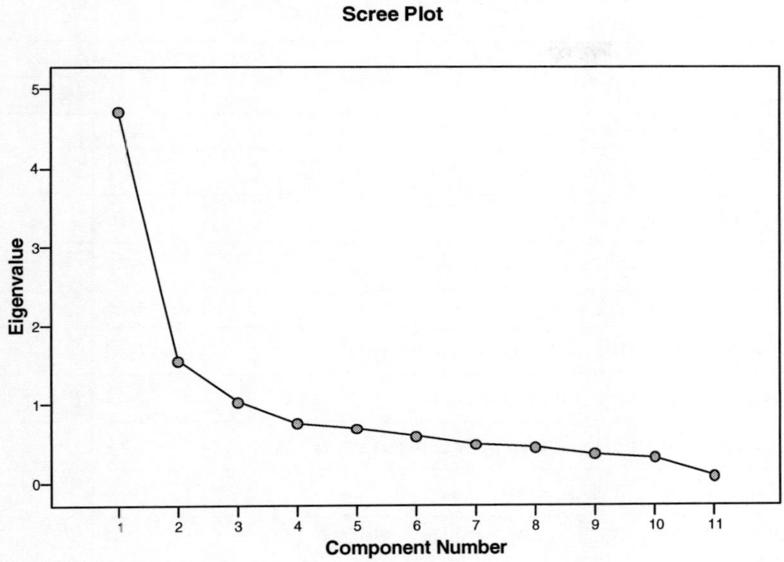

The total variance explained by the two-factor model was 56.94 per cent. This represents 9.35 per cent less than the three-factor model presented above (see Table 4-10). The first component, which accounted for 36.2 per cent of the variance, comprised the items designed to measure *Care, Openness, Fairness* and *Competence*. Once again, this could be interpreted as a general trust component. The second component accounted for 24.28 per cent of variance of the original items and was concerned with *Credibility, Reliability and Integrity*. The latter component found here reflects the sceptical view on how the Government manage the risks related to insurance companies accessing genetic test results found in the three-factor analysis.

Table 4-10 Rotated Component Matrix(a) – Trust in the Insurance Industry – Two Components

	Component	
	1	2
Insurance industry - listens to public concerns	.801	
Insurance industry - listens to ordinary people	.762	
Insurance industry – competent	.756	
Insurance industry - acts in public interest	.675	
Insurance industry – skilled	.674	
Insurance industry - fair decision-making	.670	
Insurance industry - does good job	.633	
Insurance industry - provides information	***.525***	
Insurance industry - distorts facts		.860
Insurance industry - changes policies without good reason		.782
Insurance industry - influenced by life insurance companies		.773

Extraction Method: Principal Component Analysis.
Rotation Method: Varimax with Kaiser Normalization.
a Rotation converged in 5 iterations.

This time all of the items were placed into two factors, one that measured a general trust dimension and the second a more sceptical dimension (see Table 4-10). Item 11, designed to measure *Openness,* was included as part of the first factor measuring a general trust dimension. This corresponds with Poortinga and Pidgeon's (2003) findings, although not with the above PCA on trust in the Government (see Table 4-8). It is questionable as to why this is the case. Both analyses demonstrated that item 11 is not included as a component of the condition being measured when three factors are extracted. However, when reduced to two factors, to simplify the results and to compare with previous findings, item 11 remains missing from trust in the Government, but appears as part of Component 1 in trust in the insurance industry. As a result of the above findings, it was decided to exclude item 11 from future analyses of trust in both entities, as its pattern of appearance as a condition of trust is highly irregular and can, therefore, be deemed unstable for future analyses.

4.6.2.4 Section 5 – Risk Acceptance

The fifth section of the questionnaire examined risk acceptance and the respondents' opinion of decision-making regarding life insurers' use of genetic test results to decide if they will insure an applicant (see Appendix B2). Risk acceptance was examined through analysis of agreement with the following statement; "the life insurance industry should be allowed to use genetic test results to decide if they will insure an applicant". Agreement was, once again, measured using a five-point scale ranging from strongly disagree (1) to strongly agree (5). It was not possible to adapt Siegrist's (2000) means of evaluating risk acceptance, where acceptance was dependent upon the willingness of an individual to purchase genetically modified food. Willingness to purchase life insurance is not suggestive of risk acceptance, as many individuals need life insurance in order to purchase a mortgage or to cater for the financial needs of their family, and such a necessity may overcome the risks perceived from the usage of genetic test results by life insurance companies. Hence, it was deemed more suitable to question whether or not insurers should be allowed to use genetic test results. This is more appropriate than willingness to purchase an insurance product from an insurer that demands genetic test results. By phrasing the question as 'allowance to use genetic test results' indicates, to some extent, whether or not an individual believes that constraints should be placed upon insurers' usage of genetic tests in order to manage the perceived related risks. Therefore, acceptability of the risks and management of perceived risks was questioned through the use of this statement.

The theme of risk management was pursued by other questions in the same section, including questions regarding agreement with the current rules and regulations in the Republic of Ireland in respect of this matter and evaluation of adequate regulation by the Irish Insurance Federation (IIF) and the Irish Government of insurers using genetic test results (see Appendix A.1). The level of knowledge had to be accounted for in relation to these questions, as a low level of self-assessed knowledge of the policies governing insurers using genetic test results, gauged in Section 1 of the questionnaire, may indicate inability to accurately respond to this line of questioning. However, the capabilities of the IIF to manage insurers' use of genetic test results was questioned, to understand whether respondents' judge the IIF as capable of managing the risks perceived from insurers using genetic test results (see Appendix A.1). In addition, means to hold the Irish Government and the insurance industry accountable for its decisions regarding life insurance companies' use of genetic test results were explored (see Appendix A.1). In doing so, an understanding of respondents' impression of their involvement in the risk management process surrounding this situation and their perceptions of ability to control risk managers were examined. Furthering this particular analysis, respondents were questioned as to whether or not they believed that they have a say in the content of risk management decisions and are involved in how decisions are made regarding life insurers' use of genetic test results, as control of the risk management situation may be a factor related to risk perceptions and risk acceptance (see Appendix A.1 and Figure 3-2).

4.6.2.5 Section 6 – Socio-Demographics

The sixth, and final, section of the questionnaire was dedicated to understanding the socio-demographics of the sample population (see Appendix A.1). Gender, age and occupational status were questioned; as each of these may be factors that influence risk perception (see Sections 3.7.2 and 3.7.3). Ownership of life insurance was also examined as it is plausible that individuals who own life insurance products or, who are interested in purchasing a life insurance product(s) in the future, may respond to questions regarding knowledge, social trust, risk perceptions, benefit perceptions and risk acceptance differently than those individuals who do not have life insurance and do not intend to ever purchase any life insurance products in the future (see Section 3.7.4). The dominant or national culture surrounding an individual may also influence their responses, hence several questions examined country of residence and years of residency (see Section 3.7.1). Only the responses of those individuals who were resident in the Republic of Ireland for the last five years or more when the survey was completed were analysed as such individuals were considered to be more representative of the Irish public and were exposed to similar cultural influences. Indeed, those individuals resident in the country for five years or more were more likely to have knowledge of this situation than those who may have recently entered the country as a new resident. Finally, respondents were asked if they would be interested in taking part in a further discussion of the use of genetic test results by the life insurance industry, as a means of selecting those individuals who were interested in getting involved with the interviews as part of the case study. Following this section, there was a space provided for any additional, qualitative comments of the respondent.

As a result of the two separate evaluations of trust, it was decided that general trust and scepticism should be evaluated as separate factors in this study when examining their association with knowledge, risk perception, benefit perception and risk acceptance. Figure 4-6 represents the updated objectives of the study in the form of a diagram.

Figure 4-6 Research Framework

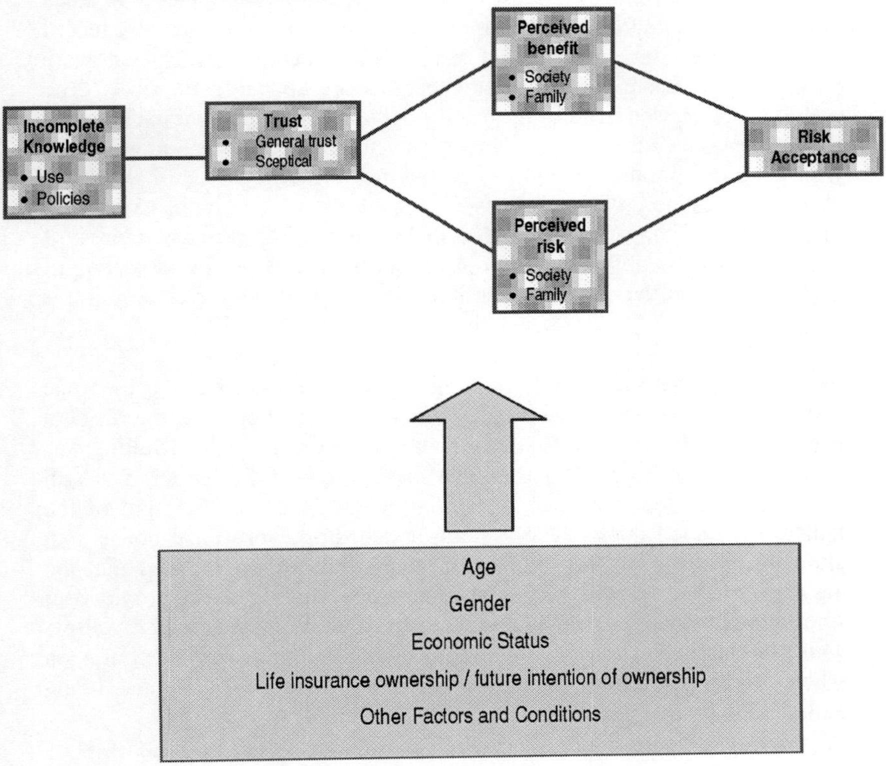

4.6.3 Sample Population

It was not the objective of this study to establish a true cross-section of the Irish population. It was intended to obtain a sample that is varied in socio-demographic terms, in order for multiple perceptions of knowledge, risk and trust from a wide variety of individuals to be studied in the analysis. It was beyond the time and financial budget of the research to complete a census of the Irish population. As it was possible for only one researcher to conduct all of the questionnaires, it was necessary that the choice of sample reflect these constraints. A total of 192 individuals living in County Limerick and County Clare successfully completed the questionnaire. The above two counties were chosen based upon their proximity to the place of residence of the interviewer and her ability to travel to these areas during the day using public transport. Those interviewed were chosen systematically (every third house) from a structured sample based on areas of different housing type, to ensure a sample population reasonably mixed in age, income, occupation and sex. The housing type chosen were residential estates in most cases where houses appeared in rows. The estates were chosen according to the population that habited each estate, for example, close to the University of Limerick, there are new estates with a young population renting or owning these houses. In the more established estates, where houses are older, larger and more expensive, there tend to be an older generation of habitants. The primary interest of the survey was to obtain perceptions of individuals from a cross-section of Irish society and the above estates were deemed suitable for obtaining this objective.

Regarding the interview system, if there was no answer from a residence visited, the house was revisited after one hour in the hope that the resident had returned. If the resident still did not answer the door, the dwelling was revisited once more after another hour had passed. If the resident was still not there after the third call, an adjoining residence on the right of the house was called upon. The system of choosing every third house and, choosing a house on the right of an unanswered door, ensured that the survey remained systematic. Two further opportunities were provided of finding a respondent after the absenteeism of an initially chosen dwelling. This systematic choice of respondents ensured that there was no confusion when choosing respondents and reduced the possibility of a house being called upon twice by mistake.

The interview was self-completed by the respondent in approximately 20 minutes. Following ethical guidelines, each individual contacted with regard to the survey was provided with information leaflets before and after the interview and an informed consent form. The context of the study was explicitly set forth in the information sheets given to the respondent and outlined in the questionnaire itself, thereby overcoming such problems as lack of specificity (see Appendix B.3 and B.4). Approximately 83 per cent of households contacted completed the survey. Upon the removal of outliers, there remained a total of 180 individuals' responses to be analysed. An outlier is an observation that can be considered 'far away' from the rest of the data. Outliers (see Section 4.6.4 – "No extreme scores") were identified through the use of visual aids whereby the researcher firstly checked the distribution of data values by levels of each categorical variable, followed by the creation of scatter plots to examine how severe the outlying observations actually were. Thirdly, box plots were used to view values that were considered extreme in relation to the rest of the data (see Appendix C.2). Twelve outliers were deleted when it was considered that the data constituted a legitimate error that could not be corrected or was too far outside the range of the remainder of the data that they distorted statistical inferences. Three outliers were accommodated where the data was considered correct and necessary to use as part of the analysis.

Appendix B.2 provides a list of the questions that were asked of participants regarding knowledge, trust, risk perception, benefit perception and risk acceptance (see Appendix A.1 for full copy of questionnaire). Data collected was assessed using the computer package SPSS. Before undertaking any of the analyses required for exploring such data, it was firstly necessary to examine whether the data was suitable for parametric or non-parametric analysis. It was decided that assumptions of parametric data are not fully satisfied, although it was considered that the parametric tests used in this study are robust enough to cater for such mild deviations from the criteria that qualify data as parametric. Consideration of the assumptions of parametric data in relation to its use in the present study is contained in the following section (see Section 4.6.4).

4.6.4 Assumptions of Parametric Data

All parametric tests have five basic assumptions that should be met for the test to be accurate (Field, 2004:37) and (Dancey & Reidy, 2002:148-149). **Normally distributed data:** The majority of variables appear normal, except for those items measuring knowledge and one item measuring Government provision of information. The Kolmogorv–Smirnov test reveals that all variables have a significant value of 0.001 (sig. less than 0.05) that indicates a deviation from normality. The reason that some of the data may be slightly skewed could be due to the subject matter under question.[17] Therefore, it is not always possible for data to be normally distributed.

Homogeneity of variance: According to this assumption, the variance should not change systematically throughout the data. In designs in which several groups of subjects are tested, variances of the populations should be approximately equal, as it is not possible to check if populations have exactly equal variances. "Approximately equal" in this study shall be taken to mean that as long as the largest variance being tested is not more than three times the smallest, there is roughly equal variances (Dancey & Reidy, 2002:148-149). In this study, variances were distributed from 0.665 to 1.343 and, therefore, satisfy this general rule of thumb. Two measures of variance, however, did not satisfy this particular requirement for parametric tests. These include measures for two variables: knowledge items 2 and 3, which were $s^2 = 0.393$ and $s^2 = 0.154$ respectively (see Appendix B.2 for list of items).

Interval data: Data should be measured at least at the interval level; this is where the data has order and can be measured in equal intervals. Counts that are considered interval include counts of income, years of education, or number of votes in an election. This study uses ordinal scales, 1-5, and it cannot be assumed that the distance between points on the scale is equal. It may be that the intervals between scale points are uneven, as, for example, the distance from the highest knowledge to the second highest may be great but the interval from the second highest to the third highest may be very close in comparison. However, it is understood that this highest self reported knowledge is higher than the second highest self reported knowledge. Nonetheless, the data used in the present study can be considered ordinal rather than interval.

---

[17] For example, if the majority of individuals do not accept the use of genetic test results by the life insurance industry, then it is likely that the data would be positively skewed toward the 'disagree' response (see Appendix C.1). Or, indeed, if the majority of participants in the survey are of the opinion that there are few benefits and many risks related to life insurance companies using genetic test results, it would be expected that distributions are positively and negatively skewed respectively (see Appendix C.1). If there is relatively low trust in the Government and/or the life insurance industry it is not likely that there will be a normal distribution, instead, the data is more likely to be positively skewed (see Appendix C.1). Finally, where knowledge of a particular subject is low and, indeed, unusual among a population, a normal curve is unlikely; whereby one side represents low knowledge and the opposite side high instead display a positively skewed distribution (see Appendix C.1).

Independence: This assumes that data from different subjects are independent. In this study, all interviews took place at separate times and separate locations. Responses to interviews were not discussed with any respondent. It can, therefore, be assumed that the behaviour of any one participant in this study did not influence the behaviour of any other.

No extreme scores: Given that the mean is sensitive to extreme scores and that parametric tests involve the calculation of the mean as central tendency, twelve outliers were removed from the data upon box plot analysis (see Appendix C.3). The presence of outliers can lead to inflated error rates and substantial distortions of parameter and statistic estimates when using either parametric or nonparametric tests (Zimmerman, 1994, 1995, 1998). Upon removing these outliers, 180 participants were left. The only outliers not to be removed from the sample were three individuals with some knowledge of the policies governing insurers' use of genetic test results. The data output displayed every individual with 'some knowledge' of the policies governing the use of genetic test results by life insurance companies as outliers. It is deemed important to include this information so that comparisons might be made between those individuals with no knowledge and those with some knowledge. Therefore, the above outliers for this one item were not removed.

4.6.4.1 MANOVA

Multivariate analysis of variance (MANOVA) was also conducted to examine whether people differed in their mean responses across trust in the two entities managing the risks related to life insurance companies using genetic test results. The assumptions underlying MANOVA are a little more complex than those already explained for parametric data.

Multivariate normality: The first assumption underlying MANOVA is to ensure that the data displays multivariate normality, that is, that each of the individuals and all linear combinations of these dependent variables are normally distributed (Dancey and Reidy, 2002:479). As it is difficult to assess multivariate normality, due to the need to look at the distributions of each dependent variable and all possible linear combinations of these to check that they are normal, it is recommended that at least each dependent variable be checked for normality (Dancey and Reidy, 2002:479). Box plots demonstrated that the data for each dependent variable in each condition of the independent variable are not normally distributed, although violations appear modest (see Appendix C.2).

Homogeneity of variance–covariance matrices: To satisfy this assumption, the univariate tests of equality of variances between groups should be met (Field, 2004: 398). There is a test of the homogeneity of variance-covariance matrices called Box's M test. However, this is a within-participants design and sample sizes are equal, therefore, Box's test is disregarded as highly unstable and impossible to use.

4.6.4.2 Regression Analysis

Regression analysis is also conducted as an extension of correlation analysis and has several assumptions attached to it that have not previously been discussed but that also require to be satisfied before performing these tests. The general purpose of multiple-regression is to learn more about the relationship between several independent or predictor variables and a dependent or criterion variable. This form of statistical analysis ensures that each predictor is controlled for the others, thereby isolating its independent effect on the criterion. This allows a fuller picture to be obtained of the effect that each of the factors measured in the study have on each other individually.

Participant numbers: It is recommended that the study use at least more than 15 participants per variable (Dancey & Reidy, 2002: 398-399). This study uses 180 participants per variable.

Variables should be linearly related to the dependent variable: Correlation analysis tests whether there is a linear relationship between variables. The correlation analysis of the present study reveals that the strength of the relationship between incomplete knowledge, social trust, risk perceptions, benefit perceptions and risk acceptance is generally weak or minimal. These results indicate that, in the majority of cases, there are imperfect linear relationships between variables; while the correlation findings also demonstrate several non-linear relationships. Given an expression of linearity between the majority of variables, it can be assumed that most variables are linearly related to the dependent variables.

Multicollinearity: No correlations correlated very highly with each other (0.8 and above) and, therefore, multicollinearity is not considered a problem.

4.6.4.3 Discussion of Violations of Assumptions

It can be observed from the above analyses of the assumptions for parametric tests that the data from the present study appears to mildly violate the majority of assumptions. However, it was considered that the parametric tests used in the study are robust enough to cope with such deviations. According to Howell (1994:394), many researchers consider the assumptions normally cited as being required of parametric tests as overly restrictive in practice, and that parametric tests are remarkably unaffected by violations of distribution assumptions. In addition, the present study has quite a large sample size thereby increasing the likelihood that the data will demonstrate significant deviations from normality (Simon, 2005). However, non-normality is considered least troublesome when using parametric tests on large samples (Simon, 2005). One can assume that the sampling distribution is normal even if the distribution of the variables in the population do not appear normal, as long as the sample is large enough, for example 100 or more observations, as is the case with the present study (Statsoft, 2003). According to the central limit theorem, when samples are large the sample means follow the normal distribution even if the respective variable is not normally distributed in the population, or is not measured very well, for example as ordinal and not interval. Thus, parametric methods, which are usually much more sensitive (i.e., have more statistical power) are in most cases appropriate for large samples. In addition, the assumption of normality is important if the results of the analysis are to be generalised beyond the sample collected. However, it is not the intention of the present study to generate universal predictions; instead, it is intended to make generalisations regarding risk acceptance in the form of a description of local mechanisms or tendencies (see Section 4.2.3 and 4.2.3.1).

With regard to the Likert scales used in the questionnaire, according to Sarle (1995) it is likely that given the tests used in the present study, it makes little difference whether the data measurement is ordinal or interval. If this statement is the case, then the use of parametric tests to analyse the parametric data in the present study should not be a cause of concern. With regard to the assumption of the homogeneity of variance, it is noted above that two measures of variance did not satisfy this particular requirement for parametric tests (see Section 4.6.4). In addition, three outliers were accommodated in the present study that serves to act as extreme scores when testing the data (see Section 4.6.4). However, given the claims for robustness of parametric tests, these violations of parametric assumptions should not affect the parametric tests undertaken on these two variables (Howell, 1994: 394). There are also minimal violations of the assumption of multivariate normality but the fact that there are equal sample sizes and a large number of participants, MANOVA should still be considered a valid test (Dancey and Reidy, 2002: 479). To conclude, the violations of assumptions underlying the tests used in the present study should not disqualify the researcher from applying each of the above parametric tests to the analysis of the questionnaire data. Section 7.7 discusses the use of non-parametric tests as opposed to the parametric tests used in the present study, and the merits and demerits of using each type of statistical test.

Linear regression was used to discover the effect of one variable on another. It is similar to correlation analysis, but while correlation analysis allows one to conclude how strongly two variables related to each other (both magnitude and direction), linear regressions answers the question "by how much will 'y' change if 'x' changes?". In other words, if x changes by a certain amount, it is possible to estimate how much y will change. Multiple-regression is able to give information regarding the ways in which the independent variables combined relate to the dependent variable, and it also gives information on how each of the variables relates to the dependent variable, separately. Logistic regression is multiple-regression but with an outcome variable that is a categorical dichotomy and predictor variables that are continuous or categorical. In other words, by using logistic regression it may have been possible to predict which of two categories ('risk acceptance' or 'non-acceptance of risk') given certain other information, for example risk perceptions, benefit perceptions, trust and knowledge evaluations. Logistic regression is a variation of ordinary regression. It is used to determine the impact of multiple independent variables presented simultaneously to produce a desired result. The main difference between the two is in ordinary regression the dependent variable is an amount (e.g., level of risk acceptance) or score, while in logistical regression the dependent variable is an event occurrence.

It is possible to argue that logistic regression may have been more suitable to use than the tests performed in the present study. It could be considered that the measurement of the dependent variables could be seen as equivalent to the measurement of an event, for example, whereby one accepts the risks related to insurers using genetic test results, or one does not. In other words, the dependent variables could be considered as binary or dichotomous. However, risk acceptance was not measured in binary form, but instead was treated as interval data whereby risk acceptance was assigned numbers such that differences between the numbers reflected differences in levels of the variable. For example risk acceptance is measured from 1 = none/low risk acceptance to 5 = high risk acceptance. Given this treatment of the variables, it can be said that logistic regression is not suitable for the study of the association between the predictors and the dependent variable in the present study. Further analysis of this decision is contained in Section 7.7.

Finally, it is important to note that the statistical approach chosen, and the choice of statistical tests used for analysing the data from the questionnaires, was corroborated by a statistician from the Statistical Consulting Unit.[18] In addition, all conclusions drawn from the analyses were reviewed and agreed with by the Executive Director of the Statistical Consulting Unit in the University of Limerick.

4.6.5 Surmounting Limitations of the Questionnaire

Four pre-tests of the questionnaire were administered to pilot groups, each containing 15-30 individuals. Feedback on the questionnaire was used to readjust and improve the questions and design of the survey, an initial attempt to overcome any misunderstanding of the questions and to ensure that the objective in asking each question was achieved. The resulting questions after the pilot testing were discussed in Section 4.6.2.

Self-reported trust and risk perception are disadvantaged in that they can induce a form of self-reflection that could bias results and require a greater number of assumptions to infer their meaning. Given that there is a need to make preliminary efforts to explore distinctions in trust and risk acceptance based on specific contexts, the study uses three items to capture an understanding of knowledge, eleven items of trust, two items to measure risk perception, two items to describe benefit perception and one item to determine risk acceptance (Cvetkovich & Lofstedt, 1999:155) (see Appendix B.2).

As the questionnaire design needed to be easy to administer by the researcher and, easily understood by the respondent, the questions tended

---

[18] All tests and the conclusions drawn from the analyses were corroborated by Dr Jean Saunders, Executive Director, Statistical Consulting Unit, Department of Mathematics and Statistics, University of Limerick

to be relatively simple. The resulting responses may be inadequate to capture complex situations or processes. Moreover, the predetermined list of questions and fixed format of a questionnaire did not allow for probing of initial responses of respondents or checking understandings of questions, making this technique inflexible for exploratory inquiry (Miller & Brewer, 2003:303). However, this limitation was overcome by the secondary stage of data collection involving semi-structured interviewing of several cases. It is necessary to obtain multiple perceptions of a single reality, therefore, respondents were initially presented with the same standardised questions, carefully worded after piloting. This ensured that each participant was asked about his or her view regarding the same aspect of reality as every other individual. The questionnaire provides a relatively simple and straightforward approach to the study of perceptions and beliefs and can be adapted to collect generalisable information from almost any group. The use of a standardised survey and, the transparency of the methods and procedures, allow relatively straightforward future assessment of the research design (Haskim, 1987).

As in previous research related to the subjects examined in this survey, Likert scales were used to rate individuals' knowledge and agreement or disagreement with the questions or statements posed. It is noted that the choice of measurement scales can affect research findings. Differences exist in the strength of the relationship between trust and risk perception when different measurement scales are used. Sjoberg (2002) demonstrates that attitude scale items (Likert scale: agree – disagree) yield a much stronger relationship between trust and risk than when trust and risk items are measured using rating scales (whereby numbers are assigned in a manner that indicates perceived size of risk and level of trust). When Likert-type items, rather than the traditional ratings of trust and risk, are used to measure trust and perceived risk, the common variance of these two dimensions doubles, from 15 per cent to 30 per cent (Viklund, 2003:730). The study used Likert-type scale items, in order to ensure the possibility of comparing results to previous research on trust and risk, as these studies also used the Likert scale in their measurement of trust and risk. However, it is noted that the use of Likert scales may affect results from the study. To reduce such effects, response alternatives were numbered so as to communicate to the respondent that the intervals between the scale points are intended to be of equal distance. In addition, by applying such Likert-type objective scales of measurement, statistical analysis of the data is more easily calculable.

It is noted that respondents may not have answered the questions truthfully, due to an unknown mixture of politeness, boredom, or a desire to be seen as agreeable (Robson, 2002:231). Data may have been affected by the characteristics of the respondents (e.g. their memory, and personality). As a self-completed questionnaire was used, it was not possible to check for misunderstandings or to inquire into a respondent's answer. Non-sampling errors, such as measurement errors and non-response errors, may also have influenced the data collected by the survey and, as a result, the research findings. Measurement error is an aspect that must be considered whenever a measurement instrument is used. Respondent characteristics, both stable (*respondent likes to disagree*) and temporary (*respondent in a bad mood*), together with imperfections of the measurement instrument (*ambiguous/misleading questions, omission of alternatives*) and problems related to the measurement process characteristics (*situational factors, interviewer effects, coding mistakes*), all combine to introduce error into measurements (Robson, 2002). Little can be done to overcome measurement problems associated with respondent characteristics; however, the survey was piloted four times to reduce measurement process characteristics.

Since some people did not answer their doors or refused to answer the survey, the respondent sample may not be representative of the population living in the areas investigated. Because of this selection bias the characteristics of those who agreed to be interviewed may be markedly different from those who declined, that is the actual sample could be considered a biased version of the population analysed. If the people who refused to answer their door to the researcher in order to complete the survey, or were never reached, have the same characteristics as the people who did answer, the final results can be considered as unbiased. However, if the people who did not answer or refused to respond to the survey have different opinions, then there is bias in the results. A high non-response rate can present several disadvantages for researchers. In the present study, the researcher attempted to overcome these problems by persuading individuals to take part in the survey face-to face. A copy of the questionnaire was left with each respondent and the researcher returned to collect the survey and to ensure that all the questions were answered. If there was no answer from a residence visited, the house was revisited twice more upon hourly intervals. If there was still no answer, an adjoining residence on the right of the house was called upon. By approaching the survey in this manner it is hoped to reduce the bias caused by non-response.

There are a number of constraints regarding resource availability, such as time, money and personnel that affect the study. Owing to such constraints, only one person conducted the surveys that contribute to the findings of the study. This posed a limitation toward the choice of sample size. It was not possible to do a census to gauge Irish public opinion on the use of genetic test results by the life insurance industry, instead, a representative sample of the population was chosen and only one researcher conducted all of the surveys. However, every effort was made to ensure that the sample chosen was representative of the cross-section of the Irish public chosen.

4.7 The Case Study

A case study is an in-depth exploration of a particular case/s (situation or subject) in the context of its environment, for the purpose of gaining depth of understanding into the issues being investigated. Yin (1993) presents Giddens' (1984) view that considers case methodology "microscopic" as it lacks a sufficient number of cases. However, Yin (1993, 1994) argues that a single case study is a satisfactory method provided it meets the established objectives of the research. The sampling principle in this intensive research design is strategic (Danermark *et al.,* 2002:170). Four types of case are used – the normal (representative) case, extreme case, extremely varied case and critical case (Danermark *et al.,* 2002:170). The cases were chosen based upon the responses of individuals to the survey. The representative case implies that the responses given reflect the average responses of the individuals surveyed. For example, if most of those surveyed have incomplete knowledge of insurers using genetic test results, this characteristic is then considered as typical for the study. The representative case is, therefore, typical of the average response pattern. The extreme case implies that the responses are odd in comparison to the representative case, although not necessarily uncommon. Extremely varied case individuals were selected based upon their differences from each other with regard to some of the dimensions considered as important to the study, such as risk perception, trust judgements and risk acceptance. The critical case was selected based upon the requirement of one particular symptom appearing that should not have done so. For example, an individual may have a high level of knowledge of life insurers using genetic test results and yet may still respond to the other questions in the same manner as the representative case.

4.7.1 Case Study using Semi-structured Interviews

Case studies involve a variety of different methods, including interviews and participant observation, where inferences can be drawn from the results without the use of 'simplifying' statistical analysis (Danermark *et al.,* 2002:171). Interviews are one of the most widely used research methods allowing researchers to generate knowledge and to attain a richer understanding of human beings (Miller & Brewer, 2003:166-167). According to Layder (1993:112), there are three types of interview. Firstly, the structured interview is similar to a survey, in that the questions are developed in advance and answered by the respondent in a particular order. This allows for ease of analysis and comparison across cases, however, it does not cater for the needs of the researcher to probe for further detail on the respondents' answers. The questions asked in the interview must be clear and unambiguous so that the respondent understands what is being asked. Unlike the structured interview, the second type, the unstructured interview, has no strict guidelines and allow the discussion to be open. However, the interviews are not necessarily concise in nature and the information provided by the respondent may not be related or; indeed, useful to the study and make the data provided difficult to analyse. Finally, the third type, the semi-structured interview, is an amalgamation of the structured interview and the unstructured interview and, thereby, overcomes the weaknesses inherent in both extremes. The semi-structured interview allows an open discussion of the subject matter and for the researcher to steer the interview to ensure that no valuable points are excluded. In addition, the use of a semi-structured structured interview ensures the collection of data at the micro level in this research, allowing individuals to tell their story and the interviewer to probe for further detail. Semi-structured interviews can also glean knowledge regarding previously unknown relationships. This may lead to a rethinking of the phenomena being studied and a richer picture of the situation (Stake, 1995:47). Therefore, to encourage the development of theory, semi-structured interviews were used.

The study examined multiple cases as these have the advantage of increasing confidence in the robustness of the theory developed from research findings. By classifying individuals into distinct cases and interviewing them separately, it was possible to identify inter-relationships between those categories. This ensured that the same questions devised from the results of the survey could be asked for each case allowing comparison between cases. Also, information and concepts gained from previous cases were used to gain further information from other cases.

Continuous interplay between deduction and induction ascertain the ongoing process of theory advancement (Parkhe, 1993:252, 256). The choice of cases is discussed in Section 4.7 and shall be further described in Chapter 6 based upon the results of the survey.

4.7.2 Sample Population

Each respondent to the survey was asked if they would like to be contacted for further interview, on the basis that such contact may or not take place dependent upon case and quota requirements. The names, contact details and responses of the participants willing to go forward for further interview were amalgamated for ease of selection for further interview. This resulted in 76 individuals available for interview, which were then sub-divided into the above groups (see Section 4.7.1), whereby individuals with similar characteristics were divided into each of the four cases. The use of these cases help to attain information surrounding the significance of various conditions and experiences for producing the results obtained by the questionnaire. By choosing four different cases, the possibility of a variational range of possible comparison is achieved, whereby mechanisms that operate under a variety of different conditions can be identified. Realism relies on triangulating multiple perceptions of a single reality to develop theoretical findings. By adopting such a means of selecting interview candidates, it was considered possible in the next stage of analysis to delineate the prevailing factors and conditions that influence acceptance of insurers using genetic test results and to build a theory regarding the reality of this situation. Further analysis of case selection is available in Chapter 6.

4.7.3 Surmounting Limitations of Case Studies using Semi-Structured Interview

A lack of a routine procedure for the collection of data and inadequate investigation during case studies may result in the collection of either too much, too little or the wrong data. To ensure the reliability of the data collected in the case study, the Case Study Protocol as designed by Yin (1994: 63-74) was followed as closely as possible given the nature and the constraints of the present study. The study does not proclaim to represent the population as a whole, however, the findings from the case study can serve to validate the findings of the survey and reveal additional factors that may affect the risk acceptance pattern outlined from the survey findings. Findings from the case study may have been influenced by biased opinions on the part of the researcher; however, by having an understanding of the subject under study, the cases and their background, this problem is overcome to a large degree. To conclude, the case study method is an advantageous research tool if used correctly and, especially when there is a need for an in-depth analysis of the cases chosen.

The interviews can also serve to confirm findings from the survey. The main drawback of semi-structured interviews is that the richness of this data is largely dependent on the interviewer, although being semi-structured, this limitation is partly overcome. Another limitation of the semi-structured interview is that they can take a long time to complete. However, if managed correctly, this disadvantage can be surmounted. According to Robson (1993:229), interviews should last over half an hour and under an hour in order to ensure the collection of valuable data and avoid taking up too much time on part of the respondent. In light of this, the interviews were restricted to above half an hour and did not stretch beyond one hour. To guarantee that all information provided during the interview was documented, all interviews were recorded having sought prior permission from each interviewee. The recording of interviews also assisted the ease of data analysis. In addition, semi-structured interviews are subject to problems of bias, poor recall and inaccurate or inadequate articulation on the part of the interviewee. The interviewee may also provide the interviewer with the information he/she believes is required or wanted by the interviewer. It is important to note that semi-structured interviews are verbal reports only and may provide an inaccurate illustration of reality (Yin, 1994:85). It was not entirely possible to overcome such problems in the case studies adopted by the present study. However, by ensuring that several sources of information were used, so that the information from only one interviewee was not solely relied upon as representative of each case, it was possible to compare and contrast responses and to verify, to a certain extent, the legitimacy of responses. In addition, the interviewee was advised that it was their individual opinion that was of interest to the interviewer. To ensure that subject bias (where a sensitive subject is broached) was minimised, the interviewer attempted to build up trust with the interviewees, whereby, a general discussion was held prior to the commencement of the interview. In addition, participants were ensured of their anonymity. Finally, effort was made to control the pace of the interview to suit the interviewee and to ensure that there was little or no interruption to the discussion.

As in the questionnaire, constraints regarding resource availability, such as time, money and personnel affected the case study. The numbers of individuals that make up each case as part of the qualitative analysis were limited as a result of the sample population size. As the cases were chosen from those individuals who agreed to complete a further interview, the original small sample resulted in the selection of only three members for three cases and one for the fourth case. Therefore, only one to three people from each case expressed their opinion on the topics under consideration, perhaps limiting the findings somewhat. In addition, given that one individual conducted all surveys and interviews, an opportunity for bias on the part of the interviewer may creep into the findings. Respondents' answers can be influenced by the interviewer's behaviour, choice of clothes, sex, accent and prompting when a respondent does not understand a question. It was not possible to reduce all interviewer effects, such as the gender of the interviewer; however, it was possible to reduce some effects. By wearing a similar style of clothes and behaving in a similar manner at each interview, it was possible to state that such interviewer affects should be relatively similar in all interview situations. The effect was also dependent upon the individual interviewee, however, as mentioned above, it is not possible to rid the research from all interviewer effects. As all of the interviews were recorded it is possible to acknowledge prompting made and the effects of such suggestions on the interviewee. By acknowledging these forms of bias and by presenting then in an open manner within the study, the researcher is not so much attempting to reduce such interviewer effects, but instead is promoting awareness of these effects so that they may be accounted for when interpreting the results. However, the researcher did attempt to eliminate as much interviewer bias as possible from the study by having an understanding of the subject area, the cases and their background, and by completing semi-structured interviews (see Section 4.7.3). Coder effects are also difficult to check when there is only a sole researcher. To reduce such effects, a colleague of the researcher scanned through the methods and codes adopted and used by the researcher and provided a detached opinion on their quality and validity. In addition, the coding system applied was presented in the Appendices for other researchers to examine (see Appendix D). In this way, the researcher was not so much minimising the coder bias introduced, but recognising that it does have an effect on the results and this effect may be made clearer by presenting the codes used in the Appendices.

Figure 4-7 Outline of Methodology

Sampling Frame

County Clare & County Limerick:

- Based upon proximity to the place of residence of the interviewer and her ability to travel to these areas during the day using public transport

Sample Drawn

192 individuals:

- Chosen systematically (every third house) from a structured sample based on areas of different housing type, in order to ensure a sample population reasonably mixed in age, income, occupation and sex.

Distribution of Non-Responses to Survey

Systematic Approach:

- Respondents persuaded to take part face-to-face. If no answer from a residence visited, the house was revisited twice more upon hourly intervals. If no answer, an adjoining residence on the right of the house was called upon.

Distribution of Non-Responses to Questionnaire

Systematic Approach:

- The researcher, to ensure that respondents completed the questionnaire in full, followed up non-responses. This resulted in all questions being answered by the 192 individuals surveyed.

Identification of Outliers

12 Outliers Removed, 3 Outliers Accommodated

Outliers identified using visual aids:

1. Checked the distribution of data values by levels of each categorical variable.
2. Scatter plot was used to examine how severe the outlying observations actually were.
3. Box plots were used to view values that were considered extreme in relation to the rest of the data

Methods used to deal with outliers:

1. Deletion: when it was considered that the outliers were legitimate errors that could not be corrected or were too far outside the range of the remainder of the data that they distorted statistical inferences.
2. Accommodation: three outliers were accommodated where the data was considered correct and necessary to use as part of the analysis.

Please Turn Over

Sub-Sample for Case Studies

Representative Case:

Representative of the view of the majority of repsondents to the survey

- Incomplete knowledge of the use of genetic test results
- High risk perceptions
- Low benefit perceptions
- Low general trust and high sceptical trust in the government and the insurance industry
- Lack of willingness to accept insurers' use of genetic test results to underwrite applicants for life insurance
- Inverse relationship between risk perception and benefit perception

Extremely Varied Case:

Displaying a lack of agreement with normality, however, on separate issues

- No inverse relationship between risk perception and benefit perception

and/or

- Incomplete knowledge of the use of genetic test results
- High risk perceptions
- Low benefit perceptions
- Low general trust and low sceptical trust in the government and the insurance industry
- Lack of willingness to accept insurers' use of genetic test results to underwrite applicants for life insurance

and/or

- Incomplete knowledge of the use of genetic test results
- High risk perceptions
- Low benefit perceptions
- Low general trust and low sceptical trust in the government and the insurance industry
- Unsure whether to accept insurers' use of genetic test results to underwrite applicants for life insurance

Critical Case:

One particular response appears that should not have typically done so

- Incomplete knowledge of the use of genetic test results
- High risk perceptions
- Low benefit perceptions
- Low general trust and high sceptical trust in the government and the insurance industry
- Lack of willingness to accept insurers' use of genetic test results to underwrite applicants for life insurance
- No inverse relationship between risk perception and benefit perception

or

- Incomplete knowledge of the use of genetic test results
- High risk perceptions
- Low benefit perceptions
- Low general trust and high sceptical trust in the government and the insurance industry
- Accepts insurers' use of genetic test results to underwrite applicants for life insurance
- Inverse relationship between risk perception and benefit perception

or

- Incomplete knowledge of the use of genetic test results
- High risk perceptions
- Low benefit perceptions
- High general trust in the government, low general trust in the insurance industry
- High sceptical trust in the government and the insurance industry
- Lack of willingness to accept insurers' use of genetic test results to underwrite applicants for life insurance
- Inverse relationship between risk perception and benefit perception

Extreme Case:

Majority of responses are beyond what is considered typical

- Incomplete knowledge of the use of genetic test results
- Low risk perceptions
- High benefit perceptions
- High general trust and low sceptical trust in the government
- Low general trust and high sceptical trust in the insurance industry
- Willingness to accept insurers' use of genetic test results to underwrite applicants for life insurance
  Inverse relationship between risk perception and benefit perception but opposite to inverse relationship of representative case

4.8 Criticism regarding the use of a Quantitative method followed by a Qualitative method

Figure 4-7 presents an outline of the methodology used in the present study. As mentioned above, combining methodologies in the study of the same phenomena enables the researcher to focus on a research question from a variety of angles (see Section 4.4.1). Given that different methods show up different aspects of reality, multi-method research is considered increasingly valuable in respect of ensuring validity and reliability, when compared to a study adopting a single method approach. A quantitative survey is a useful strategy for obtaining a quantifiable description of society's views. It is possible to classify features, count them, and construct complex statistical models in an attempt to explain what is observed. Findings can be generalised to a larger population, and direct comparisons can be made between two corpora if valid sampling and significance techniques are adopted. Case studies allow for description, understanding and explanation of a particular phenomenon or event whereby no attempt is made to assign frequencies to the linguistic features which are identified in the data. In the present study, the quantitative survey preceded the qualitative case study.

By completing the survey in advance of the case study, it was intended to formulate a basis from which further analysis of the role and relationship of trust and risk perception could be explored by intensive interviews. When developing a theory regarding the factors and conditions that affect risk acceptance, it was considered valid to test the application of a previous theory to the present study and to then develop this theory in the context of the study. The first requirement is clearly theory-testing whilst the second is theory-building. Several academics state that quantitative methods are more appropriate for answering *what* and *how many* (theory-testing) questions, whilst qualitative methods are useful for finding information on *how* and *why* (theory-building) things occur (Yin, 1994; Robson, 1993:6; Layder, 1996:3). Theory-testing methods include survey by questionnaire and a popular theory-building method includes in-depth interviews as part of a case study. Hence, in an effort to test and develop a theory surrounding Irish society's perceptions of insurers using genetic test results it was decided to proceed with a survey in advance of a case study. An analogy can be formulated which communicates the intention of the survey and the case study. The survey acted as a pencil sketch formulating the borders and angles of the study, while the case study was used to develop the findings from the survey by filling in and adding colour to the pencil drawing. Thereby together they provided a complete canvas of the reality surrounding the relationship between risk perception and trust within the confines of the present study.

However, Schmied (1993) notes that a stage of qualitative research is often a precursor for quantitative analysis, since before phenomena can be classified and counted, the categories for classification must first be identified. Qualitative analysis allows for fine distinctions to be drawn because, unlike quantitative analysis, it is not necessary to shoehorn the data into a finite number of classifications. Ambiguities and omissions that may appear in the quantitative survey can be recognised in the qualitative stage of analysis (see Section 7.7). For example, the quantitative analysis assumed that, based upon previous research, trust is composed of several factors, whereby an evaluation of each factor contributes to an overall view of trust. Hence, each of these factors was adopted in the survey to measure trust. However, if the Irish public were to identify more or less, or even a different composition of factors that delineate trust, it is not possible to incorporate such dimensions if the survey is completed prior to the in-depth interviews. It was the intention of the present study to overcome the disadvantages associated with completing the interviews subsequent to the survey by applying an extensive review of literature in the area of risk perception and trust to minimise the appearance of ambiguities and/or omissions.

4.9 Time Horizons

The time horizon of the study is particularly important. The study is a contemporary study and is cross-sectional by design. It is important to recognise the significance of this time horizon for the results of the study to be understood fully and practically. The survey was completed during the months of October and November 2004 and the interviews for the case studies were conducted four months later, based upon results of analyses completed on the survey. During this time frame or; indeed previously, there was little media attention given to the use of genetic test results by life insurance companies. The Disability Bill 2004 had not yet been enacted and the IIF Code of Practice was the single risk management strategy in place. The time frame of the study, therefore, influences the results of the study and all findings and conclusions observed.

4.10 Summary

This chapter attempted to justify and explain the methodological approach adopted by the study. It explained how and why the critical realist paradigm and the nature of the research problem determined the research approach and design. By combining a deductive and inductive approach to research, the 'textured' or interwoven nature of the different dimensions of social reality, as depicted by a realist philosophy, is conveyed and, thereby, has the ability to capture real-world complexity. The objectives of this research required the researcher to explore and describe the role that knowledge, social trust evaluations, risk perceptions and benefit perception have in determining acceptance of the risks related to insurers using genetic test results. The first stage of research involves investigating whether an influential link exists between incomplete knowledge, social trust, risk perception, benefit perception and risk acceptance in relation to the use of genetic test results by the life insurance industry in the Republic of Ireland. The second objective aims to identify the factors and conditions that lead to trust and risk related attitudes and evaluations in order to understand the risk acceptance pattern suggested by the findings from the research behind the first objective.

Yin (1994) states that quantitative methods are more appropriate for answering *what* and *how many* (theory –testing) questions, whilst qualitative methods are useful for finding information on *how* and *why* (theory-building) things occur (Robson, 1993:6; Layder, 1996:3). A survey is a useful strategy for obtaining a quantifiable description of society's views. According to Yin (1994:6), a case study is particularly appropriate for describing, understanding and explaining an area of study. Based on such arguments, it was decided to use a survey followed by a case study to gain the information required for this research. It was demonstrated how the researcher attempts to improve the quality of the survey and interviews. Having decided upon the strategies, the choice of research instruments, a questionnaire and semi-structured interview, were justified. Finally, the influence that the time horizon has upon the study was discussed.

Principal components analysis for each of the items designed to measure knowledge, trust, risk perception, benefit perception and risk acceptance were provided. From this initial data analysis, the resulting research framework was presented. Following this framework, the relationship between incomplete knowledge, general trust, sceptical trust, risk perception, benefit perception and risk acceptance shall be examined. In addition, the influence that age, gender, economic status and life insurance ownership have upon each of the above items shall be explored. Differences in trust shall also be examined between trust in the Government and trust in the insurance industry, and the results discussed.

CHAPTER FIVE

SURVEY FINDINGS

5.1 Introduction

The previous chapter justified and explained the methodological approach to be adopted by the present study. Reasons as to why a survey is considered as suitable for obtaining a quantifiable description of the relationship between risk perception and risk acceptance were explained. This chapter examines the results of the survey undertaken and analyses the nature of the data obtained. In doing so, it shall be determined whether respondents lacking detailed knowledge of the use of genetic test results by the life insurance industry apply social trust in the life insurance industry and the Government to formulate their assumptions of the risks and benefits related to life insurers using genetic information in underwriting, thereby reducing the complexity of risk management decisions. It is predicted that trust in those managing the risks related to such usage, that is, trust in the life insurance industry and the Government, results in a positive evaluation of the benefits related to the use of this technology in life insurance underwriting and the perception of fewer risks, an assumption based on previous research (see Chapter 3). If this is the case, it is expected that public risk and benefit perceptions influence society to accept the current management of the associated risks. In other words, if individuals associate the use of genetic test results by life insurance companies with few risks and many benefits, they are more likely to be satisfied with the current management of the hazard.

The opposite of the above assumption asserts that relatively little trust in entities managing these risks results in a negative appraisal of the associated benefits and an increase in awareness and concern with regard to the risks. The expected result is rejection or a lack of agreement with current management of the use of genetic test results in underwriting practices. The following section shall examine the main statistical findings from the survey analysis followed by a discussion of the process by which the findings were arrived at. The influence that socio-demographics have upon the predictor and dependent variables shall also be explored.

5.2 Main Findings

The main findings from the study demonstrate a weak correlation between incomplete knowledge, general trust, sceptical trust, risk perceptions, benefit perceptions and risk acceptance. Figure 5-1 demonstrates the expected results from the correlation analysis according to findings from previous research, while Table 5-1, Figure 5-2 and Figure 5-3 indicate the results from the correlation analysis of the present study.

Figure 5-1 Conceptual Model of the Relationship between Incomplete Knowledge, Social Trust, Risk Perception, Benefit Perception and Risk Acceptance

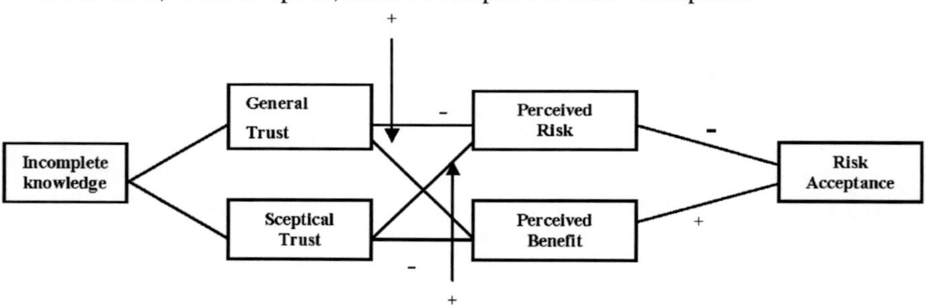

Source: Model adapted from Siegrist (Siegrist, 2000:197; Siegrist and Cvetkovich, 2000:713-719)

Table 5-1 Correlation between Incomplete Knowledge, Social Trust, Risk Perception, Benefit Perception and Risk Acceptance

	Risk Acceptance	Risk Perception	Benefit Perception	Ins. General Trust	Ins. Sceptical Trust	Govt. General Trust	Govt. Sceptical trust	Knowledge
Risk Perception	-.314**							
Benefit Perception	.469**	-.229**						
Ins. General Trust	.221**	-.024	.211**					
Ins. Sceptical Trust	-.132	.25**	-.092	-.356**				
Govt. General Trust	-.065	.265**	.018	.243**	.066			
Govt. Sceptical trust	-.065	.098	-.042	-.279**	.409**	-.195**		
Knowledge	.056	-.047	.032	.095	-.028	.03	.07	

* Correlation is significant at the 0.05 level (1-tailed).
** Correlation is significant at the 0.01 level (1-tailed).

Figure 5-2 Model of the Correlation between Incomplete Knowledge, Social Trust in the Government, Risk Perception, Benefit Perception and Risk Acceptance

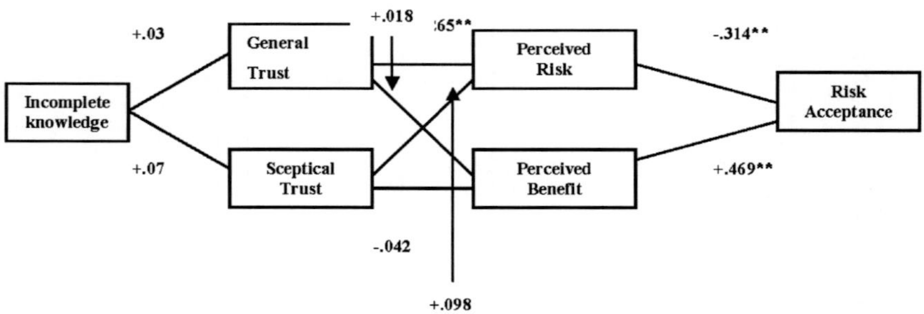

* Correlation is significant at the 0.05 level (1-tailed)
** Correlation is significant at the 0.01 level (1-tailed)

Figure 5-3 Model of the Correlation between Incomplete Knowledge, Social Trust in the Insurance Industry, Risk Perception, Benefit Perception and Risk Acceptance

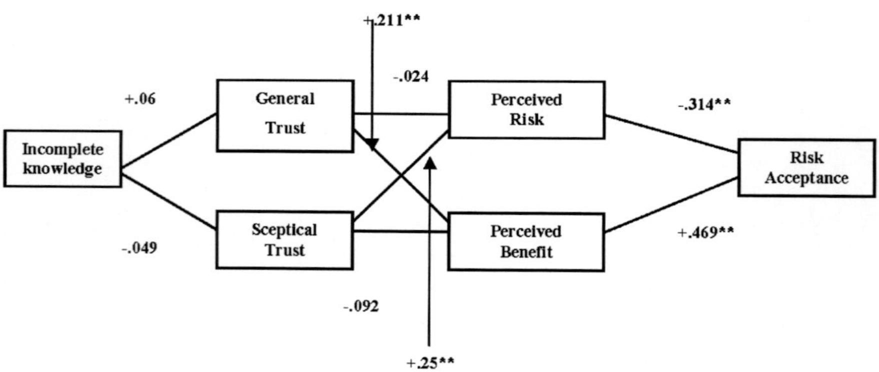

* Correlation is significant at the 0.05 level (1-tailed).
** Correlation is significant at the 0.01 level (1-tailed).

As expected from previous research, there is a positive correlation between mean benefit perception and risk acceptance (r = 0.469, df = 177-180, p = 0.001) which is statistically significant at the 0.01 level. There is a negative correlation between risk perception and risk acceptance (r = -0.314, df, 177-180, p = 0.036) which may also be considered statistically significant at the 0.01 level. The direction of these relationships corresponds with previous research in the area of risk, although the strength of the correlation can be considered weaker than recent research indicates. With regard to trust in the insurance industry, analysis of the survey reveals that general trust is positively correlated with perceived benefit (r = 0.211, df = 176, p = 0.002) and sceptical trust is positively correlated with perceived risk (r = 0.25, df = 176, p = 0.111), both significant at the 0.01 level. Although these findings can be considered as demonstrating weak relationships, they are more compelling as an indication of a relationship than the statistical findings for general trust and perceived risk (r = -0.024, df = 176, p = 0.374) and sceptical trust and perceived benefit (r = -0.092, df = 176, p = 0.111).

Interestingly though, Figure 5-2 demonstrates that the direction of the relationship between general trust in the Government and perceived benefit is opposite to what was expected from previous literature. Indeed, the findings from the present study indicate that general trust is positively correlated with perceived risk, whereby the correlation is significant at the 0.01 level (r = 0.265, df = 176, p = 0.001) which represents a departure from the theory that risk perception is negatively correlated with benefit perception. The correlation (r = 0.265) is relatively weak; however, it would be erroneous to disregard the finding altogether due to its weak nature and it will be interesting to note any reasons why this may be so in the case studies. Other findings regarding trust demonstrate a weak correlation between general trust in the Government and perceived benefit (r = 0.018, df = 176, p = 0.403) and between sceptical trust in the Government and perceived benefit (r = -0.042, df = 176, p = 0.288) and between sceptical trust in the government and perceived risk (r = 0.098, df = 176, p = 0.096) with none of the findings demonstrating statistical significance. Figure 5.2 and Figure 5.3 demonstrate an insignificant relationship between incomplete knowledge and general trust and incomplete knowledge and sceptical trust in both the insurance industry and the Government (see also Table 5-1). Overall correlation analysis reveals relatively little association between incomplete knowledge, social trust, risk perceptions, benefit perceptions and risk acceptance in the case of the risks related to insurers using genetic test results. Indeed, the effect that each of the factors have upon each other in the majority of cases is so weak that it may even indicate a non-existent relationship between incomplete knowledge, social trust, risk perceptions, benefit perceptions and risk acceptance.

The inverse relationship between perceived risks and perceived benefits found in previous studies (Siegrist and Cvetkovich, 2000; Alhakami and

Slovic, 1994; Frewer *et al.,* 1998; Gregory and Mendelsohn, 1993) is replicated in the present study. Respondents that perceived the use of genetic test results by life insurance companies as risky ascribed less benefits than people perceiving same as not risky. According to Frewer *et al.* (1998), perceived risks can be reduced by emphasising the benefits associated with a technology. However, results of the present study suggest that this may not work in a straightforward way. A partial correlation controlling for the general trust dimension and the sceptical trust dimension in the Government and the insurance industry reduces the association between perceived-risk means and perceived-benefit means, $r = -0.203$, $df = 174$, $p = 0.003$. These findings are in line with other studies in which trust in gene technology was investigated (Siegrist and Cvetkovich, 2000; Siegrist, 2000; Siegrist, 1999). However, the difference in correlations when trust is controlled is minimal, suggesting that the correlation between perceived risk and perceived benefit is not strongly influenced by social trust evaluations.

Regression analysis was also used in addition to correlation analysis to explore the results from the survey. Regression analysis is an extension of correlation analysis. While correlation analysis can indicate how strongly two variables relate to each other (both magnitude and direction), regression analysis reveals by how much one variable will change in response to a change in another variable(s). Regression analysis demonstrates only low associations between incomplete knowledge, social trust, risk perceptions, benefit perceptions and risk acceptance. Incomplete knowledge, general and sceptical trust in the Government and the insurance industry and benefit perception account for only 12 per cent of the variation in risk perception. The association between benefit perception and the predictors mean knowledge, general and sceptical trust in the Government and the insurance industry and risk perception is moderate with the predictors accounting for only 10 per cent of the variation in benefit perception. Incomplete knowledge, general and sceptical trust in the Government and the insurance industry, risk perception and benefit perception account for 27 per cent of the variation in risk acceptance. Based on the findings from correlation and regression analysis, it is possible to state that the relationship between incomplete knowledge, social trust, risk perceptions, benefit perceptions and acceptance of the risks related to insurers using genetic test results is weak, with the results indicating that the influence that each of the factors have upon each other and risk acceptance is minimal.

5.3 Discussion of Findings: Process of Correlation Analysis

This section provides an overview of the process by which the above correlation results were obtained (see Section 5.2). The findings are divided according to the predictors and dependents tested and tables are provided to illustrate the data.

5.3.1 Correlation: Knowledge and Trust

Only knowledge items 2 and 3 are used in the correlations analysis as an indicator of people's knowledge of the use of genetic test results by life insurance companies, as these two items provide a truer or more accurate gauge for individuals' knowledge of the use of genetic test results by life insurance companies (see Section 4.6.2). Given the majority of individuals' low knowledge of insurers' use of genetic test results, it is impossible to test if/how incomplete knowledge versus good/complete knowledge affects the relationship between trust and risk perception and trust and benefit perceptions (see Section 5.4.1). Due to the irregularity of the appearance of trust item 11, measuring *Openness*, this item was excluded from those factors designed to measure general trust in the insurance industry and general trust in the government (see Section 4.6.2). It was previously explained how there appears to be two separate evaluations of trust, with the first consisting of items which encompass a general trust factor, while the second category consists of negative or sceptical items (see Section 4.6.2). As a result of these two separate evaluations of trust, it was decided that general trust and scepticism should be evaluated as separate factors in this study when examining their association with knowledge, risk perception, benefit perception and risk acceptance (see Table 5-2 for the two evaluations of trust and their associated measurement items):

Table 5-2 Descriptions of General Trust and Sceptical Trust

General Trust	Sceptical Trust
Item 1: Doing a does good job	Item 4: Distorts facts in its favour
Item 2: Competent	Item 5: Changes policies without good reasons
Item 3: Skilled	Item 6: Influenced by life insurance companies
Item 7: Acting in public interest	
Item 8: Listens to concerns raised by ordinary people	
Item 9: Listens to what ordinary people think	
Item 10: Provides all relevant information	

Table 5-3 and Table 5-4 show the results from partial correlations between knowledge and trust items, controlling for the knowledge and trust items not included in the immediate analysis.

Table 5-3 Correlations – Knowledge and Trust in Government

	Trust Item 1	Trust Item 2	Trust Item 3	Trust Item 4	Trust Item 5	Trust Item 6	Trust Item 7	Trust Item 8	Trust Item 9	Trust Item 10	General Trust	Sceptical Trust
Know Item 2	-0.63	0.182	-0.133	0.044	-0.066	-0.04	-0.098	0.017	0.047	-0.026	0.027	-0.051
Sig.	*0.42*	*0.018*	*0.086*	*0.571*	*0.397*	*0.605*	*0.204*	*0.832*	*0.549*	*0.742*	*0.723*	*0.505*
Know Item 3	0.096	-0.166	0.136	-0.143	0.211	0.012	0.067	0.023	-0.058	-0.012	-0.004	0.097
Sig.	*0.215*	*0.032*	*0.08*	*0.065*	*0.006*	*0.88*	*0.388*	*0.766*	*0.456*	*0.88*	*0.96*	*0.201*
Know Mean	0.057	-0.147	0.091	0.033	0.088	0.053	0.174	0.104	-0.128	-0.082	0.03	0.07
Sig.	*0.464*	*0.058*	*0.243*	*0.675*	*0.256*	*0.498*	*0.024*	*0.181*	*0.1*	*0.292*	*0.65*	*0.324*

(df = 166: across row Know. Item 2 & Item 3)
(df = 165: across row Know. Mean)
two-tailed

Table 5-4 Correlations – Knowledge and Trust in Insurance Industry

	Trust Item 1	**Trust Item 2**	**Trust Item 3**	**Trust Item 4**	**Trust Item 5**	**Trust Item 6**	**Trust Item 7**	**Trust Item 8**	**Trust Item 9**	**Trust Item 10**	**General Trust**	**Sceptical Trust**
Know Item 1	0.037	-0.008	0.082	-0.027	-0.034	0.186	0.084	0.118	-0.071	0.062	0.23	0.093
Sig.	*0.633*	*0.919*	*0.288*	*0.732*	*0.666*	*0.0.16*	*0.276*	*0.126*	*0.357*	*0.425*	*0.002*	*0.22*
Know Item 2	-0.09	0.011	-0.06	-0.094	0.04	-0.129	0.02	-0.13	0.102	-0.033	-0.139	-0.158
Sig.	*0.244*	*0.885*	*0.439*	*0.227*	*0.609*	*0.094*	*0.798*	*0.094*	*0.19*	*0.675*	*0.066*	*0.036*
Know Mean	-0.016	-0.025	0.034	-0.119	0.04	0.072	0.098	0.005	0.002	0.024	0.095	-0.028
Sig.	*0.832*	*0.752*	*0.664*	*0.124*	*0.606*	*0.353*	*0.205*	*0.852*	*0.778*	*0.752*	*0.21*	*0.715*

(df = 166: across row Know. Item 2 & Item 3)
(df = 167: across Know. Mean)
two-tailed

The correlation between the two knowledge items and each of the trust items is so low that it could be considered non-existent, given that r is close to 0 in all cases. However, it can be noted that correlations between knowledge and trust are higher in the Government than in the insurance industry. The highest correlations occur mostly between self-assessed knowledge and the items measuring scepticism in trust with regard to all cases except for mean knowledge and general trust, which is higher than mean sceptical trust. However, even these statistics indicate a minimal or non-existent relationship between incomplete knowledge and trust.

5.3.2 Correlation: Knowledge and Perceived Risk and Perceived Benefit

The results of correlation analysis between self-assessed knowledge and perceived risk and perceived benefit demonstrate a non-existent relationship between these factors (see Table 5-5 and Table 5-6). These tests each control for item 1 of knowledge and each of the items not being tested in the correlations. Looking at the direction of the results, it appears that mean knowledge and mean risk perception are negatively correlated r = -0.047, df = 177, p = 0.533, while mean knowledge and mean benefit perception are positively correlated r = 0.032, df = 177, p = 0.669.

Table 5-5 Correlations – Knowledge and Risk Perception

	Risk Perception Society	**Risk Perception Family**	**Risk Perception Mean**
Knowledge Item 2	0.05	-0.017	0.047
Sig.	*0.511 (df = 175)*	*0.818 (df = 175)*	*0.533 (df = 176)*
Knowledge Item 3	-0.016	-0.072	-0.159
Sig.	*0.835 (df = 175)*	*0.339 (df = 175)*	*0.034 (df = 176)*
Knowledge Mean Items 2 & 3	0.064	-0.081	-0.047
Sig.	*0.398 (df = 176)*	*0.281 (df = 176)*	*0.533 (df = 177)*

two-tailed

Table 5-6 Correlations – Knowledge and Benefit Perception

	Benefit Perception Society	Benefit Perception Family	Benefit Perception Mean
Knowledge Item 2	0.065	-0.028	0.052
Sig.	*0.392 (df = 175)*	*0.707 (df = 175)*	*0.492 (df = 176)*
Knowledge Item 3	-0.111	0.093	-0.031
Sig.	*0.141 (df = 175)*	*0.218 (df = 175)*	*0.684 (df = 176)*
Knowledge Mean Items 2 & 3	0.033	-0.009	0.032
Sig.	*0.661 (df = 176)*	*0.908 (df = 176)*	*0.669 (df = 177)*

two-tailed

5.3.3 Correlation – Perceived Risk, Perceived Benefit and Trust

This research questions if, when lacking knowledge about risks and benefits, people have to rely on trust to make judgements regarding risk acceptance. The results below indicate that the speculated direction of the relationship is generally the case, although individual trust items appear to go against the theory in particular cases but in no identifiable pattern.[1] Mean risk perception is positively correlated with general trust in the Government ($r = 0.265$, $df = 176$, $p = 0.001$) and with sceptical trust in the government ($r = 0.098$, $df = 176$, $p = 0.096$) (see Table 5-7). However, the findings illustrate a minimal association between risk perception and trust in the Government; indeed, the results are so weak that it could be

[1] Given the absence of good/complete knowledge, it is expected that there will be strong negative correlations between trust and perceived risks, as well as positive correlations between trust and perceived benefits. In addition, it is anticipated that there will be negative correlations between general trust and perceived risk and positive correlations between sceptical trust and perceived risk. It is also predicted that there will be positive correlations between general trust and perceived benefit and negative correlations between the sceptical dimension of trust and perceived benefits.

determined that no relationship exists between these factors. The direction of the relationship between the sceptical or negative dimension of trust corresponds with the previous research; however, the positive correlation found between risk perception and general trust goes against previous research results. Mean risk perception is negatively correlated with general trust in the insurance industry ($r = -0.024$, $df = 176$, $p = 0.374$) and positively correlated with sceptical trust in the insurance industry ($r = 0.025$, $df = 176$, $p = 0.001$) (see Table 5-8). However, once again, these correlations can be considered weak and indicative of a non-existent relationship between these factors. Therefore, although the inverse relationship between trust in the insurance industry and risk perception corresponds with previous research, the findings illustrate that there is no foundation upon which to demarcate a relationship between these factors.

It is predicted that there will be positive correlations between general trust and perceived benefit and negative correlations between sceptical dimension of trust and perceived benefits. The results in Table 5-9 indicate that this appears to be the case, although some individual trust items appear to go against the prediction. Mean benefit perception is positively correlated with general trust in the Government ($r = 0.018$, $df = 176$, $p = 0.403$) and negatively correlated with sceptical trust in the government ($r = -0.042$, $df = 176$, $p = 0.288$) (see Table 5-9). Once again, the relationship between risk perception and trust is minimal and indicative that no relationship exists between these factors. The above association between general trust, the positive dimension of trust, and the sceptical, negative dimension of trust, corresponds with previous research. Mean benefit perception is positively correlated with general trust in the insurance industry ($r = 0.211$, $df = 176$, $p = 0.002$) and negatively correlated with sceptical trust in the insurance industry ($r = -0.092$, $df = 176$, $p = 0.111$) (see Table 5-10). The inverse relationship between trust in the insurance industry and benefit perception corresponds with previous research. However, it is important to note that each of the above correlations can be considered very weak and, although they indicate the direction of the relationship between trust and risk perception, the correlation results provide no foundation upon which to illustrate any association between risk perception and trust in the insurance industry ($r < 0.25$).

Table 5-7 Correlations – Risk Perception and Trust in Government

	Trust Item 1	Trust Item 2	Trust Item 3	Trust Item 4	Trust Item 5	Trust Item 6	Trust Item 7	Trust Item 8	Trust Item 9	Trust Item 10	General Trust	Sceptical Trust
R.P. Society Item 1	0.066	0.035	-0.079	-0.028	0.079	0.029	-0.037	-0.028	0.128	0.071	0.156	0.014
Sig. *df*	*0.198* *df=167*	*0.324* *df=167*	*0.155* *df=167*	*0.361* *df=167*	*0.152* *df=167*	*0.356* *df=167*	*0.319* *df=167*	*0.359* *df=167*	*0.049* *df=167*	*0.181* *df=167*	*0.019* *Df=175*	*0.427* *df=175*
R.P. Family Item 2	-0.036	-0.031	0.114	0.071	-0.073	0.038	0.03	0.021	-0.052	-0.072	0.009	0.055
Sig. *df*	*0.319* *df=167*	*0.343* *df=167*	*0.07* *df=167*	*0.179* *df=167*	*0.174* *df=167*	*0.311* *df=167*	*0.348* *df=167*	*0.392* *df=167*	*0.249* *df=167*	*0.175* *df=167*	*0.453* *Df=175*	*0.232* *df=175*
R.P. Mean	0.024	0.012	0.059	0.082	-0.019	0.099	0.001	-0.005	0.101	0.009	**0.265**	**0.098**
Sig. *df*	*0.379* *df=168*	*0.437* *df=168*	*0.223* *df=168*	*0.145* *df=168*	*0.401* *df=168*	*0.1* *df=168*	*0.496* *df=168*	*0.477* *df=168*	*0.095* *df=168*	*0.452* *df=168*	***0.001*** *Df=176*	***0.096*** *df=176*

One-tailed

Table 5-8 Correlations – Risk Perception and Trust in Insurance Industry

	Trust Item 1	**Trust Item 2**	**Trust Item 3**	**Trust Item 4**	**Trust Item 5**	**Trust Item 6**	**Trust Item 7**	**Trust Item 8**	**Trust Item 9**	**Trust Item 10**	**General Trust**	**Sceptical Trust**
R.P. Society Item 1	0.128	0.068	0.015	0.015	0.093	0.051	0.123	-0.09	0.041	-0.069	0.089	0.162
Sig. df	*0.049 df=167*	*0.192 df=167*	*0.412 df=167*	*0.422 df=167*	*0.115 df=167*	*0.256 df=167*	*0.056 df=167*	*0.123 df=167*	*0.3 df=167*	*0.188 df=167*	*0.12 df=175*	*0.015 df=175*
R.P. Family Item 2	-0.151	-0.163	0.051	0.005	-0.012	-0.016	-0.099	0.142	-0.085	0.068	-0.102	-0.01
Sig. df	*0.025 df=167*	*0.017 df=167*	*0.256 df=167*	*0.474 df=167*	*0.437 df=167*	*0.42 df=167*	*0.1 df=167*	*0.032 df=167*	*0.135 df=167*	*0.189 df=167*	*0.089 df=175*	*0.449 df=175*
R.P. Mean	-0.063	-0.137	0.095	0.038	0.137	0.054	0.027	0.103	-0.089	0.009	**-0.024**	**0.25**
Sig. df	*0.208 df=168*	*0.037 df=168*	*0.108 df=168*	*0.309 df=168*	*0.037 df=168*	*0.253 df=168*	*0.362 df=168*	*0.091 df=168*	*0.125 df=168*	*0.455 df=168*	***0.374** df=176*	***0.001** df=176*

one-tailed

Table 5-9 Correlations – Benefit Perception and Trust in Government

	Trust Item 1	Trust Item 2	Trust Item 3	Trust Item 4	Trust Item 5	Trust Item 6	Trust Item 7	Trust Item 8	Trust Item 9	Trust Item 10	General Trust	Sceptical Trust
B.P. Society Item 1	-0.082	-0.073	0.065	0.117	-0.018	-0.053	0.134	-0.035	0.13	-0.108	0.078	0.02
Sig.	*0.143 df=167*	*0.173 df=167*	*0.2 df=167*	*0.065 df=167*	*0.409 df=167*	*0.245 df=167*	*0.041 df=167*	*0.324 df=167*	*0.046 df=167*	*0.082 df=167*	*0.15 df=175*	*0.397 df=175*
B.P. Family Item 2	0.049	0.117	-0.063	-0.063	-0.029	0.017	-0.193	0.01	-0.112	0.149	-0.074	-0.043
Sig.	*0.262 df=167*	*0.064 df=167*	*0.208 df=167*	*0.21 df=167*	*0.355 df=167*	*0.411 df=167*	*0.006 df=167*	*0.448 df=167*	*0.073 df=167*	*0.027 df=167*	*0.163 df=175*	*0.287 df=175*
B.P. Mean	-0.033	0.072	0.008	0.088	-0.091	-0.05	-0.096	-0.037	0.029	0.049	**0.018**	**-0.042**
Sig.	*0.335 df=168*	*0.176 df=168*	*0.457 df=168*	*0.127 df=168*	*0.119 df=168*	*0.259 df=168*	*0.106 df=168*	*0.317 df=168*	*0.355 df=168*	*0.262 df=168*	***0.403*** *df=176*	***0.288*** *df=176*

(df = 167-176) one-tailed

Table 5-10 Correlations – Benefit Perception and Trust in Insurance Industry

	Trust Item 1	Trust Item 2	Trust Item 3	Trust Item 4	Trust Item 5	Trust Item 6	Trust Item 7	Trust Item 8	Trust Item 9	Trust Item 10	General Trust	Sceptical Trust
B.P. Society Item 1	0.175	0.078	-0.011	0.113	-0.112	-0.076	-0.087	-0.09	-0.009	-0.009	-0.057	-0.036
Sig. *df*	*0.012* *df=167*	*0.157* *df=167*	*0.444* *df=167*	*0.071* *df=167*	*0.073* *df=167*	*0.162* *df=167*	*0.131* *df=167*	*0.112* *df=167*	*0.49* *df=167*	*0.456* *df=167*	*0.225* *df=175*	*0.316* *df=175*
B.P. Family Item 2	-0.094	0.07	-0.012	-0.1	0.092	0.037	0.074	-0.045	0.094	0.105	0.201	-0.017
Sig. *df*	*0.11* *df=167*	*0.181* *df=167*	*0.437* *df=167*	*0.098* *df=167*	*0.118* *df=167*	*0.317* *df=167*	*0.171* *df=167*	*0.28* *df=167*	*0.111* *df=167*	*0.086* *df=167*	*0.004* *df=175*	*0.413* *df=175*
B.P. Mean	0.112	0.228	-0.043	0.004	-0.019	-0.047	-0.034	-0.208	0.145	0.162	**0.211**	**-0.092**
Sig. *df*	*0.073* *df=168*	*0.001* *df=168*	*0.287* *df=168*	*0.477* *df=168*	*0.402* *df=168*	*0.271* *df=168*	*0.329* *df=168*	*0.003* *df=168*	*0.029* *df=168*	*0.017* *df=168*	***0.002*** *df=176*	***0.111*** *df=176*

(df = 167-175) one-tailed

The results of the individual items demonstrate that individually the two benefit items and the ten trust items do not define trust and they do not satisfy the assumption made in this study and the results of previous research. However, when the two benefit items are amalgamated and taken to indicate overall benefit perception, and the individual trust items are divided in terms of those that measure a general trust dimension and a sceptical trust dimension, the overall results indicate that the assumption made regarding a link between trust and perceived risk and benefit is satisfied in agreement with previous research on this area. However, it has to be noted that the correlations are distinctively weak, in that there is no correlation higher than r = 0.265, df = 176, p = 0.001, and even this correlation is in the opposite direction as indicated by the assumption. This is the only overall correlation which goes against the predicted direction of the relationships between perceived risk and benefit and trust and it is questionable why this is so. It is plausible that the differences arise from a disparity between trust in the government and trust in the insurance industry, with the latter being lower or perhaps it may be due to the influence of an unmeasured factor.

Table 5-11 Correlations – Benefit Perception and Risk Perception

	R.P. Society	**R.P. Family**	**R.P. Mean**
B.P. Society Item 1	0.007	0.068	0.123
Sig. *df*	*0.464* *df=176*	*0.182* *df=176*	*0.05* *df=177*
B.P. Family Item 2	0.011	-0.188	-0.279
Sig. *df*	*0.441* *df=176*	*0.006* *df=176*	*0.001* *df=177*
B.P. Mean	0.027	-0.174	**-0.229**
Sig. *df*	*0.358* *df=177*	*0.01* *df=177*	***0.001*** *df=180*

(df = 176-180) one-tailed

A weak association between perceived risk means and benefit means is observed, ($r = -0.229$, df = 180, $p = 0.001$) (see Table 5-11). The inverse relationship between perceived risks and perceived benefits found in previous studies (Siegrist and Cvetkovich, 2000; Alhakami and Slovic, 1994; Frewer *et al.,* 1998; Gregory and Mendelsohn, 1993) is replicated in the present study. Respondents that perceived the use of genetic test results by life insurance companies as risky ascribed less benefits than people perceiving same use as not risky. According to Frewer *et al.* (1998), perceived risks can be reduced by emphasising the benefits associated with a technology. However, results of the present study suggest that this may not work in a straightforward way. A partial correlation controlling for the general trust dimension and the sceptical trust dimension in the Government and the insurance industry reduces the association between perceived-risk means and perceived-benefit means, $r = -0.203$, df = 174, $p = 0.003$. These findings are in line with other studies in which the trust on gene technology (Siegrist and Cvetkovich, 2000; Siegrist, 2000; Siegrist, 1999) was investigated. However, the difference in correlations when trust is controlled is minimal, suggesting that the correlation between perceived risk and perceived benefit is not strongly influenced by social trust evaluations.

No significant correlation was found between knowledge and any other items. It is impossible to note the correlations between perceived risk and trust for the possible risks and benefits perceived by participants who possessed a high level of knowledge, as none of the 180 respondents believed themselves to possess such knowledge. Therefore, in this particular study, it is not possible to note correlations between lower and higher self-assessed knowledge and any other variable. This is a significant drawback in this study, although it can also be perceived as a significant finding. All of the individuals randomly chosen in this sample do not possess detailed knowledge of the use of genetic test results by life insurance companies. Due to the homogenous nature of the participants' level of knowledge, it is not possible to sub-divide the sample population into groups (no knowledge/much knowledge) and to compare separate correlations for each (see Section 5.4.1). Therefore, it is not possible for this study to state whether or not individuals with low knowledge depend upon social trust more so than individuals with high knowledge, in order to make their risk and benefit decisions. However, the findings do indicate that although there was incomplete knowledge of insurers use of genetic test results, trust did not appear to influence risk or benefit perceptions, a finding that is associated with a high level of knowledge (see Siegrist and

Cvetkovich, 2000:716-717). Given the statistical findings from the present study, it does not seem that trust is a significant determinant in judgements of risks in the absence of knowledge.

5.3.4 Correlation – Incomplete Knowledge, Social Trust, Risk Perception, Benefit Perception and Risk Acceptance

Correlations between general trust in the Government and risk acceptance are unusually negative (although this corresponds with the unexpected result earlier) ($r = -0.065$, df = 168-177, $p = 0.196$). General trust in the insurance industry is positively correlated with risk acceptance ($r = 0.221$, df = 168-177, $p = 0.002$). However, the direction of the relationship is less significant, given the low correlations between the factors. In particular the association between general trust in the Government and risk acceptance is so low that r could be either side of zero and representative of either a positive or a negative relationship. Further results indicate that the sceptical dimension of trust in the Government and the insurance industry is correlated negatively with risk acceptance ($r = -0.065$, df = 168-177, $p = 0.196$; $r = -0.132$, df = 168-177, $p = 0.039$ respectively). However, these correlations are once again very low and indicate that there is little or no relationship between trust and risk acceptance. Therefore it cannot be said that high general trust is associated with high risk acceptance and high sceptical trust is associated with low risk acceptance, as was initially speculated.

There is a negative correlation between mean risk perception and risk acceptance but this is relatively low ($r = -0.314$, df, 177-180, $p = 0.036$), although the positive correlation between mean benefit perception and risk acceptance is higher ($r = 0.469$, df = 177-180, $p = 0.001$). These results suggest that high risk perception is associated with low risk acceptance and high benefit perception is associated with high risk acceptance. Nonetheless, both correlations can be considered rather weak, and perhaps even reflect a lack of association between these factors, given that $r<0.05$ in all tests.

5.4 Discussion of Findings: Process of Regression Analysis

This section pertains to the process by which findings from the regressions analysis were attained (see Section 5.2). Similar to the correlation analysis, the findings are divided according to the predictors and dependents tested and tables are provided to illustrate the results.

5.4.1 Risk Perception

The association between risk perception and the predictors (mean knowledge, general and sceptical trust in the Government and the insurance industry and benefit perception) is moderate (Multiple R = 0.385). Together, the predictors accounted for only 12 per cent of the variation in risk perception (adjusted R2). In this case, the F-value is 5.014, with an associated probability of <0.001. This means that such a result is highly unlikely to have arisen by sampling error, assuming the null hypothesis to be true. Both mean knowledge and benefit perception negatively relate to risk perception (see Table 5-12). The regression coefficient for mean knowledge is 0.012 (95 per cent CI = -0.293 – 0.268); and for benefit perception it is –0.236 (95 per cent CI = -0.407 - -0.065). This means that, for every one unit increase in risk perception, knowledge drops by –0.012 units. At this stage of the analysis, it is important to note that this interpretation of the regression analysis findings is rather exact and precise, while, bearing in mind that the data for the present study is ordinal, such findings over-imply a sense of accuracy that is not the case. Therefore, while the results indicate a decrease in knowledge of 0.012 units for every one unit increase in risk perception, it would seem more correct to interpret this finding as a slight drop in knowledge as risk perception increases. Such a loose interpretation of the findings should apply to all of the statistical results of the present study.

As for benefit perception, with every one unit increase in risk perception there is a decrease in benefit perception of –0.236 units, once again demonstrating a minimal effect between factors. The other predictors are positively related to risk perception. The regression coefficient for general trust in the Government is 0.336 (95 per cent CI = 0.125 – 0.547); general trust in the insurance industry is 0.001 (95 per cent CI = -0.226 – 0.227); for sceptical trust in the Government the correlation coefficient is 0.051 (95 per cent CI = -0.219 – 0.321); and sceptical trust in the insurance industry is 0.051 (95 per cent CI = 0.024 – 0.486). These predictors determine that as risk perception increases, they also increase. However, the above relations do seem relatively low. Since the confidence limits do not encompass a negative value for general trust in the Government and sceptical trust in the insurance industry, it can be concluded that the population regression coefficients for general trust in the Government and sceptical trust in the insurance industry are positive (General Trust: t = 3.142; p = 0.002 / Sceptical Trust: t = 2.183; p = 0.03). The population regression coefficients are also positive for general trust in the insurance industry (t = 0.003; p = 0.998), sceptical trust in the Government (t =

3.142; p = 0.002), although they are negative for knowledge (t = -0.085, p = 0.993) and benefit perception (t = -2.723; p = 0.007).

Table 5-12 Risk Perception – Coefficients

	Unstandardised Coefficients		Standardized Coefficients	t	Sig.	95% Confidence Interval for B	
	B	Std. Error	Beta			Lower Bound	Upper Bound
(Constant)	2.119	.712		2.977	.003	.714	3.524
Mean Knowledge	-.012	.142	-.006	-.085	.933	-.293	.268
Insurance industry - general trust	.000	.115	.000	.003	.998	-.226	.227
Insurance industry – sceptical trust	.255	.117	.181	2.183	.030	.024	.486
Government – skeptical trust	.051	.137	.030	.372	.711	-.219	.321
Government - general trust	.336	.107	.236	3.142	.002	.125	.547
Benefit Perception	-.236	.087	-.201	- 2.723	.007	-.407	-.065

a Dependent Variable: Risk Perception

5.4.2 Benefit Perception

The association between benefit perception and the predictors (mean knowledge, general and sceptical trust in the Government and the insurance industry and risk perception) is moderate (Multiple R = 0.363). Together, the predictors accounted for only 10 per cent of the variation in benefit perception (adjusted R^2). In this case the F-value is 4.378, with an associated probability of <0.001. Mean knowledge, sceptical trust in the insurance industry, and risk perception negatively relate to benefit perception (see Table 5-13). The regression coefficient for mean knowledge is -0.038 (95 per cent CI = -0.279 - 0.203); for sceptical trust in the insurance industry it is –0.097 (95 per cent CI = -0.298 – 0.103) and for risk perception it is –0.174 (95 per cent CI = -0.3 - -0.048). This means that for every one unit increase in benefit perception knowledge drops by –0.038 units, a precise statistic that, loosely interpreted, indicates a minimal association between benefit perception and knowledge. As for risk perception, with every one unit increase in benefit perception there is a decrease in risk perception of –0.174 units, a finding which is indicative of a minimal association between benefit and risk perception. The other predictors are positively related to risk perception. The regression coefficient for general trust in the Government is 0.048 (95 per cent CI = -0.138 – 0.234); general trust in the insurance industry is 0.305 (95 per cent CI = 0.116 – 0.494); and for sceptical trust in the Government the correlation coefficient is 0.106 (95 per cent CI = -0.125 – 0.338). These predictors determine that as risk perception increases, they also increase.

Table 5-13 Benefit Perception - Coefficients

	Unstandardised Coefficients		Standardized Coefficients	t	Sig.	95% Confidence Interval for B	
	B	Std. Error	Beta			Lower Bound	Upper Bound
(Constant)	2.156	.605		3.565	.000	.963	3.350
Mean Knowledge	-.038	.122	-.022	-.311	.756	-.279	.203
Insurance industry - general trust	.305	.096	.254	3.185	.002	.116	.494
Insurance industry - sceptical trust	-.097	.101	-.081	-.960	.338	-.298	.103
Government - sceptical trust	.106	.117	.073	.906	.366	-.125	.338
Government - general trust	.048	.094	.040	.510	.611	-.138	.234
Risk Perception	-.174	.064	-.205	-2.723	.007	-.300	-.048

a Dependent Variable: Benefit Perception

However, the above relations appear relatively low also. The population regression coefficients for general trust in the Government and the insurance industry and sceptical trust in the Government are positive (General Trust Government - $t = 0.51$; $p = 0.611$ / General Trust insurance industry – $t = 3.185$, $p = 0.494$ / Sceptical Trust Government – $t = 0.906$; $p = 0.366$). The population regression coefficients are negative for mean knowledge ($t = -0.311$; $p = 0.756$), sceptical trust in the insurance industry ($t = -0.96$; $p = 0.338$) and risk perception ($t = -2.723$, $p = 0.007$).

5.4.3 Risk Acceptance

The association between risk acceptance and the predictors (mean knowledge, general and sceptical trust in the Government and the insurance industry, risk perception and benefit perception) is moderate (Multiple R = 0.522). Together, the predictors accounted for 27 per cent of the variation in benefit perception (adjusted R^2). In this case, the F-value is 9.22 with an associated probability of <0.001. This means that such a result is highly unlikely to have arisen by sampling error, assuming the null hypothesis to be true. Sceptical trust in the insurance industry and general trust in the Government are negatively related to risk perception (see Table 5-14). The regression coefficient for sceptical trust in the insurance industry is -0.137 (95 per cent CI = -0.361 - 0.087) and for general trust in the Government it is –0.102 (95 per cent CI = -0.310 – 0.105). Given the nature of the data under analysis and the exact nature of statistical analysis, the above findings should be more loosely interpreted as indicating a weak influential relationship between risk acceptance and trust. The other predictors are positively related to risk acceptance. The regression coefficient for mean knowledge is 0.084 (95 per cent CI = -0.184 – 0.353); general trust in the insurance industry is 0.273 (95 per cent CI = -0.056 – 0.49); for sceptical trust in the Government the correlation coefficient is 0.001 (95 per cent CI = -0.259 – 0.258); risk perception is 0.012 (95 per cent CI = -0.132 – 0.156); and finally benefit perception is 0.489 (95 per cent C.I. = 0.322 – 0.656). These predictors determine that as risk perception increases, they also increase. However, the above relations do seem relatively low also and they do not correspond with previous correlation results. The population regression coefficients for mean knowledge, risk perception and benefit perception are positive (Mean knowledge - $t = 0.62$; $p = 0.536$ / General Trust insurance industry – $t = 2.482$, $p = 0.014$/ Risk Perception – $t = 0.164$; $p = 0.87$ / Benefit Perception – $t = 5.772$, $p = 0.001$). The population regression coefficients are negative for general trust in the Government ($t = -0.973$; $p = 0.332$), sceptical trust in the insurance industry (t = -1.208; p = 0.229), and sceptical trust in the Government (t = -0.002, p = 0.999), respectively.

Table 5-14 Risk Acceptance – Coefficients

	Unstandardised Coefficients		Standardised Coefficients	t	Sig.	95% Confidence Interval for B	
	B	Std. Error	Beta			Lower Bound	Upper Bound
(Constant)	.994	.699		1.423	.157	-.385	2.373
Mean Knowledge	.084	.136	.041	.620	.536	-.184	.353
Insurance industry - general trust	.273	.110	.187	2.482	.014	.056	.490
Insurance industry - sceptical trust	-.137	.113	-.094	-1.208	.229	-.361	.087
Government - sceptical trust	.000	.131	.000	-.002	.999	-.259	.258
Government - general trust	-.102	.105	-.070	-.973	.332	-.310	.105
Risk Perception	.012	.073	.012	.164	.870	-.132	.156
Benefit Perception	.489	.085	.403	5.772	.000	.322	.656

a Dependent Variable: Agreement with life insurers use of genetic test results

5.5 Descriptive Statistics

This section further explores the findings from the survey to describe the data obtained. Descriptive statistics are provided that measure average knowledge, trust, risk perceptions, benefit perceptions and risk acceptance evaluations in order to provide a holistic view of the views of the group surveyed.

5.5.1 Knowledge

A considerable 70 per cent of individuals feel that they have no knowledge about the use of genetic test results by life insurance companies to assess if they will insure an applicant (see Appendix E.2, Table E-1). This corresponds with 81 per cent of those surveyed lacking any knowledge of current policies guiding life insurance companies' use of such results. Approximately 74 per cent of respondents believe that the Government and the insurance industry have not provided all relevant information about life insurers' use of genetic test results or the management of this situation (see Appendix E.2, Tables E-6 and E-7).

Table 5-15 contains some basic descriptive statistics for knowledge and trust. Knowledge items 2 and 3 demonstrate respondents' very low knowledge of life insurance companies' use of genetic test results and the policies governing such usage. If it is taken that these two items best represent mean knowledge of life insurance companies' use of genetic test results and the risk management of same, the mean knowledge can be said to be very low indeed. On a scale of 1 (no knowledge) to 5 (perfect knowledge), most respondents self-assessed knowledge of the above is non-existent, with some respondents possessing some knowledge (mean = 1.36).

Table 5-15 Descriptive Statistics for knowledge across subjects

	Mean	Std. Deviation
Knowledge of use of genetic test results by life insurers	1.38	.627
Knowledge of policies governing use of genetic test results	1.19	.393
Mean Knowledge	1.36	.557

(n = 180)

5.5.2 Risk Perception and Benefit Perception

Given the incomplete knowledge of life insurers using genetic test results, 63.7 per cent of respondents to the survey agree that life insurance companies using genetic test results to decide if they will insure an applicant expose society and their families to many risks, with 69.8 per cent of people disagreeing that there are many benefits for society or their families if life insurance companies were to use genetic test results (see Appendix E.2, Table E-3). Although 20.6 per cent of respondents believe that there may be benefits for society associated with life insurance companies using genetic test results, 73.4 per cent think that life insurers should not be allowed to use genetic test results to decide if they will insure an applicant (see Appendix E.2, Tables E-3 and E-4). Out of those individuals that are of the opinion that there are more benefits than risks in this scenario, 41 per cent believe that insurers should still not be allowed to use genetic test results (see Appendix E.2, Tables E-3 and E-4).

5.5.3 Trust in the Government

Trust in the Government across all items is relatively low (see Table 5-16). Positive trust items 1-3 and 7-10 demonstrate respondents' replies to the related questions as falling between 2 and 3 (2 represents 'disagree' or low trust; 3 represents 'don't know' or not sure if trust or don't trust). The mean of general trust, which represents the above items, is 2.85. The mean of items 4, 5 and 6, negative items is 3.22. These items represent a skeptical dimension of trust, with items falling between 3 (represents 'don't know' or not sure if trust or don't trust) and 4 ('agree' or low trust). Therefore, general trust is low, with people disagreeing that the Government is competent, fair, or caring for the public or individuals with regard to the management of life insurance companies' use of genetic test results. Sceptical trust is high, with people agreeing that the government could distort facts in its own favour, that it changes policies without good reasons and, is too influenced by life insurance companies regarding this situation. According to the survey respondents, the Irish Government lacks credibility, reliability and integrity with regard to managing life insurance companies' use of genetic test results. Although, it is notable that the average scores of individuals for general trust and skeptical trust lie in and around 3, which corresponds with respondents not knowing whether to trust or not to trust the Government on this issue. This may be due to peoples' general lack of knowledge of life insurance companies' use of genetic test results and individuals' low or, non-existent, knowledge of any risk management policies. Given this low knowledge, in the majority of cases, respondents are unsure whether or not to rely on, or to trust, the Government to manage this situation. Descriptive statistics in the form of percentage population providing responses to each of the items measured with regard to trust in the Government can be observed in Appendix E.2 (see Tables E-5 and E-8 to E-28).

Table 5-16 Descriptive Statistics of Items measuring Trust in the Government across subjects

	Mean	Std. Deviation
Government - does good job	2.72	.786
Government – competent	2.80	1.111
Government – skilled	2.86	1.159
Government - distorts facts	3.24	.925
Government - changes policies without good reason	3.16	.958
Government - influenced by life insurance companies	3.26	.861
Government - acts in public interest	2.96	.905
Government - listens to public concerns	2.97	1.019
Government - listens to ordinary people	2.81	1.020
Government - fair decision-making	2.82	.916
Government – general trust (positive items)	2.85	0.787
Government – skeptical trust (negative items)	3.22	0.656

(n = 180)

5.5.4 Trust in the Insurance Industry

Trust in the insurance industry is lower yet again, with positive items ranging between 2 and 3 (2 represents 'disagree' or low trust; 3 represents 'don't know' or not sure if trust or not trust) (see Table 5-17). Approximately 45 per cent of respondents disagree that the insurance industry is competent to manage insurance companies' use of genetic test results, while 42.6 per cent believe that the Government is incompetent to manage this situation (average of the three statements designed to measure competence – does good job, competent and skilled) (see Appendix E.2, Table E-6). The mean of general trust, which represents the above items, is 2.38. The mean of items 4, 5 and 6, negative items is 3.64. These items represent a skeptical dimension of trust, with items falling between 3 (represents 'don't know' or not sure if trust or not) and closer to 4 ('agree' or low trust). Therefore, general trust in the insurance industry falls less toward 'don't know' than trust in the Government, with the majority of respondents disagreeing that the insurance industry is competent, fair, or caring for the public or individuals with regard to managing life insurance companies' use of genetic test results to decide if they will insure an individual.

Sceptical trust is high, with most participants agreeing that the insurance industry could distort facts in its own favour, that it changes policies without good reasons and is too influenced by life insurance companies regarding the above. This demonstrates that the greater part of respondents were of the opinion that the insurance industry lacks credibility, reliability and integrity with regard to managing life insurance companies' use of genetic test results to decide if they will insure an individual.

Table 5-17 Descriptive Statistics for Items measuring Trust in Insurance Industry across subjects

	Mean	**Std. Deviation**
Insurance industry - does good job	2.65	.815
Insurance industry – competent	2.51	1.022
Insurance industry – skilled	2.79	1.082
Insurance industry - distorts facts	3.69	.993
Insurance industry - changes policies without good reason	3.47	.960
Insurance industry - influenced by life insurance companies	3.79	.897
Insurance industry - acts in public interest	2.29	.942
Insurance industry - listens to public concerns	2.29	.930
Insurance industry - listens to ordinary people	2.24	.912
Insurance industry – fair decision-making	2.38	.879
Insurance industry – general trust (positive items)	2.45	.793
Insurance industry – skeptical trust (negative items)	3.64	.795

(n = 180)

5.5.5 Differences between Trust in the Government and Trust in the Insurance Industry

Multivariate analysis of variance (MANOVA) is conducted to examine whether people differed in their mean responses across trust in the two entities managing the risks related to life insurance companies using genetic test results. The Irish Government and the insurance industry are, therefore, investigated with respect to the 11 items evaluating trust in both entities.

Error bar charts demonstrate that there is quite a large difference, with no overlap of the 95 per cent confidence intervals, between items 4, 5, 6, 7, 8, 9 and 10 (see Appendix C.3). However, there are little or no differences in terms of items 1 and 3, with only slight overlap between item 11 and, even less so, item 2. MANOVA is used because of intercorrelations observed between the dependent variables. Consistent with Tabachnick and Fidell (1997), Wilk's lambda is reported. This is the most commonly reported of the four tests that calculate the F-value.[1] A repeated-measures MANOVA with one within-participants factor of trust (trust in the Government and trust in the insurance industry) is conducted with measures of trust as the dependent variables. This reveals that there is a multivariate difference between the two trust conditions that is not attributable to sampling error ($F(11,169) = 8.36$, $p \leq 0.001$, Wilk's lambda = 0.603).

It can be seen from the analysis that there is an effect of the within-participants independent variable on the linear combination of the dependent variables, which is unlikely to have arisen by sampling error (see Table 5-18). There is a multivariate difference between the conditions. Consequently, it is assumed that if the dependent variables measure trust, it could be concluded that there is a difference between trust in the insurance industry and trust in the Government.

[1] Several multivariate tests can be used to calculate the F-value: Wilks' lambda, Pillai's Trace, Hotelling's Route and Roy's largest root

Table 5-18 Descriptive Statistics

	Mean	Std. Deviation
Government - does good job	2.72	.786
Insurance industry - does good job	2.65	.815
Government - competent	2.80	1.111
Insurance industry - competent	2.51	1.022
Government - skilled	2.86	1.159
Insurance industry - skilled	2.79	1.082
Government - distorts facts	3.24	.925
Insurance industry - distorts facts	3.69	.993
Government - changes policies without good reason	3.16	.958
Insurance industry - changes policies without good reason	3.47	.960
Government - influenced by life insurance companies	3.26	.861
Insurance industry - influenced by life insurance companies	3.79	.897
Government - acts in public interest	2.96	.905
Insurance industry - acts in public interest	2.29	.942
Government - listens to public concerns	2.97	1.019
Insurance industry - listens to public concerns	2.29	.930
Government - listens to ordinary people	2.81	1.020
Insurance industry - listens to ordinary people	2.24	.912
Government - fair decision-making	2.82	.916
Insurance industry - fair decision-making	2.38	.879
Government - provides information	2.07	1.003
Insurance industry - provides information	1.96	.956

(n = 180)

What is immediately striking about Table 5-18 is the number of real differences in people's evaluation of trust with respect to the Government and the insurance industry. The mean scores not only differ between different questions, they also slightly differ within related questions. Some may find this high level of across-entity disagreement surprising, given that each trust entity is evaluated by the same sample of respondents. However, the two entities, i.e. the Government and the insurance industry, have contrasting facets and histories providing them with different views on the management of the use of genetic test results by life insurance companies. The mean results demonstrate a lack of trust in both the Government and the insurance industry, with the sample population trusting the insurance industry less than the Government with regard to the management of the use of genetic test results. Mean responses are below the scale mid-point for most of the 'positive' items, such as competence, care and fairness, whereas, the mean responses are above the scale mid-point for 'negative' items 4, 5 and 6. Interestingly, the item with the lowest average agreement is the one designed to measure the concept of 'openness'. This corresponds with the results of Poortinga and Pidgeon (2003).

It is apparent from the MANOVA Table (Table 5-18) that there is a considerable multivariate difference between trust in the two entities (the Government and the insurance industry). In the multivariate analyses, where there is more than one dependent variable, once a multivariate difference is found it is necessary to find out which dependent variables are contributing to this difference. It is likely, especially if there are many dependent variables, that not every dependent variable will contribute to the overall difference observed. To evaluate each dependent variable, it is necessary to see whether they are correlated. Having partialled out all other items designed to measure trust, the Pearson's Product Moment Correlations for each item are shown below (see Table 5-19).

Table 5-19 Pearson's Product Moment Correlations (Trust in Government and Trust in Insurance Industry)

Item	Correlation
Government – does good job Insurance industry – does good job	r = 0.136 p ≤ 0.086
Government – competent Insurance industry – competent	r = 0.289 p ≤ 0.001
Government – skilled Insurance industry – skilled	r = 0.253 p ≤ 0.001
Government – distorts facts Insurance industry – distorts facts	r = 0.044 p ≤ 0.582
Government – changes policies Insurance industry – changes policies	r = 0.026 p ≤ 0.746
Government – influenced by life insurance companies Insurance industry – influenced by life insurance companies	r = 0.048 p ≤ 0.545
Government – acts in public interest Insurance industry – acts in public interest	r = -0.082 p ≤ 0.301
Government – listens to concerns raised by public Insurance industry – listens to concerns raised by public	r = 0.067 p ≤ 0.396
Government – listens to what ordinary people think Insurance industry – listens to what ordinary people think	r = 0.067 p ≤ 0.399
Government – fair decision-making Insurance industry – fair decision-making	r = 0.176 p ≤ 0.026
Government – provides information Insurance industry – provides information	r = 0.316 p ≤ 0.001

It is evident from the above correlation coefficients that there is no correlation between the majority of the dependent variables. The highest correlation occurs in item 11 ($r = 0.316$, $p \leq 0.001$) (see Table 5-19). However, this correlation is still relatively weak and is considered too weak to influence the evaluation of the dependent variables (see Dancey & Reidy, 2002:166). It is, therefore, possible to investigate the relative contribution of each dependent variable to the linear combination of the dependent variables. To establish which dependent variables are contributing to this effect, a post-hoc univariate analysis is conducted on the individual dependent variables. There are a variety of ways in which to proceed with post-hoc analysis, and different researchers prefer different techniques. The most popular techniques include conducting related t-tests or by using the F-values from the univariate test. In this particular analysis, there is a multivariate difference for a two group independent variable design, therefore eleven paired univariate t-tests are used to establish which dependent variables are contributing to differences in trust, with the alpha set at 0.0046 (see below). The ∂ is adjusted to take account of the number of analysis to be conducted. Such an adjustment of ∂ is necessary to control the family wise error rate. The approach recommended by Stevens (1997) is to set a level for the overall ∂ (the ∂ in this study is set at 5 per cent) and then divide this by the number of comparisons to be made. In this particular analysis there are eleven dependent variables, therefore the alpha is adjusted as follows:

$$0.05 / 11 = 0.0046$$

The t-test analysis of all of the items measuring trust are examined separately, with trust in the Government and trust in the insurance industry as the independent variable in each case, 'setting' the alpha for each at 0.0046 (0.46 per cent). By doing so, it is increasingly likely that any findings are unlikely to have arisen due to sampling error. The following table illustrates the t-test analysis results (see Table 5-20).

Table 5-20 Paired Samples Statistics

		95% Confidence Interval of the Difference			
		Lower	**Upper**	**t**	**Sig. (2-tailed)**
Pair 1	Government - does good job Insurance industry – does good job	-.080	.213	**.899**	**.370**
Pair 2	Government - competent Insurance industry – competent	.095	.483	**2.941**	**.004**
Pair 3	Government - skilled Insurance industry – skilled	-.131	.253	**.629**	**.530**
Pair 4	Government - distorts facts Insurance industry - distorts facts	-.616	-.273	**-5.124**	**.000**
Pair 5	Government - changes policies without good reason Insurance industry - changes policies without good reason	-.476	-.135	**-3.539**	**.001**

Pair 6	Government - influenced by life insurance companies Insurance industry - influenced by life insurance companies	-.687	-.369	**-6.555**	**.000**
Pair 7	Government - acts in public interest Insurance industry - acts in public interest	.474	.871	**6.686**	**.000**
Pair 8	Government - listens to public concerns Insurance industry - listens to public concerns	.495	.860	**7.332**	**.000**
Pair 9	Government - listens to ordinary people Insurance industry - listens to ordinary people	.381	.753	**6.013**	**.000**
Pair 10	Government - fair decision-making Insurance industry - fair decision-making	.276	.613	**5.210**	**.000**
Pair 11	Government - provides information Insurance industry - provides information	-.037	.271	**1.494**	**.137**

(95 per cent Confidence Interval, df = 179)

Results from the t-test analysis reveal that item 10 ($t = 5.21$, $p \leq 0.001$), item 9 ($t = 6.013$, $p \leq 0.001$), item 8 ($t = 7.332$, $p \leq 0.001$), item 7 ($t = 6.686$, $p \leq 0.001$), item 6 ($t = -6.555$, $p \leq 0.001$), item 5 ($t = -3.539$, $p \leq 0.002$), item 4 ($t = -5.124$, $p \leq 0.001$) and item 2 ($t = 2.941$, $p \leq 0.004$) dependent variables have an associated probability of being due to sampling error that is less that the 0.46 per cent level of alpha. It is, therefore, possible to conclude that trust in the Government and trust in the insurance industry differ only in terms of their evaluation of the above variables. It is assumed that this is a valid measure of trust, therefore it appears that there is a difference between the evaluations of these two entities in terms of trust. If it is considered that the above named variables on their own are not a sufficient indicator of trust, which is the assumption undertaken in the present study, then this analysis would seem to suggest that overall there is no difference between trust in the two entities beyond that which is attributable to the differences in the items designed to measure credibility, integrity, reliability, care, fairness and one item of competence. There is no contribution to the multivariate difference between the items designed to measure two items of competence and openness: item 1 ($t = 0.899$, $p \leq 0.37$), item 3 ($t = 0.629$, $p \leq 0.53$) and item 11 ($t = 1.494$, $p \leq 0.137$). The confidence interval for the difference between trust in the Government and trust in the insurance industry demonstrates that the population mean difference is likely (95 per cent).

5.5.6 Risk Acceptance

Given that a mere 10.6 per cent believe that the current rules and regulations in the Republic of Ireland are sufficient to control the use of genetic test results by life insurance companies, it can be said that consumers lack confidence in the insurance industry and the Government to adequately regulate the use of genetic test results by life insurance companies (see Appendix E.2, Tables E-6 and E-7). Approximately 74 per cent of individuals surveyed believe that there is a need for the Irish Government to create legislation to prohibit life insurers from using genetic test results to decide if they will insure an applicant (see Appendix E.2, Tables E-4 and E-5). When making decisions regarding life insurers' access to, and use of, genetic test results, 91.7 per cent of the population believe that the public should participate more in any decisions made regarding life insurers' use of genetic test results (see Appendix E.2, Table E-9). Given such enthusiasm for involvement in the decision-making processes, only 18.4 per cent of those surveyed believe that they currently have a say in the content of these decisions and only 20 per cent of the population agree that they have a say in how decisions regarding life insurers' use of genetic test results are made (see Appendix E.2, Tables E-10 and E-11).

5.6 Socio-Demographic Differences Across Knowledge, Trust, Risk Perceptions, Benefit Perceptions and Risk Acceptance

This section explores differences across gender, age, economic status and ownership of life insurance in evaluations of knowledge, trust, risk perception, benefit perception and risk acceptance. Such an analysis is pursued in order to understand if and/or how these factors serve to influence the views of the group surveyed.

5.6.1 Gender

The study records 106 female respondents and 74 male remaining respondents to the survey.[2] Therefore, 59 per cent of the population surveyed are female, while the remaining 41 per cent are male.[3] Given this difference in the proportion of respondents, knowledge of insurers' use of genetic test results and the policies relating to such usage is similarly low (mean = 1.35/1.37; SD = 0.535/0.574) across both genders. Female respondents perceive more risks and more benefits for society, their families and themselves, than male respondents (mean = 3.36/3.76; SD = 1.2/1.01). Several studies demonstrate that women express higher concerns about environmental and technological risks than men (Davidson and Freudenburg, 1996; Siegrist, 2000). In addition, Bord and O'Connor (1997:832) found that women consistently exhibit much stronger perceptions of vulnerability to illness and physical debilitation. Given the technological and health related nature of genetic testing, the risks perceived by women from insurers gaining access to genetic test results may be higher than their male counterparts. The mean difference between risk perceptions is 0.4, and between benefit perceptions it is even less (mean difference = 0.09; mean = 2.3/2.39; SD = 0.975/0.942) (see Table 5-21). The 95 per cent confidence interval for the estimated population mean difference for risk perception is between 0.068 and 0.73. The difference between the perceptions of men and women is significant (t = 2.38, DF = 178, $p < 0.018$). It is, therefore, possible to conclude that females perceive more risks than males in relation to insurers using genetic test results. The 95 per cent confidence interval for the estimated population mean difference for benefit perception is between 0.196 and 0.377. The difference between the perceptions of men and women is not significant (t = 0.618, DF = 178, $p < 0.537$). It is, therefore, concluded that females perceive a similar number of benefits as males in relation to insurers using genetic test results.

Overall trust in the Government and the insurance industry does not differ greatly between genders (see Table 5-21). The 95 per cent confidence interval for the estimated population mean difference for trust in the

[2] This calculation is based upon the total population after removal of outliers

[3] This slightly disproportionate gender classification may be seen as a result of the times during which the survey was conducted. The researcher began door-to-door visits at 4pm between Monday and Friday. It may be that there are more women at home during the early hours of the evening as a result of child minding or working at home.

Government is between 0.134 and 0.148 (see Table 5-21). It can be concluded that there is little difference between gender and trust judgements in the Government (t = 0.093, DF – 178, p = 0.926). This does not fully correspond with the findings of Siegrist (2000), who indicates that a different level of trust in institutions determines the different risk perceptions between males and females. Although, the findings do indicate that trust does not explain the difference in perceived benefits and, hence, does not explain the observed differences in risk acceptance.

Table 5-21 Gender Independent Samples Test

	F	Sig.	t	Sig (2-tailed)	Mean Difference	95 per cent Confidence Interval of the Difference	
						Lower	Upper
Knowledge	.009	.923	.011	.991	.00	-.187	.189
Risk Perception	10.512	.001	-2.380	.018	-.40	-.730	-.068
Benefit Perception	.001	.973	-.618	.537	-.09	-.375	.196
Government Overall Trust	3.073	.081	-.093	.926	-.01	-.148	.134
Insurance industry Overall trust	.069	.793	.792	.429	.05	-.079	.184
Risk Acceptance	2.136	.146	.371	.711	.07	-.282	.412

(Degrees of Freedom: 178, n = 180)

Similarly, overall trust in the insurance industry does not differ greatly between genders (see Table 5-21). The mean difference between the sexes is 0.05, while the 95 per cent confidence interval for the estimated population mean difference for trust in the insurance industry is between 0.079 and 0.184 (see Table 5-21). It can also be concluded that there is little difference of trust in the insurance industry between gender (t = 0.792, DF = 178, p = 0.429).

Risk acceptance is slightly higher for males than females, with women less likely to accept the risks related to life insurance companies using genetic test results to underwrite individuals than their male counterparts (mean difference = 0.065). However, the mean difference is relatively low and the confidence interval is 0.282 - 0.412. There is a not a significant difference of risk acceptance between genders (t = 0.371, DF = 178, p = 0.711).

Correlation analysis reveals that the strength of the relationship between gender and each of the variables knowledge, trust, risk perceptions, benefit perceptions and risk acceptance is weak (see Table 5-22). The highest correlation occurs between gender and risk perception (r = 0.176), suggesting that the influence of gender upon risk perception is stronger than the effect that gender may have upon any other factor.

Table 5-22 Gender, Paired Samples Correlations

	Correlation	**Sig.**
Mean Knowledge & Gender	.015	.845
Risk Perception & Gender	.176	.018
Benefit Perception & Gender	.046	.537
Government - Overall Trust & Gender	.007	.926
Insurance industry - overall trust & Gender	-.059	.429
Agreement with life insurers use of genetic test results & Gender	-.028	.711

(n = 180)

Overall, given that female respondents perceive more risks and more benefits, have similar risk acceptance and, yet, similar trust judgements in the risk managers as the male participants, it is possible to say that the findings of the present study do not mirror exactly the theory put forward by Siegrist (2000) and Flynn *et al.* (1994). This theory suggests that men perceive more benefits than women and have a higher risk acceptance because they have more power and control in society than women. It appears, in this case, that there may be another generative mechanism, other than trust, affecting the results of the analysis. Although the differences illustrated above are not significant, the findings illustrate a need for future studies to better understand the reasoning for gender differences in risk and benefit perception and risk acceptance of technologies.

5.6.2 Age

The majority of individuals surveyed (74 per cent) fall between the age brackets of 26 – 56 years of age (see Table 5-23). Given that 67.5 per cent of the Irish population is aged between 15 and 64 years of age, while those individuals occupying the age category of 65 years and over make up approximately 11.5 per cent of the Irish population (Central Statistics Office, 2005 estimate), it would seem that the population sampled in the present study is not representative of the age cross section of the Irish public. There are a variety of reasons that may influence the above finding; for example, some of those individuals occupying the older categories may be wary about answering their door to strangers or filling out the survey. Younger individuals, under the age of 25, were more likely to be out of the home during the hours the survey was conducted (4pm to 8.30pm).

Table 5-23 Age Group Percentage

	Valid Percent
18-25	8.3
26-35	27.8
36-45	26.1
46-55	20.0
56-65	10.0
66-75	5.0
76-85	2.2
86-95	.6

(n = 180)

Table 5-24 represents the percentage agreement of individuals in each age group with life insurers using genetic test results to decide whether or not to insure an individual. It is noteworthy that it is the 26-35 year old age group, those most likely to purchase life insurance for a mortgage, that have the highest percentage disagreement with insurers accessing genetic test results (84 per cent). This disagreement may be seen as a reaction against insurers using genetic test results for fear of the risks related to such usage. Indeed, this statement is qualified by the fact that 62 per cent of the respondents from this category agreed that insurers using genetic test results expose society, themselves and their families to many risks, while 76 per cent disagreed that there were many benefits.

Table 5-24 Age Group and Risk Acceptance

	Agreement with life insurers' use of genetic test results					
	N	Strongly disagree	Disagree	Don't know	Agree	Strongly Agree
18-25	15	20.0	46.7	13.3	20.0	
26-35	50	38.0	46.0	2.0	12.0	2.0
36-45	47	29.8	46.8	4.3	17.0	2.1
46-55	36	36.1	33.3	8.3	16.7	5.6
>56	32	31.3	28.1	6.3	31.3	3.1

(n = 180)

The group most likely to disagree that there are risks related to insurers using genetic test results are those respondents in the 36-45 age groups, while the 18-25 age groups follows closely behind (29.8 per cent and 26.7 per cent respectively). The 18-25 age group are also the most likely to agree that there are benefits relating to such usage of genetic test results (26.7 per cent), while the 26-35 age group perceived the least benefits related to this situation (see Appendix E.2, Table E-20 – 76 per cent). Those respondents aged 18-25 and over the age of 46 are the most likely to agree that insurers should be allowed to use genetic test results. In addition, individuals aged 18-25 and over 36 are the groups that express the highest percentage disagreement that there are risks related to insurers using genetic test results while the same groups are the most likely to agree that there are benefits related to this.

Turning to trust as an explanation for the risk perception of individuals, given the uneven proportion of respondents in each group, the 26-35 age group are the most likely to disagree that the Government can be trusted (see Appendix E.2, Table E-29). Those aged between 56 and 65 have the highest percentage disagreement that the insurance industry can be trusted to manage insurers' use of genetic test results. Interestingly, this group may have had the most experience dealing with insurers given their age. Although the over 56 years of age categories are the least represented by the survey, they are the groups who declare most agreement with insurers accessing genetic test results and indicate overall agreement with trust in the Government and the insurance industry to manage insurers' use of genetic test results (see Appendix E.2, Table E-31). Such a finding may be due to the disproportionate number of respondents in the older categories as compared to the younger age groups. Knowledge is low amongst all age groups, with no age group dominating higher or lower knowledge.

Correlation analysis reveals that there is relatively little or no correlation between age and any of the variables, with the highest correlation existing between age and trust in the Government (r = 0.14, which is very weak indeed) (see Table 5-25). This suggests that the relationship between age and trust in the Government is strongest for this pair of variables in comparison with all others.

Table 5-25 Age - Paired Samples Correlations

	Correlation	**Sig.**
Mean Knowledge & Age	-.044	.554
Risk Perception & Age	.038	.616
Benefit Perception & Age	.052	.486
Government - Overall Trust & Age	.140	.060
Insurance industry - overall trust & Age	.070	.351
Agreement with life insurers use of genetic test results & Age	.112	.133

(n = 180)

5.6.3 Economic Status

The majority of individuals (56.1 per cent) surveyed are employed (see Figure 5-4). The minority are students or self-employed who, together, make up 16.6 per cent of the sample.

Figure 5-4 Graphical Distribution of Occupational Status

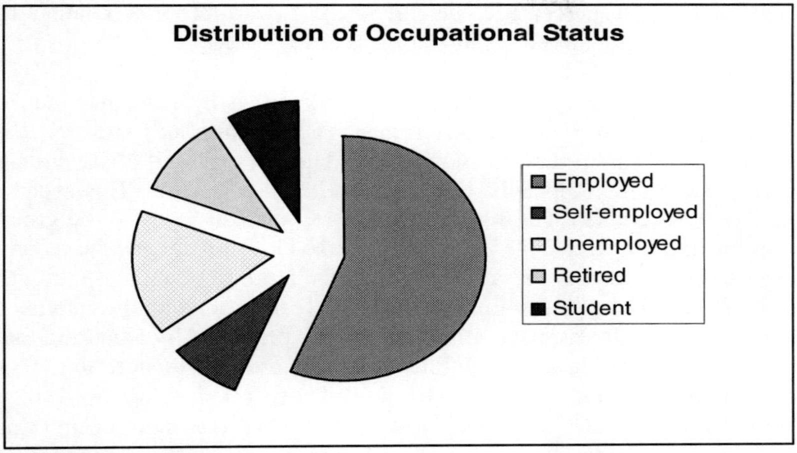

(n = 180)

Individuals who are unemployed feel the strongest regarding not allowing insurers the use of genetic test results, with 80 per cent against and only 10 per cent for allowing insurers access to this information (see Appendix E.2, Table E-22). On the other hand, retired respondents are the least likely to disagree with insurers using genetic test results (57.9 per cent disagree), while 36.8 per cent agree that insurers should be allowed to use such data when deciding whether or not to insure an applicant for insurance. What this initial data suggests is that those individuals who are most vulnerable with regard to being able to afford life insurance are those most likely to disagree with genetic stipulations on access to insurance products. And those who are least likely to apply for life insurance, respondents who are retired, are the group with the highest declaration of agreement with insurers using genetic test results.

Although retired respondents were the most likely to accept insurers using genetic test results, they displayed the highest percentage agreement that there were many risks related to insurers using genetic test results. However, contrary to these findings, they were also the group which, percentage wise, had the highest rate of agreement that there were also many benefits for society (see Appendix E.2, Table E-23). Approximately 63.4 per cent of individuals who are unemployed believed that there were many risks related to insurers using genetic test results, while only 30 per cent agree that there were benefits regarding this situation. Those individuals who occupy the employed categories, (employed / self-employed), as a percentage, perceived the lowest benefits relating to insurers using genetic test results.

The highest percentage of individuals as classified by economic status, which displayed trust in the Government and the insurance industry, are those respondents in the retired category (31.6 per cent and 26.3 per cent respectively) (see Appendix E.2, Tables E-25 and E-26). However, it should also be noted that this category of respondents are also the group with the highest percentage lack of trust in the Government and the second highest with regard to the insurance industry (15.8 per cent and 18.8 per cent respectively). Students are the least likely to trust either the insurance industry or the Government, with only 6.7 per cent actually admitting that they felt trust towards both entities. It is interesting to note the large number of individuals present in the 'don't know' category, whereby, they are unsure as to their trust in the two entities. The reasoning behind this may relate to the number of dimensions of trust examined that lead to the trust evaluations indicated in the tables. The average response is taken as each individual's evaluation of trust in each entity and, in some cases, the number of 'don't know' answers outweigh other responses provided. It is perhaps a lack of familiarity of each entity's management of insurers' use of genetic test results that generate so many 'unsure' responses. All categories of individuals display low knowledge of insurers' use of genetic test results or the policies guiding such usage.

Correlation analysis reveals that there is little or no association between occupational status and knowledge, risk perceptions, benefit perceptions, trust or risk acceptance (see Table 5-26).

Table 5-26 Occupational Status, Paired Samples Correlations

	Correlation	Sig.
Knowledge & Current occupational status	-.140	.060
Risk Perception & Current occupational status	-.080	.285
Benefit Perception & Current occupational status	.182	.015
Trust in Government & Current occupational status	.030	.691
Trust in Insurance industry & Current occupational status	-.033	.659
Risk Acceptance & Current occupational status	.020	.785

(n = 180)

5.6.4 Ownership of Life Insurance

A slightly higher percentage of those respondents who declare that they own life insurance are more willing to agree that life insurance companies should be allowed to use the results of genetic tests than those who don't own life insurance (22.5 per cent and 18.3 per cent respectively) (see Appendix E.2, Table E-27). Relatively the same percentage of respondents who own and don't own life insurance products replied that they disagree that insurers should be allowed to assess individuals based on their genetic test results (72.5 – 75 per cent). Out of the respondents that believe that they would purchase life insurance in the future, only 13.3 per cent agree that life insurance companies should be allowed to use genetic test results, compared to 28 per cent of respondents who have no intention of purchasing life insurance with the same belief (see Appendix E.2, Table E-28). While 80 per cent of those interviewed who wish to purchase life insurance in the future disagree that life insurance companies should be allowed to use genetic test results, only 68 per cent of those who do not intend to do so disagree. These findings seem to suggest that it is those individuals in the life insurance purchasing category that are most against insurers using genetic test results, whilst it is those respondents who already have life insurance or, do not intend purchasing it, that feel the least against insurers using genetic test results.

Given the difference in the proportion of respondents currently owning life insurance, knowledge of insurers' use of genetic test results and the policies relating to such usage is similarly low (mean = 1.44/1.20; SD = 0.605/0.403) across those who currently own and who do not own life insurance. There appears to be no difference in risk perception between the two groups (mean = 3.60/3.60; SD = 1.115/1.061). There is no mean difference between risk perceptions (0.00); while the mean difference between benefit perceptions is slight (0.13) (see Table 5-27). The 95 per cent confidence interval for the estimated population mean difference for risk perception is between -0.351 and 0.351. There is no difference between the risk perceptions dependent on life insurance ownership ($t = 0.00$, DF = 178, $p < 1.0$). The 95 per cent confidence interval for the estimated population mean difference for benefit perception is between -0.423 and 0.173. The difference between the benefit perceptions of the two groups is not significant ($t = -0.828$, DF = 178, $p < 0.409$). It is possible to conclude that there is very little difference between risk and benefit perceptions based upon current life insurance ownership. Overall trust in the Government and the insurance industry does not differ greatly between those who own life insurance and those who don't. Upon examination of overall trust in the Government, the mean responses between groups were similar (mean = 3.01/3.02; SD = 0.494/0.415). The 95 per cent confidence interval for the estimated population mean difference for trust in the Government is between -0.255 and 0.038 (see Table 5-27). Overall trust in the insurance industry is also similar between groups (mean = 3.04/3.07; SD = 0.474/0.362) and ($t = 0.359$, DF = 178, $p < 0.72$). Finally, risk acceptance does not differ between those who own and those who don't own life insurance (mean = 2.18/2.17; SD = 1.233/1.028) and ($t = 0.091$, DF = 178, $p < 0.928$). Table 5-27 displays how the mean difference between life insurance ownership is very low for all variables.).

Overall trust in the insurance industry is also similar between groups (mean = 3.04/3.07; SD = 0.474/0.362) and ($t = 0.359$, DF = 178, $p < 0.72$). Finally, risk acceptance does not differ between those who own and those who don't own life insurance (mean = 2.18/2.17; SD = 1.233/1.028) and ($t = 0.091$, DF = 178, $p < 0.928$).

The strength of the relationship between current ownership of life insurance and the variables designed to measure overall knowledge, trust, risk perceptions, benefit perceptions and risk acceptance is rather weak (see Table 5-27). The highest correlation occurs between mean knowledge and current ownership of life insurance ($r = -0.205$), suggesting a negative relationship between these two factors. Overall, correlation analysis suggests that life insurance ownership plays a minimal or, perhaps even non-existent, role in determining evaluations regarding the above variables.

Table 5-28 displays how the mean difference between life insurance ownership is very low for all variables.

The strength of the relationship between current ownership of life insurance and the variables designed to measure overall knowledge, trust, risk perceptions, benefit perceptions and risk acceptance is rather weak (see Table 5-28). The highest correlation occurs between mean knowledge and current ownership of life insurance (r = -0.205), suggesting a negative relationship between these two factors. Overall, correlation analysis suggests that life insurance ownership plays a minimal or, perhaps even non-existent, role in determining evaluations regarding the above variables.

Table 5-27 Life Insurance Ownership Independent Samples Test

	F	Sig.	t	Sig. (2-tailed)	Mean Difference	95 per cent Confidence Interval of the Difference	
						Lower	Upper
Knowledge	31.42	.000	2.796	.006	.24	.071	.412
Risk Perception	.402	.527	.000	1.00	.00	-.351	.351
Benefit Perception	.000	.987	-.828	.409	-.13	-.423	.173
Government Overall Trust	.282	.596	-1.46	.146	-.11	-.255	.038
Insurance Industry Overall Trust	1.374	.243	-.359	.72	-.03	-.162	.112
Risk Acceptance	3.222	.074	.091	.928	.02	-.346	.379

(Degrees of Freedom: 178, n = 180)

Table 5-28 Current Ownership of Life Insurance, Paired Samples Correlations

	Correlation	Sig.
Knowledge & Currently own life insurance	-.205	.006
Risk Perception & Currently own life insurance	.000	1.000
Benefits for society & Currently own life insurance	.109	.144
Trust in Government & Currently own life insurance	.109	.146
Trust in Insurance industry & Currently own life insurance	.027	.720
Risk Acceptance & Currently own life insurance	-.007	.928

(n = 180)

Turning to future ownership of life insurance as an explanation for these factors, 30 respondents replied that they would like to purchase life insurance in the future, while 25 replied that they have no intention of doing so. The remaining 125 individuals responded that they were unsure whether or not they would purchase insurance in the future. The following analysis examines only those individuals who responded positively or negatively to these questions, so that it is possible to note any differences in the variables based upon future intention to purchase life insurance. Both groups appear to have similarly low knowledge of insurers' use of genetic test results and the associated policies (mean = 1.17/1.28; SD = 0.379/.458). There appears to be relatively little difference across risk perceptions and benefit perceptions (mean risk perceptions = 3.47/3.60; SD = 1.106/1.041; mean benefit perceptions = 2.43/2.36; SD = 0.744/1.075). The mean difference between risk perceptions is -0.13, while the mean difference between benefit perceptions is even less (0.07) (see Table 5-29). The difference between the two groups and their risk perceptions is not significant ($t = -0.457$, $DF = 53$, $p < 0.649$). Neither is the difference in benefit perceptions significant ($t = -0.293$, $DF = 53$, $p < 0.77$). These findings demonstrate that, statistically, there is very little difference between risk and benefit perceptions based upon intention to buy life insurance in the future.

Overall trust in the Government and the insurance industry does not differ greatly between those who intend to purchase life insurance and those who don't. The mean responses between groups for trust in the Government were similar (mean = 3.03/3.16; SD = 0.32/0.473). The 95 per cent confidence interval for the estimated population mean difference for trust in the Government is between -0.342 and 0.089 (see Table 5-29). Overall trust in the insurance industry is also similar between groups (mean = 3.03/3.08; SD = 0.32/0.4) and (t = -0.481, DF = 53, p < 0.633). Finally, risk acceptance does not appear to differ greatly between those who intend and those who do not intend to purchase life insurance in the future (mean = 2.03/2.32; SD = 0.964/1.18) and (t = -0.992, DF =53, p < .326).

Table 5-29 Future Ownership of Life Insurance, Independent Samples Test

	F	**Sig.**	**t**	**Sig. (2-tailed)**	**Mean Difference**	**95 per cent Confidence Interval of the Difference**	
						Lower	**Upper**
Knowledge	4.021	.05	-1.004	.32	-.11	-.34	.113
Risk Perception	.543	.464	-.457	.649	-.13	-.718	.452
Benefit Perception	4.195	.046	.293	.77	.07	-.428	.575
Government Overall Trust	6.193	.016	-1.18	.243	-.13	-.342	.089
Insurance Industry Overall Trust	1.2	.278	-.481	.633	-.05	-.241	.148
Risk Acceptance	3.698	.06	-.992	.326	-.29	-.867	.293

(Degrees of Freedom: 53, n = 180)

The magnitude of the relationship between future purchase of life insurance and knowledge, trust, risk perceptions, benefit perceptions and risk acceptance is small, with most correlations displaying no association between future purchase of life insurance and any other variable (see Table 5-30). This indicates that future purchase of life insurance does not play a significant role in evaluations regarding these variables.

Given that respondents who own life insurance and those who don't, display similar knowledge evaluations, trust judgements, risk perceptions, benefit perceptions and risk acceptance, it is possible to state that ownership of life insurance does not appear to be a factor that affects these variables to any great extent. In addition, findings from the present study also demonstrate that future intention to purchase life insurance does not influence these variables either.

Table 5-30 Future Purchase of Life Insurance, Paired Samples Correlations

	Correlation	Sig.
Mean Knowledge & Future purchase of life insurance	-.137	.068
Risk Perception & Future purchase of life insurance	-.027	.715
Benefit Perception & Future purchase of life insurance	.024	.747
Trust in Government & Future purchase of life insurance	.089	.235
Trust in Insurance industry & Future purchase of life insurance	.018	.816
Risk Acceptance & Future purchase of life insurance	.018	.806

(n = 180)

The majority of individuals surveyed (74 per cent) fall between the age brackets of 26 – 56. Such an age group consists of those individuals most likely to be in the life insurance purchasing or ownership category (73 per cent of those who answered Qu.6 Part C (ii) that "yes", they think they will purchase a life insurance product in the future, belong to the 26-35 age group). A total of 84 per cent of the respondents who already own life insurance products belong to the >36 age group categories. In addition, the 26-35 age group are the most likely to disagree that the insurance industry can be trusted. This may be because they have not yet had many experiences dealing with the life insurance industry, being the group that is most likely to not yet have purchased insurance, and the group with the highest percentage of intention to purchase insurance in the future. The findings from Section 5.6.2 demonstrate that the age groups most likely to be in the purchasing bracket for life insurance in Ireland are those that most strongly disagree with insurers using genetic test results, perceive the more risks, and the least benefits. Correlation analysis between age and previous and current ownership of life insurance reveal a Pearson's correlation of ($r = -0.209$, $p \leq 0.005$) (two tailed), demonstrating a weak negative relationship between the two factors. Correlation analysis between age and future ownership of life insurance is significantly weaker ($r = 0.017$, $p \leq 0.821$). Therefore, the statistical analysis demonstrates a weak to non-existent relationship between age and life insurance ownership.

Results from the analysis of economic status also suggest that those individuals who are most vulnerable with regard to being able to afford life insurance are those most likely to disagree with genetic stipulations on access to insurance products. Those who are least likely to apply for life insurance, respondents who are retired, are the group with the highest declaration of agreement with insurers using genetic test results (see Section 5.6.2). Therefore, there appears to be a link between economic status and life insurance ownership; however, upon statistical analysis, the correlation between occupational status and life insurance ownership is rather weak ($r = 0.238$, $p \leq 0.001$), perhaps placing less importance upon this theory. Nonetheless, both are representative of factors that may affect knowledge, trust, risk perceptions, benefit perception and risk acceptance within the subject confines of the present study.

To conclude this section, preliminary analysis of the results from the survey suggested that gender, age, economics status and life insurance ownership may be factors that affect the relationship between knowledge, trust evaluations, risk perceptions, benefit perceptions and acceptance of the risks related to insurers using genetic test results. It was revealed how there is a difference between risk perceptions and risk acceptance based upon gender that cannot be explained by differing trust judgements. Initial analysis also suggested that risk perceptions and acceptance of the risks related to insurers using genetic test results is dependent upon age and economic status. It was assumed that these differences may be related to intention to purchase life insurance; however, correlation analysis revealed that there is no relationship between these factors. Furthermore, evaluations of knowledge, trust, risk perceptions, benefit perceptions or risk acceptance did not appear to be dictated by current life insurance ownership or intention to purchase life insurance. Correlation analysis indicated that gender, age, economic status and life insurance ownership play a minimal or non-existent role in influencing knowledge, trust, risk perceptions, benefit perceptions or risk acceptance.

5.7 Summary

Correlation analysis revealed relatively little association between incomplete knowledge, social trust, risk perceptions, benefit perceptions and risk acceptance in the case of the risks related to insurers using genetic test results. Indeed, the effect that each of the factors have upon each other in the majority of cases was so weak that it may even indicate a non-existent relationship between incomplete knowledge, social trust, risk perceptions, benefit perceptions and risk acceptance. Regression analysis was used in addition to correlation analysis and revealed only low associations between incomplete knowledge, social trust, risk perceptions, benefit perceptions and risk acceptance. Based on the findings from correlation and regression analysis, it is possible to state that the relationship between incomplete knowledge, social trust, risk perceptions, benefit perceptions and acceptance of the risks related to insurers using genetic test results is weak, with the results indicating that the influence that each of the factors have upon each other and risk acceptance is minimal.

Findings from the survey demonstrated that respondents have incomplete knowledge of the use of genetic test results, high risk perceptions, low benefit perceptions, low trust in the government and the insurance industry and a lack of willingness to accept insurers' use of genetic test results to underwrite applicants for life insurance. Differences across gender, age, economic status and ownership of life insurance in the means of knowledge, trust, risk perception, benefit perception and risk acceptance were examined.

Further study of the above findings shall be completed by interview in the following chapter in an effort to explicate the reasoning behind the statistical results. According to Sayer (2000:17) and Holmwood (2000:950-951), the lack of association found between knowledge, trust, risk perceptions, benefit perceptions and risk acceptance could indicate either of two things. Firstly, the non-occurrence of events could be an indication that there are countervailing tendencies generated by the operation of other relevant mechanisms that have interfered with the causal risk acceptance pattern. Or indeed, such lack of association between trust, risk perceptions and risk acceptance could indicate that the claimed mechanism between these objects is false (Siegrist, 2000; Siegrist and Cvetkovich, 2000; Poortinga and Pidgeon, 2005). The following chapter shall explore which of the above conclusions appears to be most appropriate in the present study. Through the analysis of several cases, it is intended to explore the factors that may have had a role to play in people's responses to the questions relating to knowledge, social trust, risk and benefit perception and risk acceptance and to further our knowledge regarding their association, or indeed, any misrecognition of causality.

CHAPTER SIX

INTERVIEW FINDINGS

6.1 Introduction

The previous chapter examined the results of the survey and explored the effect of socio-demographic factors and ownership of life insurance upon knowledge, trust, risk perception, benefit perception and risk acceptance. The chapter analysed the relationship between knowledge, trust, risk perception, benefit perception and risk acceptance and explained whether people differed in their responses across trust in two entities managing the risks related to life insurance companies using genetic test results, the Irish Government and the insurance industry. In addition, the effects that these variables have on each other were discussed. This chapter examines the responses to the semi-structured interviews conducted as part of the study in an effort to understand what factors or conditions generated the results from the statistical analysis as outlined in Chapter 5. The relationship between knowledge, trust, risk perception, benefit perception and risk acceptance shall briefly be examined based upon the results of the interviews. Finally, several risk management perspectives shall be discussed based upon the analysis of interview discussions, allowing for further insight into the conditions whereby individuals agree to tolerate or to accept the risks perceived by insurers using genetic test results.

6.2 Case Description

The original 76 individuals available for interview were sub-divided into groups whereby individuals with similar characteristics were divided into four cases: representative case, extremely varied case, critical case and an extreme case (see Sections 4.7.1 and 4.7.2); (Danermark, 2002:170). It was attempted to select three individuals representative of each case, however, this method of assortment resulted in a total of only ten individuals, as there was only one individual who satisfied the selection requirements of the extreme case. All ten individuals were subsequently interviewed as part of the qualitative analysis of the study.

To distinguish between cases it is first necessary to define what is accepted as representative of the sample population. Section 5.5.1 revealed how the findings from the survey demonstrated that the majority, approximately 70 per cent, of all respondents have incomplete knowledge of the use of genetic test results, high risk perceptions, low benefit perceptions, low general trust and high sceptical trust in the government and the insurance industry, and a lack of willingness to accept insurers' use of genetic test results to underwrite applicants for life insurance. Such characteristics are considered as typical of the sample population, and allow for a template from which it is possible to select the cases to be analysed. Based upon the above characteristics, three representative cases were chosen for further interview (Subjects 4, 5 and 10) (see Table 6-1). Upon identification of the representative case, it was possible to identify all other cases in accordance with such features.

Three individuals were chosen as part of the study of extremely varied cases, based on their responses displaying a lack of agreement with normality, however, on separate issues. All three demonstrated the lack of an inverse relationship between risk perception and benefit perception and placed general trust in the Government. One of these individuals, beyond these two specifications satisfied the representative case parameters (Subject 6) (see Table 6-1). Another individual did not have sceptical trust in the Government or in the insurance industry, and judging by the majority of his answers he typically should have agreed with insurers using genetic test results, yet did not (Subject 9) (see Table 6-1). The final respondent to be interviewed from this group had general trust in the Government and the insurance industry but was unsure as to whether or not he agreed that life insurance companies should be allowed to use genetic test results (Subject 1) (see Table 6-1).

Table 6-1 Case Selection Based on Survey Analysis

Case	Subject	Know	Risk	Benefit	Gvt. General	Gvt. Sceptical	Ins. General	Ins. Sceptical	Acceptance
Representative	4	Some	Some	Some	Disagree	Don't know	Disagree	Don't know	Not accept
Representative	5	Some	Some	Low	Don't know	Agree	Don't know	Don't know	Not accept
Representative	10	None	Some	Low	Disagree	Agree	Disagree	Agree	Not accept
Extremely Varied	1	Some	Some	Low	Agree	Don't know	Agree	Agree	Don't know
Extremely Varied	6	None	Low	Low	Agree	Don't know	Disagree	Don't know	Strongly not accept
Extremely Varied	9	None	Low	Low	Agree	Disagree	Disagree	Disagree	Not accept
Critical	2	None	High	Low	Don't know	Don't know	Agree	Agree	Strongly accepts
Critical	7	Some	Some	Low	Agree	Don't know	Disagree	Don't know	Not accept
Critical	8	None	Low	Low	Don't know	Don't know	Don't know	Don't know	Not accept
Extreme	3	Some	Low	Some	Agree	Disagree	Disagree	Agree	Accepts

The third group of individuals to be selected are identified as critical cases, whereby a particular response appears that should not have typically done so. In one scenario, the individual displayed a non-inverse relationship between risk perception and benefit perception, and yet satisfied all other factors considered 'typical' for the purpose of this analysis (Subject 8) (see Table 6-1). Another individual satisfied most of what was accepted as typical responses and yet still agreed that life insurance companies should be allowed to use genetic test results (Subject 2) (see Table 6-1). One other individual who agreed for further interview complied with all typical responses, except with general trust in the Government, in that he had general trust in the Government, unlike the majority of respondents (Subject 7) (see Table 6-1). All of the above three individuals were considered as part of a critical case for the purpose of this study.

None of the respondents available for interview responded to every single question in the survey beyond what was considered typical. However, one male disagreed that there were any risks related to insurers using genetic test results and agreed that there were benefits, thereby displaying an inverse relationship between risk and trust however, contrary to what was accepted as typical for this particular study. In addition, this individual had general trust in the Government, did not have sceptical trust for the Government and agreed that insurers should be allowed to use genetic test results for underwriting. As this individual's responses were considered atypical as compared to the representative case, this individual satisfied the requirements for analysis of an extreme case. Although, it is not necessary for the extreme case to be uncommon, it is noteworthy that this individual was the most atypical case found within the 76 individuals who agreed for further interview (Subject 3) (see Table 6-1).

Table 6-2 notes the characteristics of the cases chosen. It is notable that the socio-demographics of the case populations are somewhat imbalanced. For example, 70 per cent of those chosen for further interview were male, occupied the under 35 years of age categories, were employed and did not own life insurance. The original sample of individuals for selection contained 43.5 per cent male and 56.5 per cent female, while 36.2 per cent of the population were under the age of 35, 51 per cent, were employed (with the remaining either self-employed, unemployed, students or retired), and 63.5 per cent owned insurance (see Section 5.6). Therefore, the sample chosen for the cases were not representative of the total population available for interview. The reasons for this phenomenon are due to a variety of factors. All of the survey participants available for interview were sorted according to the characteristics of four cases and were subsequently contacted via telephone. Given the small population occupying each case, it was decided not to further limit the choice of respondent by considering factors such as gender age, employment status and life insurance ownership as further criteria for the selection of interview candidates. Instead, individuals were randomly selected for further contact from a list of the divided cases. Some individuals contacted no longer wished to partake in further interview, a factor that limited the choice of interviewees somewhat. These factors and chance each contribute to the imbalanced case portfolio. The limitations of the socio-demographics of those interviewed may affect the interview results. Although findings from Section 5.6 illustrate that these factors did not have a strong influence upon incomplete knowledge, social trust, risk perceptions, benefit perceptions and risk acceptance. However, given the low case population, the lack of socio-demographic representativeness was not considered to affect the overall findings too greatly, although for future studies involving a larger number of individuals, socio-demographic criteria should be taken into account in case selection.

Table 6-2 Characteristics of Cases

Case	Subject	Gender	Age	Employment Status	Insurance Ownership
Representative	4	F	26-35	Employed	No
Representative	5	M	18-25	Self-employed	No
Representative	10	M	56-65	Employed	Yes
Extremely Varied	1	M	26-35	Employed	Yes
Extremely Varied	6	M	18-25	Student	No
Extremely Varied	9	M	26-35	Employed	Yes
Critical	2	F	46-55	Employed	No
Critical	7	M	26-35	Student	No
Critical	8	F	26-35	Employed	No
Extreme	3	M	46-55	Employed	No

Source: Survey Analysis

6.3 Knowledge

The following section examines what factors may cause society's lack of knowledge of genetic testing and its use by life insurance companies.

According to the results of the survey, 70 per cent of individuals felt that they had no knowledge about the use of genetic test results by life insurance companies to assess if they will insure an applicant (see Section 5.5.1) Upon initial investigation, it seemed that this incomplete knowledge may be related to a lack of understanding of genetic testing, given that all ten interviewees when asked how much they know about genetic testing in the questionnaire responded that they had none or very little knowledge of this test (see Appendix D.1, Table D-1). However, upon further probing, the interviews revealed otherwise. Eight respondents demonstrated that they knew what genetic testing is, in that it tests for diseases that one may be "predisposed" or "susceptible" to genetically (see Appendix D.1, Table D-1). Most individuals were able to mention some diseases that are hereditary, the only two that did not were not asked by the interviewer to do so (see Appendix D.1, Table D-2). Interviewees' knowledge of genetic diseases, illustrates that although they initially emphasised their lack of knowledge of genetic testing, they may indeed have some knowledge of the subject. Given the original response of a lack of knowledge of what genetic testing is in both the questionnaire and the interview and subsequent good understanding of genetic tests, it is questionable if the majority of respondents to the questionnaire also possessed general knowledge of genetic testing. It is important to note, however, that all respondents, upon completion of the questionnaire, were provided with an information sheet that explained genetic testing and its use by life insurance companies (see Appendix B.4). Therefore, it may be likely that interviewees recalled such information, however, were reluctant to say so for fear of further questioning regarding their knowledge.

This point may be supported by the fact that questioning individuals' knowledge of genetic tests commenced at the beginning of the questionnaire and the interview, whereby, the respondents were unaware of the next line of questions, perhaps reinforcing their reluctance to declare general or good knowledge of this subject. Only one respondent (Subject 4) was questioned as to why they answered differently on the two points and declared that they "wouldn't have known too much" but "would have known a little about genetic testing itself…" and stated that she "probably misread" the question. The interviewee became flustered and embarrassed. As a result, further candidates were not asked why any of their interview answers differed to their questionnaire responses in order to avoid interviewee discomfort.

Although, individuals may have some or good knowledge of genetic testing, few were aware of insurers' interest in gaining access to genetic test results with some declaring that "during the survey was the first time" they heard about this issue and that they knew nothing about this until they were approached regarding the survey (see Appendix D.1, Table D-2). Such remarks further illustrate why responses regarding knowledge of genetic testing and, indeed, its use by life insurance companies differ between the questionnaire and the interviews. It is interesting to note that the only individual interviewed with knowledge of insurers wishing to use genetic information prior to the survey being completed is Subject 10, a man, aged 58, who worked in insurance for several years before entering his current occupation. He understood that "genetic testing may have an impact in relation to where and how and under what conditions life insurance may be issued in respect of clients" (see Appendix D.1, Table D-2). It is also noteworthy that this individual, although he had knowledge of insurers using genetic test results, analysis of his survey responses indicates that he satisfies all other criteria deemed to qualify him as a Representative case. This raises the question as to whether or not knowledge has an effect upon one's judgement of risks and benefits.

The apparent knowledge demonstrated by the majority (80 per cent) of interviewees with regard to insurers' use of genetic test results does not detract from the results of the survey which illustrates that 70 per cent of individuals felt that they had no knowledge of the use of genetic test results by life insurance companies to assess if they will insure an applicant (see Section 5.5.1). It may be the case that unless individuals work in the insurance industry or a related body, the vast majority of individuals are unaware or uninformed of insurers' use of genetic test results, a factor which may influence the statistical finding that there is incomplete knowledge of insurers' use of genetic test results in Irish society.

6.4 Risk Perception

According to the majority of interviewees, insurers using genetic test results to underwrite applicants, expose society, their families and themselves, to some level of risk. Results from the survey, as outlined in Section 5.5.2, indicate that 63.7 per cent of those surveyed agree that life insurance companies using genetic test results to decide if they will insure an applicant expose society and their families to many risks. Responses to the survey did not differ greatly from discussion during interviews and, indeed, all individuals developed their beliefs concerning the risks related to insurers using genetic test results to a greater extent, thereby greatly increasing an understanding as to why society perceives many risks arising from this situation. The following section analyses and describes the findings from the interviews and categorises the risks viewed as a result from insurers using genetic test results.

A total of eight interviewees believed that there were risks related to insurers using genetic test results (see Appendix D.1, Table D-3, Subjects, 1, 2, 4, 5, 7, 8, 9 and 10). Two of the four individuals who disagreed in the survey that there were risks, changed their replies in the interview to indicate a belief that they did have concerns related to such usage of genetic test results (see Appendix D.1, Table D-3, Subjects 8 and 9). Two individuals remained steadfast that there are no risks related to insurers using genetic test results, with one individual outlining that he does not "see any risk for society as a whole", or indeed, "any family risks" while the second individual was of the same opinion, believing that there would not "be a major risk from it" (see Appendix D.1, Table D-3, Subjects 3 and 6). The two interviewees are representative of the extreme case and the extremely varied case. Both respondents, however, did acknowledge some risk, regarding a higher premium being charged to individuals with positive genetic test results as a risk and, the possibility of unfair discrimination (See Appendix D.1, Table D-3, Subject 3 and 6).

It is possible to categorise the risks mentioned to some extent, as the risks recalled could be divided into common recurring themes or factors that manifest themselves as risks in the minds of those interviewed. The first factor that is regarded as a risk is related to the privacy of genetic test results, with other groups following such as the 'risks' related to the denial of insurance application, the charging of high or unaffordable premiums, unfair discrimination, and finally, the selection of only low risk individuals. It is interesting to note that the two most commonly discussed categories of risk that interviewees were concerned about are those related to privacy and the possibility of insurers denying applicants access to life insurance.

Looking at the former, those interviewed felt that "your information about whether you're going to be healthy in a few years or not is "very "personal information, and it shouldn't be given to an insurance company" (see Appendix D.1, Table D-3, Subject 5). This reaction is reflected by others commenting that genetic test results "are nobody else's business" except for medics with "your health in mind", with some interviewees reiterating this belief stating that "it's very confidential information" and "outside of a family member or outside of…a medical profession… it delves too much into the person… it's too personal" (see Appendix D.1, Table D-3, Subject 2 and 8). It can be said that for the majority of interviewees who, described the provision of information regarding the results of genetic tests to an insurer as "too personal" or "too private" feel that doing so threatens or poses a risk toward their inherent right to privacy (see Appendix D.1, Table D-3, Subject 8). This view raises the question as to whether or not an individual's genetic data is private property to be shared with others only at the individual's discretion.

Following on from feelings of risks against individual privacy, lie concerns related to access to life insurance and that insurers may "increase the payments" of those individuals whose genetic test results indicate an increased predisposition to disease (see Appendix D.1, Table D-3, Subject 3). This fear amplifies to one that indicates that insurers may use genetic test results to "exclude certain individuals from cover altogether" (see Appendix D.1, Table D-3, Subject 1). Therefore, if a life insurance company receives a positive test result from an applicant, this may result in that particular individual "not being allowed to take out" life insurance with that company, or being given a "very, very high policy quote" that may be deemed unaffordable by the individual concerned (see Appendix D.1, Table D-3, Subject 4). Looking more closely at the latter concern, some individuals consider it unfair that if they reveal a positive genetic test result to an insurer this will mean that they are "going to have to pay a lot more for life insurance than everyone else" in that insurers are "going to load you" (see Appendix D.1, Table D-3, Subjects 7 and 10). In addition, there is concern that only low risk individuals will be selected for life insurance, in that insurers may use genetic test results "to be selective about the people they insure and the kind of cover they provide for individuals and, may even use it to exclude certain individuals from cover altogether" (see Appendix D.1, Table D-3, Subject 1).

Due to the workings of life insurance, an individual who is at high risk of death during the term of the policy will have to pay a higher premium. However, judging from the responses of interviewees, this aspect of insurance should not apply when it comes to the analysis of the information provided by the results of genetic tests. It can be said that genetic test results are perceived as different to the results of other medical tests, and their use by life insurance companies to assess the premium to charge an individual is considered unfair. Hence, the viewpoint that an increase in the costs of life insurance for those individuals predisposed to genetic disease, as indicated by the results of a genetic test, is unfair and is considered a risk of insurers using such data. It may be argued that although such an outcome to some may be seen as unfair, it is questionable whether or not it can be considered a risk, given that using genetic test results may be explained as simply an increasingly accurate way of making those people who are more likely to die pay for the higher risk, in other words, a more accurate way of calculating contribution to the fund.

As discussed in Chapter 2, the term 'risk' implies a possible future negative and uncertain outcome of an event or situation caused by nature or human activities where insufficient knowledge was available to prevent such an outcome (see Section 2.4.2). The results of a genetic test may be considered in either a positive or a negative light. If one's results indicate a healthy future, one will not be charged a higher premium for life insurance than that indicated by other underwriting factors. However, the risk, as perceived by the interviewees, is portrayed as negative, in that if the genetic test result determines that one has a high probability of developing a genetic disease, the life insurance company will charge that individual a higher premium that reflects this higher probability of ill health or a shortened life span. The result of the genetic test is uncertain, as is the increase in premium charged, given insurance companies earned reputation for a lack of underwriting transparency. In the case of many genetic diseases, such as Alzheimer's and Huntington's disease, there is insufficient knowledge to prevent the negative outcome of such a disease. Therefore, the belief that being charged a higher premium because of a genetic test result can be considered a risk, as is perceived by 50 per cent of those interviewed regarding this matter.

Perhaps underlying much of this argument and perception of risk, lies a deeper concern, one that is more closely related to the workings of life insurance and the lack of transparency regarding underwriting methods and calculations. One individual states that insurers are "obviously going to hype up the price if you have, you know say, anything different with your genes or even anything that is kind of, how would you say, anything that would be alarming, let's say" (see Appendix D.1, Table D-3, Subject 9), which appears to indicate a concern that an insurer may draw unsubstantiated conclusions from the result of a genetic test due to lack of experience and/or knowledge of genetics. Such beliefs give rise to the statement that life insurance companies "have all the cards, they can say we will or we won't, no we can or we can't, or we're going to give it to you at such a price or indeed make you not want it, or it's too expensive…morally and ethically and every which way you're on a hiding to nothing, you're caught between a rock and a hard place, they've got you by the short and curlies" (see Appendix D.1, Table D-3, Subject 10). What is most noteworthy regarding this category of risk is the negative view of life insurance companies and their underwriting methods, which leads to the notion of the existence of risk in their practices when it comes to their inclusion of the results of genetic tests in their analysis of premium to be charged. This disagreement with the workings of life insurance shall be reviewed in more detail in the following chapter (see Section 7.2.5.1 and 7.2.9).

Three of the ten individuals interviewed mention risks with regard to unfair discrimination believing genetic test results "would be used unethically and that people would be discriminated against because of their genetic make-up" (see Appendix D.1, Table D-3, Subjects 4, 6 and 8). This may lead to "discrimination against…families who may wish to take out life insurance but who probably have quite a few genetic disorders in their family" (see Appendix D.1, Table D-3, Subject 4). This concern with discrimination may stem from the fact that it is "discriminating against people for something that's not their fault", "a person should not be "…discriminated against for something that they can't control" (see Appendix D.1, Table D-3, Subject 8 and Table D-4, Subject 6). Several interviewees took into account the faultless nature of genetic disease, stating that, "if you're born with a family genetic history of disease, it's not your fault" and consider it wrong for insurers to decide whether or not to insure an individual based upon the result of a genetic test. To do so, would be "…discriminating against people for something that's not their fault" and given that a persons' "…genetic make-up…is…something that they cannot prevent" (it) "should not be a major issue in discerning insurance…" (see Appendix D.1, Table D-4, Subjects 7 and 6). It is believed to be wrong to decide upon the premium charged to an applicant based upon the results of a genetic test as "a genetic disease…is not something that a person brings on themselves, they inherit, therefore it isn't something they should have to pay extra premium because they cannot prevent having a genetic disease, therefore I do not believe that insurance companies should be allowed to use them to determine their premiums" (see Appendix D.1, Table D-4, Subject 6). As previously discussed in Chapter 2, it is probable that no person is genetically 'perfect' and that all humans have a genetic predisposition to diseases, even though, at this stage many of them are not amenable to testing. Those interviewed above seem to be questioning whether or not it is fair to use genetic tests that may reveal adverse factors for one population but not to use those that reveal risk factors for other populations. Given this opinion, everybody is open to the risk of being 'unfairly discriminated' against and being 'punished' by life insurance companies if they are allowed to use genetic test results.

The concept that "none of us know, but we could have something genetic, but it just hasn't come about yet" (see Appendix D.1, Table D-4, Subject 2) emulates the quintessential feature of genetic discrimination, in that it is the use of information about an asymptomatic person. If the disorder related to a genetic characteristic has occurred, discrimination based on the disorder may be unfair but is not customarily considered genetic discrimination (Macer, 2000:107-121). The probability factor of genetic test results appears to lead people to believe that it is unfair for insurers to use this information to calculate premium or to decide whether or not to insure an applicant as "...genetic testing just delves too much into...something that might not happen..." and this would not "be fair at all" (see Appendix D.1, Table D-4, Subjects 8 and 4). In support of this, one individual who is at risk from ovarian cancer states that "...there's no definite answer that I would ever get ovarian cancer, I'm just at a more slight risk than another person, but then again another person, another woman, could develop this in a year's time who took out an insurance policy and, in a year's time, she could actually get ovarian cancer, you know, as in it could start a lot quicker" (see Appendix D.1, Table D-4, Subject 4).

Looking at the above arguments a little further, it seems that there is an attitude that genetic test results are not accurate enough or wide ranging across diseases or people for the results of tests to be used in underwriting, as genetic test results merely produce a probability of developing a particular disease for one individual, whereas, another who may not have taken the test, or who may not have a genetic predisposition to a disease, might become ill and die before the individual who has revealed the result of a genetic test. Therefore, all individuals should be charged similar premiums, as a genetic test result reveals a predisposition "to such a thing that you may never get", similar to the probability of dying from "getting knocked down", a risk that is not singled out as an underwriting factor at present and instead is included in the overall mortality statistics of an insurer (see Appendix D.1, Table D-4, Subject 8). Altogether, this suggests that people believe that as genetic test results reveal only the probability of developing a particular disease and, as there are many other contributing factors that lead to a claim being made upon a death policy, genetic test results are fallible regarding future mortality. Therefore, they should not be used as evidence in assessing insurability or indeed, in deciding the premium to be charged. The opinion that genetic test results are far too limited and unreliable to be used as an evidence basis from which to make underwriting decisions, may be linked to the risks of being charged a high premium or being denied insurance. This 'probability' factor could guide people's opinion that there are too many risks related to insurers using genetic test results in underwriting.

It is likely that the risks related to insurers using genetic test results are exacerbated by the perceived need for life insurance. According to one individual, "most people want insurance for themselves or their family" or an "...employer wants them to get" life insurance as a requirement for their job. Although, as the interview progresses, the wording of 'want' becomes stronger and develops to 'must have' and 'need' in that "an employer or family (is) waiting out there saying, well, you *must have* this for the next job, or we really *need* this" (see Appendix D.1, Table D-4, Subject 10). According to another respondent, "the reason you take out a policy in the first place is to look after...who's left behind" (see Appendix D.1, Table D-4, Subject 2). In other words, life insurance is a requirement should you wish to financially "protect your family" in the event of death or accident and "look after...kids, send them to college" (see Appendix D.1, Table D-4, Subject 10 and 5). The fear that one's family "may be left without anything from me, to leave to them...if I have been told that I can't take out life insurance because of that" (positive genetic test results) perhaps leads to the majority of responses to the survey to believe that life insurers using genetic test results to decide if they will insure an applicant expose society and families to many risks (see Appendix D.1, Table D-4, Subject 4). It appears that "...everybody... just wants to do their best for their family" and in order to do one's best, it is necessary to be able to take out a life insurance policy (see Appendix D.1, Table D-4, Subject 2).

There is also a fear that the risks of being denied insurance or charged a higher premium may have the ability to affect future generations. There is a belief that for individuals who have revealed a positive genetic test to an insurer, if "...their children are going for life insurance later on, well then they'll be blacklisted because they'll have found a trace of something" (see Appendix D.1, Table D-4, Subject 5). In addition, the belief that allowing insurers to use genetic test results now could "affect my family, affect, if I have kids, it could affect them when they're a bit older". It is not only the risks for the present generation, indeed, this perception of future risk is linked to the ideology that it is necessary to protect future generations from decisions made now (see Appendix D.1, Table D-4, Subject 4). The value gained by insurers using genetic test results now, when they were not asked for such information previously, is also questioned, with one individual believing that to use genetic test results now is unfair to the present generation as compared to previous applicants: "I think that it would be very unfair to an individual who wanted to take out insurance, and the fact that years ago there wouldn't have been any such thing as… things being hereditary, or genetic, or anything like that" (see Appendix D.1, Table D-4, Subject 2). This suggests that concerns about the present and future welfare of society feed into peoples' decision-making regarding the magnitude of the risks related to insurers using genetic test results.

The privacy of information regarding the results of a genetic test is also queried with respect to who has the right to access genetic test results presented to life insurance companies. One individual states that "there's nothing stopping that information from being passed on elsewhere" (to other companies) and this "would be totally unfair to the individual" (see Appendix D.1, Table D-4, Subject 2). This fear of lax privacy laws and lenient controls on access to applicants' information provided to life insurance companies may be a source of peoples' concerns about the risk to their right to privacy. It is questionable that if people felt that information provided to a life insurance company would be kept with confidence and not provided to any other source, this may reduce their perception of the privacy risk (see Appendix D.1, Table D-3). Such confidence in the life insurance industry brings into question people's trust of insurers, a matter which shall be looked at in further detail later on in the chapter (see Section 6.9).

Following on from this issue of confidence, lies a concern that insurers may try to 'force' individuals to provide genetic data, in that on applying for a life insurance policy "…the company says, hold up now, we can't (provide a policy) without genetic tests, and the client goes out and gets a genetic test" (see Appendix D.1, Table D-4, Subject 10). This may seem like forceful persuasion, however, the company is "not forcing you, but we're asking you if you want insurance" you must provide the result of a certain genetic test or multiple genetic tests. Such fear that an insurer may manipulate their allowance to use genetic test results to their advantage leads to a further concern that "…if I go and do these tests… and then I find out that… that there's a good chance that when you hit 40 years of age… you're going to suffer from this and you're going to die", in other words, one could receive unwanted knowledge regarding their future health. This individual questions "do you need to know that?" and suggests that one may lead "a happier life" by not knowing the results of such tests (see Appendix D.1, Table D-4, Subject 5). The revelation of knowledge about future ill health may have a significant negative impact upon a person's psychological well being. In addition, "in terms of actual physical risks, they have to take a blood sample" which may cause physical harm "if it was done incorrectly" (see Appendix D.1, Table D-4, Subject 10). These fears or concerns of risks against one's mental and physical well-being are dictated by the financial interests of the insurer. It can be argued that individuals have the right not-to-know the results of genetic tests and, therefore, the inherent right not to have to take a genetic test as part of an application process for life insurance. However, there appears to be a fear that through the forceful persuasion of insurers, individuals may end up finding out such unwanted information, given that it may not be possible to oversee all life insurance dealings with consumers to ensure that potential clients are not pressured into taking a genetic test.

In addition, one can extend the following statement that "…maybe if people knew that" (insurers use genetic test results) "they wouldn't go get tested" to a further mental and physical effect that insurers, by using genetic test results, may have upon society. If people are greatly influenced by the risks related to insurers using genetic test results, and decide to mitigate such risks by not taking a genetic test, this may lead to many people not taking the measures required to alleviate the symptoms or the onset of hereditary diseases. According to one interviewee, "genetic testing in a positive light would be to give an individual an indication of something in their genetic history" allowing them to "foresee the future and put in…progressive safeguards. Therefore you would put in breakers… positive medical interventions" in order to ameliorate one's predisposition to the disease or the severity of the disease (see Appendix D.1, Table D-6, Subject 10). Therefore, if people decline to take genetic tests as a result of insurers being allowed to access the results and penalise them financially, people may not be able to take precautions or preventative measures against genetic diseases. Therefore, the ability and manner in which insurers ask for genetic test results could present risks to the psychological and physical well being of an individual, as well as risks to their financial security with regard to their ability to obtain insurance and to do so at an affordable rate. This suggests that concerns that people may have about the appropriate manner in which insurers can ask for genetic test results could lead to their perception of psychological, physical and financial risks of insurers using genetic test results.

Finally, it is noteworthy that not only do individuals take into account the risks related to insurers using genetic test results, but also other risks related to genetic testing, with concern that "it is dabbling in something that… maybe we shouldn't" (see Appendix D.1, Table D-4, Subject 5). "It's pretty cool kind of thing, it's just as long as it's being used for good rather than being used for…bad things, that's probably the worry… that this kind of information can be extended to other aspects that might affect your life" (see Appendix D.1, Table D-4, Subject 5). This general fear of genetic testing may influence individuals' perceptions of the risks related to insurers using genetic test results.

6.5 Benefit Perception

Moving on to individuals' perceptions of benefits, based on the results of the questionnaire, when asked about agreement that there are many benefits for Irish society, their families or themselves if life insurers use genetic test results to decide if they will insure an applicant, eight out of the ten individuals chosen for further interview disagreed that there are many benefits (see Appendix D.1, Table D-5). Two individuals agreed that there were benefits (see Appendix D.1, Table D-5, Subjects 3 and 4). Upon interview, four of the original eight that disagreed that there were benefits described how they believed that there were no benefits for society at all (see Appendix D.1, Table D-5, Subjects 2, 6, 7 and 9). For the most part, these individuals believed that insurers, by using genetic test results when underwriting individuals, would be "very much on the insurance side", in other words, such usage would be for the sole benefit of the insurers and would not have a positive, beneficial effect upon society (see Appendix D.1, Table D-5, Subject 10). According to one individual "there can't be an advantage…'cos they're using it for their own benefit" believing that the provision of genetic test results to insurers benefits only the insurers and not the insured and "the only reason they'd be doing it is to avoid the risks for them…(selves)" (see Appendix D.1, Table D-5, Subject 2). This statement is further reinforced by other respondents, stating that "…it's only going to benefit the insurers…" and "its only going to really help the insurance companies, 'cos they're going to know who to insure... they're not going to really help society" (see Appendix D.1, Table D-5, Subject 9 and 7). The main sentiment coming back from the above interviews is that insurers "wouldn't be looking for" the results of genetic tests for the benefit of society, the main reason is for "lining their own pockets". Such statements present a highly negative view of insurance companies and, indeed, their willingness to use genetic test results in underwriting practice.

Although, there were four more individuals who disagreed that there were benefits for society from insurers using genetic test results, when interviewed, they did note that there was one benefit, in that a fairer premium could be charged to insured individuals if such knowledge is introduced to the underwriting process (see Appendix D.1, Table D-5, Subject 1, 2, 5 and 8). To explain this, one individual states that you could "benefit... from being assessed and... in getting the right premium and not being overcharged, so I suppose in that sense it's... a good thing... for people having to disclose that kind of information", however, "other than that" he was unaware "of any other benefits to society or... personally" (see Appendix D.1, Table D-5, Subject 8). Out of the remaining three respondents, one individual states that "the advantages may be" (that) "they'll find that this is an appropriate bracket for you and it's realistic" (see Appendix D.1, Table D-5, Subject 5). However, it is recognised that this benefit is one-sided in that it is "for the benefit of those that don't actually carry genetic disorders" (see Appendix D.1, Table D-5, Subject 2).

Two individuals out of the ten interviewed agreed that there would be benefits for society, their families and themselves should insurers use genetic test results of applicants when assessing whether or not to insure them (see Appendix D.1, Table D-5, Subjects 3 and 4). The same benefit as discussed above was mentioned, that of a fairer premium being charged to applicants, ensuring "a fairer assessment" (see Appendix D.1, Table D-5, Subject 3). One individual believes that if insurers are able to assess individuals genetic test results, this will prevent a cause of adverse selection, in that "...people who would be maybe more likely to suffer cancer, or something like that, if they would see tell-tale signs of it, they might immediately take out a life insurance policy and, therefore, putting other peoples premiums up". Therefore, it would be of "benefit for honest people who are coming along, taking out their own insurance policies" as genetic test results may reduce the number of people hiding predisposition to diseases (see Appendix D.1, Table D-5, Subject 4). According to the second individual who agreed that there were benefits, if insurers are better able to assess the probability of applicants claiming by using genetic test results, then "payments from insurance companies are generally reduced, then the premiums can be reduced" and this can, therefore, be considered a "benefit for society" (see Appendix D.1, Table D-5, Subject 3). Although, in summation, this respondent acknowledges that this is purely a "financial benefit" and is unaware of any "moral benefits" (see Appendix D.1, Table D-5, Subject 3).

Given the majority of respondents' perceptions of few benefits related to insurers using genetic test results for society, it is noteworthy, that all of those interviewed were aware that insurers using genetic test results, while it was not overly perceived to benefit society, it was considered to have a beneficial effect for insurers. It is understood how insurers would use the information gained from the results of a genetic test judging from the risks mentioned by interviewees, and other statements such as "insurance companies are collecting genetic information about people" in order "to segregate their population and charge more to people who are prone to a genetic disease or who have a genetic disease" and to know "who to insure" (see Appendix D.1, Table D-6, Subjects 6 and 7). Several interviewees also explain that they understand why insurers would look for genetic test results, explaining that "you can kind of see where they're coming from", given that to "know the risks that would be involved for somebody in the future would be an advantage" in that there would be an "improved accuracy" in underwriting calculations (see Appendix D.1, Table D-6, Subjects 4 and 3). The requirement for insurance companies to be risk averse is also noted: "Insurance companies have to guard themselves in relation to risk, and in relation to the premium… because… they eventually will have some sort of payout at some point in time" and "the only reason they'd be doing it is to avoid the risks for themselves" (see Appendix D.1, Table D-6, Subjects 10 and 2). One individual states that insurers may not only want this information to protect themselves against risk, but they may even "need it, obviously, 'cos it's their business… to know whether someone's going to have a long life or not" (see Appendix D.1, Table D-6, Subject 5).

However, it is this requirement of insurers to use genetic test results that leads people to believe that the use of genetic test results is "just a benefit to the insurance companies mainly than to anyone else if they have that kind of information", therefore, it is "only going to benefit the insurers" (see Appendix D.1, Table D-6, Subjects 8 and 9). It's "a business kind of thing for them" (insurers), "like the more information they get out of it, the higher premiums they can charge" (see Appendix D.1, Table D-6, Subject 8). There is a belief that "insurance companies would inevitably use it for their own financial gain first, and the welfare of the insured second, because" they are simply "companies which are there to make a profit, to maximise their profits" (see Appendix D.1, Table D-6, Subject 1). The beneficial nature that genetic test results would have upon the profits of insurers and their ability to avert risks causes one individual to believe that this leaves insurers at an advantage as compared to the applicant, in that "...it's all very much on the insurance side..." given that "insurance companies are in the business of risk and risk aversion means looking at all the potential claims areas, angles, eventualities and everything in general whether they be car insurance house insurance, life insurance... they look at all the risks, that means getting a test or an engineering assessment done or in our case looking at a genetic test, that gives the company all the cards again, they have all the information, they're at no loss at all, I know I understand that's the basis of insurance, risk balance, risk aversion, risk re-insurance, risk... versus payment, so I understand, but they seem to have all the cards and more and more people are not necessarily happy at the power at the information and how that is used by the companies" (see Appendix D.1, Table D-6, Subject 10). Once again, it is possible to note concern about the way insurance companies use information in their underwriting practices, developing from a lack of transparency regarding this issue, leading to a negative view of life insurance companies.

When discussing the benefits related to insurers using genetic test results, a few interviewees also took into account the scientific benefits of genetic testing, taking the view that genetic testing is "a scientific advantage", "it's furthering science, ...it can enhance peoples lives" and related to this there are "many benefits" (see Appendix D.1, Table D-6, Subjects 9, 3 and 10). One interviewee discusses how "genetic testing in a positive light would be to give an individual an indication of something in their genetic history" allowing oneself to "foresee the future and put in...progressive safeguards" (see Appendix D.1, Table D-6, Subject 10). This positive general view of the benefits of genetic testing, in that it can be used "for good" (see Appendix D.1, Table D-6, Subject 5) may feed into a more positive view of the benefits related to insurers using genetic test results (see Appendix D.1, Table D-6 and Table D-4, Subject 3).

6.6 Inverse Relationship between Risk Perception and Benefit Perception

Examining the nature of the relationship between risk perception and benefit perception of insurers using genetic test results, four individuals disagreed that there were many risks related to insurers using the results of genetic tests for society or their families (see Appendix D.1, Table D-3, Subjects 3, 6, 8 and 9). When interviewed, eight out of ten individuals perceived some risk or high risks from insurers using genetic test results (see Appendix D.1, Table D-3, Subjects 1, 2, 4, 5, 7, 8, 9 and 10), while two interviewees (see Appendix D.1, Table D-3, Subjects 3 and 6) stated that there were no risks related to insurers using genetic test results. Regarding the perception of benefits, two individuals stated in the questionnaire that they agreed that there are many benefits related to insurers using genetic test results to decide if they will insure an applicant (see Appendix D.1, Table D-5, Subjects 3 and 4). Upon interview, the same two individuals felt that there were benefits from insurers using genetic test results (Subjects 3 and 4). It should be noted that individuals that mentioned benefits for society but at some point stated that there were 'no benefits', are excluded from the above calculation. Therefore, from the point of view of the survey there are six inverse relationships between risk perception and benefit perception (see Appendix D.1, Table D-3 and Table D-5, Subjects 1, 2, 3, 5, 7 and 10). Judging from the responses provided during interview, there are eight out of ten instances of an inverse relationship between risk perception and benefit perception (see Appendix D.1, Table D-3 and Table D-5, Subjects 1, 2, 3, 5, 7, 8, 9 and 10). In both cases, that is, in the survey and on interview, the majority of the relationships between risk perceptions and benefit perceptions are inverse with more inverse relationships identified during interview (see Sections 5.2 and 5.3.3)

6.7 Definition of Trust

Before examining individuals trust in the Government and the insurance industry, it is relevant to note how individuals define trust. Eight interviewees mentioned how they would define trust if requested to do so for a dictionary. Five individuals included 'confidence' in their interpretation of trust (see Appendix D.1, Table D-7, Subjects 1, 3, 7, 8 and 9). According to one respondent, trust can be defined as "my level of confidence in another party…my level of confidence in that party to make certain decisions or perform certain actions on my behalf and to do so with my best interests at heart". Another states that trust means "you're confident that they'll do the right things and you don't need to worry about it" (see Appendix D.1, Table D-7, Subjects 1 and 7). In both these cases, trust is given to another in the hope or expectation that the party endowed with one's trust shall not renegade upon their position of trust and the provider of trust can feel peace of mind. Given these definitions, when questioning if individuals trust the Government and the insurance industry to appropriately manage insurers' use of genetic test results, one is really questioning people's belief as to whether or not the two authorities shall make decisions on the behalf of the public or their customers with society's or their clienteles best interest in mind. One of the factors examined in the questionnaire and the interview were people's assumptions of the Government and the insurance industry caring for society by acting in the public interest, thereby covering this definition of trust. The concept of 'care' is also discussed by another individual when he states that trust is "being able to place either money, or, or a family member or an idea with someone to look after it properly, to protect it" (see Appendix D.1, Table D-7, Subject 3).

In addition, another interviewee defines trust as the ability to "tell" a third party "something in confidence" and "you would hope that they wouldn't repeat it" (see Appendix D.1, Table D-7, Subject 8). A separate individual reiterates this definition by describing trust "as being able to confide in somebody" knowing that they won't "tell anyone else" (see Appendix D.1, Table D-7, Subject 9). The above two statements display trust in the light of privacy, and given the situation in question, trust could, therefore, be understood as the Government and the insurance industry ensuring that privacy rights are protected when data is being provided or received. Such an understanding of trust can be related to issues of care and integrity, both of which are examined in the survey and the interview.

'Reliability' and 'competence' are two concepts that define trust according to another interviewee who states that trust means that "you rely on somebody to carry out something that they said they would do and you hope that it's done to the best that they can do it" (see Appendix D.1, Table D-7, Subject 8). Reliability as a trust factor is also discussed by a second individual who asserts that trust "could be taken as to rely on somebody" for example to hold "a certain amount of money for me" for a certain amount of time and to return that money when requested (see Appendix D.1, Table D-7, Subject 9).

Another individual mentions 'fairness', 'openness' and 'credibility' as several further concepts descriptive of trust when he states "trust is my understanding that the individual or institution or entity that I am dealing with will treat me fairly" and "honestly" (see Appendix D.1, Table D-7, Subject 10). Openness and credibility are also mentioned by a second individual who asserts that when you trust another body "you hope that they are true to their word" (see Appendix D.1, Table D-7, Subject 8). These concepts are also questioned in the survey and the interview regarding individuals' trust in the Government and the insurance industry.

Trust, as a positive concept, is reinforced by the same individual who asserts that trust means that another body "will not penalise me financially, socially or any other way that might impact on me living within that society, in such a way that it is going to adversely affect what I have in my pocket or the way I have to act or interact within that society" (see Appendix D.1, Table D-7, Subject 10). Therefore, trust is a positive concept and there should not be any negative connotations attached to a trusting relationship. All of the above factors satisfy those that are examined by risk statements during the questionnaire and interview in order to understand society's trust in the insurance industry and the Government.

6.8 Trust in Government

Five of the ten individuals surveyed display general trust in the Government, a sentiment that is not mirrored by the majority of those surveyed (see Appendix D.1, Table D-10, Subjects 3, 4, 6, 7 and 9). These responses can be compared with answers provided during interview in Appendix D.1, Table D-8. It is interesting to note that out of the five individuals that displayed general trust in the Government with regard to their responses in the questionnaire, upon interview, only three of the five individuals continued to display such a sentiment (Appendix D.1, Table D-10, See Subjects 3, 6 and 7). The remaining two individuals, (Subjects 4 and 9) when asked in the interview if they had trust in the Irish Government replied that "no, I wouldn't really, to be honest" (Subject 4) and "not a lot" (Subject 9). Only one of the three individuals who responded that they were unsure whether or not they trusted the Government in the questionnaire remained steadfast to this position (Subject 5). The remaining individuals interviewed believed that they "would not necessarily" trust the Irish Government (Appendix D.1, Table D-10, Subject 10).

During the questionnaire phase of the research, individuals were also presented with a number of statements that measured sceptical trust in the Irish Government with regard to their management of life insurance companies' use of genetic test results to decide if they will insure an applicant. The majority of those surveyed agreed or were unsure as to whether or not they trusted the Government. Two of the ten individuals' surveyed displayed sceptical trust in the Government (see Appendix D.1, Table D-8, Subjects 5 and 10), two disagreed that they had sceptical trust in the Government (see Appendix D.1, Table D-8, Subjects 3 and 9); while the remaining individuals were unsure as to whether or not they had sceptical trust in the Government. These responses can once again be compared with answers provided during interview (see Appendix D.1, Table D-10). Two individuals (Subjects 1 and 3) did not have sceptical trust in the Irish Government, and one interviewee was unsure as to whether or not they had sceptical trust, believing that the Government would not grossly distort facts, but may change policies without good reasons. The other 7 respondents agreed that they had sceptical trust in the Government, in other words, they were veering toward a lack of trust in the Government due to their beliefs that the Government may distort facts, change policies without good reasons and, be overly influenced by life insurance companies.

6.8.1 Competence

Three people (Subjects 4, 5 and 6) were unsure as to whether or not the Government is doing a good job of managing insurers' use of genetic test results at present, as compared to seven of the interviewees who declared they were unsure based on the results of their questionnaires (see Appendix D.1, Table D-10, Subjects 2, 4, 5, 6, 7, 8 and 9). This may be due to the fact that the majority of individuals "know very little about" the management of insurers' use of genetic test results, even more so when filling out the questionnaire, as upon completion of same, information sheets were provided to respondents (see Appendix D.1, Table D-8, Subjects 4, 5 and 6). Upon interview, two individuals thought the Irish Government are not doing a good job of managing this issue. The first individual also disagreed on this issue when surveyed, based their response on previous history of Governmental management of insurers: "I think that they're not doing a great job regulating certain parts of the insurance industry… such as the car insurance sector…I think that reflects badly on the Government's management of the insurance industry in Ireland". According to this individual, such a lack of competence on the

Government's part lowers their level of trust in the Government "to manage the insurance companies and their use of genetic information" (see Appendix D.1, Table D-10, Subject 1). The second individual, answered that they thought the Government was doing a good job in the survey, changed his mind to believe that the Government are not doing a good job of managing insurers' use of genetic test results based upon the fact that the Code of Practice regarding this matter was designed by insurers and is similar to regulatory stipulations in the UK: "they're only beginning to look at it really… the insurance companies at the moment are the ones that are governing practice, and that's conforming to practice in England rather than Ireland… so they don't seem to be doing it at the moment" (see Appendix D.1, Table D-10, Subject 3). This change of attitude may be as a result of learning more about the situation upon reading the information sheet provided after the completion of the questionnaire. However, it is noteworthy that this individual, although not impressed by the Government's job of managing insurers' use of genetic test results, still trusts the Government. Therefore, it is fifty-fifty as to whether or not the Government doing a good job has a role to play in society's perception of trust in this case.

Those interviewed were very much divided as to their view of the competency of the Government to manage this situation. Five interviewees (Subjects 3, 4, 5, 6 and 9) believed the Government to be competent, while four did not (Subjects 1, 7, 8 and 10). This compares to six, who agreed that the Government was competent when surveyed (Subjects 2, 3, 6, 7, 8 and 9), three who disagreed (Subjects 1, 4 and 10) and one who was unsure (Subject 5). Once again, the change in attitudes could be related to the subsequent provision of information upon completion of the questionnaire. Upon analysis, of those who believed the Government to be competent enough to deal with the management of this situation, two individuals emphasise that this competence could be assisted by the contracting of a third party to analyse the situation and to advise the Government, "because it's such a new thing…it'd be great if the Government got experts in" (see Appendix D.1, Table D-10, Subject 5). Given that the Government should be able to "recruit anyone they want 'cos they have the money…if they were to recruit a special team to do that, I'm sure they would be capable" (see Appendix D.1, Table D-10, Subject 9).

The emphasis upon the recruitment of a third party to aid in the research required to formulate decisions is reiterated by those interviewed who do not believe the Government to be competent: "you need experts in the field, politicians - there's only so much they know, they need to get the right people"(see Appendix D.1, Table D-10, Subject 7). This body of experts should be "upfront, efficient, well managed, well headed up by very competent trustworthy" individuals (see Appendix D.1, Table D-10, Subject 10). Here, openness and competency are linked with views of trust. However, on a more cynical note, it is questioned if the Government would "bother to employ people… and pay those people to… monitor the situation… and… report on it" (see Appendix D.1, Table D-10, Subject 8). This attitude may be based upon the perceived previous history of Governmental management, with one individual stating that he did not see the Government as competent enough to manage this issue due "to a lot of stuff that we've seen in the last 10-15 years… from the management of the blood bank, even this morning on the radio, we hear of the Government's management of properties… and also in the papers in the last number of days looking at, especially today, all the cobwebs that have happened, these are only the ones that get to the newspapers, how many more are out there?" (see 7.5 D1 Table, Subject 10). News of the Government's lack of competence in areas other than insurance management also influences society's views of trust in the Government to manage the use of genetic test results by the life insurance companies.

Three individuals assert that the Government has the skills to manage insurers' use of genetic test results (Subject 3, 4 and 5); these beliefs do not vary considerably from survey responses (see Appendix D.1, Table D-10). According to one interviewee, once the Government employ internal or external experts to research the issues at stake, the Government are then able "to write legislation" and "they have the capabilities to write the legislation" (see Appendix D.1, Table D-10, Subject 3). The above statements seem to reflect an opinion that the Government can be trusted to manage the use of genetic test results by life insurance companies if they employ experts to research the issues at stake.

6.8.2 Credibility

Five individuals when first questioned in the survey thought that the Irish Government distorts facts in its favour with regard to insurers using genetic test results (Subjects 2, 4, 5, 9 and 10), while only two disagreed that the Government would do such a thing (Subjects 3 and 6). During the interview, two individuals (Subjects 2 and 6) changed their minds and answered differently than their questionnaire responses. Three of those interviewed (Subjects 1, 2 and 3) answered that the Government would not distort facts while seven (Subjects 4, 5, 6, 7, 8, 9 and 10) thought that they would or could.

One individual believes that the Government would not distort facts in its favour because "the Government has a duty to the people who elected it and they have to give proper information" (see Appendix D.1, Table D-10, Subject 3). This is perhaps a little too idealistic a reason for others who believe that the Government could distort information. However, it is considered that the Government is not likely to distort information as, based upon previous cases, it was punished for such an act: "they'd be found out sooner or later, because it all comes out in the wash, especially after all the enquiries that there's been lately… they would be a bit scared of doing anything underhanded". Another interviewee agrees with this statement, declaring the Government have distorted facts "in the past… however, a lot of other cases like that have been brought to light in a public forum…in recent years, (therefore) they are less likely to do that going forward" (see Appendix D.1, Table D-10, Subjects 1 and 2). As a result of such previous cases, whereby the Government distorted facts and was subsequently punished, at present "when it comes to informing the public on legalities and laws, there are constraints put upon the Government to ensure that we get all of the information, the pros and cons of everything no matter what's under control" (see Appendix D.1, Table D-10, Subject 3). Hence, the controls placed upon Government and the fear of punishment ensures that the Government would not distort facts in its own favour.

However, the majority of those interviewed are of an alternative attitude than above, believing that the Government is "run like a business" and the fear of reprisal is not sufficient to prevent the Government from distorting information (see Appendix D.1, Table D-10, Subject 4). Indeed, most of those interviewed "believe that the Government is perfectly capable" of distorting facts and "wouldn't be surprised if they did" (see Appendix D.1, Table D-10, Subjects 6 and 8). It is perceived that the Government "do it all of the time really when you think about it, the tribunals and everything that's gone on, so… if they wanted to… they could" "distort facts to suit themselves and to make themselves… sound like they're going to look after everyone else" (see Appendix D.1, Table D-10, Subjects 8 and 4).

However, it may not be so much the distortion of facts or even distortion to a "large degree" but "they might more (so) hold back on facts, rather than distort them". Indeed, it is believed by one individual that the "insurance industry and the Government, are perfectly happy to allow the public to remain in that ignorant position; as opposed to distorting facts, they are just not providing the facts" (see Appendix D.1, Table D-10, Subject 6). They may not even "withhold" information or formulate a "complete cover up", instead, "they just highlight things that suit them, kind of, and…don't tend to show you…" so one sees only "the things they actually make known" (see Appendix D.1, Table D-10, Subjects 7 and 6). However, if "you dig deep enough in the freedom of information you are going to find things they don't want you to know about" (see Appendix D.1, Table D-10, Subject 7). Therefore, according to the majority of those interviewed, the Government may distort facts or withhold information, however, it may be possible to find the truth if one searches hard enough.

6.8.3 Reliability

There may be some problems with the wording of the question designed to measure reliability, as commented upon by one individual, stating that the response "depends on what you mean by good reasons". Good reasons may be deemed differently depending on whether it is from the Government's point of view, given that "they could distort the facts and that necessarily wouldn't be for the public good" or from the public's point of view. The question does not state "whose good" is being questioned (see Appendix D.1, Table D-10, Subject 6). Therefore, this question may have some inherent problems, however, if taken to mean the good of the public, when interviewed one person (Subject 1) thought that good reasons would be required before the Government changed its policies on insurers' use of genetic test results, however, two individuals thought that the Government may change such policies without good reasons (see Appendix D.1, Table D-10, Subjects 4 and 6). Other interviewees were not asked and did not comment on this matter.

6.8.4 Integrity

The responses to the interviewees' questionnaires demonstrate that two people (Subjects 8 and 10) thought that the Government is too influenced by life insurance companies regarding their management of insurers' use of genetic test results, while two people did not (Subjects 3 and 9) (see Appendix D.1, Table D-10). When interviewed, six people thought that the Government could be influenced by insurers (Subjects 2, 5, 6, 7, 8 and

9) and three people thought otherwise (Subject 3, 4 and 9) (see Appendix D.1, Table D-10).

Perhaps one of the most contentious issues in the argument as to whether or not the Government could be influenced by insurers, is the nature of the insurance industry as "a powerful lobby", given that "they're a major commercial interest" (see Appendix D.1, Table D-10, Subject 10). The majority of interviewees recognise this factor, and see it as a reason for the Government to be influenced by insurers, however, one interviewee states that although "they're a powerful lobby…I think that any government will step back and assess the situation with the general population in mind" (see Appendix D.1, Table D-10, Subject 3). Most individuals believe that the Government could, or has no choice but to be influenced by insurers, as "insurance is such a huge industry now… for all types of businesses…it's huge…you're talking about huge money…and a lot of jobs and companies that are set up here and that they kind of help the economy so much that… maybe they (the Government) would play ball with them as well… they might… have to play ball with them" (see Appendix D.1, Table D-10, Subject 8).

Therefore, the size, wealth and the economic value of the insurance industry in Ireland may influence Governmental management of insurers' use of genetic test results. According to three individuals, such influential patterns may not always be entirely legitimate or transparent, it may be in the form of "a backhander here and a backhander there" (see Appendix D.1, Table D-10, Subject 2). In other words, insurance companies "could pay off certain members of the Government" in order to influence the management of this issue in their own favour (see Appendix D.1, Table D-10, Subject 5). As another individual states, if "insurance companies wanted something…they'd just donate a lot of money…there are ways… to influence politicians" (see Appendix D.1, Table D-10, Subject 7). This is a very cynical view of Governmental operations, although according to one individual "the Government are not all bad, but it just… would really depend on who was chairing the committee overlooking this (and) how much power the insurance companies had" (see Appendix D.1, Table D-10, Subject 7). Therefore, one's trust in the Government's management of insurers' use of genetic test results may depend upon the character of the individuals that make up the Government.

6.8.5 Care

Overall only two of the individuals interviewed believed that the Government cares for society with regard to its management of insurers' use of genetic test results (Subjects 3 and 5). The first interviewee kept the same opinion as that held at the time of the questionnaire, while the second took a more definitive attitude in the interview as compared to the survey (see Appendix D.1, Table D-10). Both individuals believe that it is necessary to have "faith" that the Government makes decisions based upon the public interest (Subjects 5 and 3). Public interest is linked with trust, in that it is a matter that the public "have to trust the Government" and the public "shouldn't think of them as who's in power at the moment... or who was in power 20 years ago. The Government is who people choose, and the Government normally operates with its constituents in mind, and... that's the way it should be" (see Appendix D.1, Table D-10, Subject 3). According to this interviewee "any government will step back and assess the situation with the general population in mind" (see Appendix D.1, Table D-10, Subject 3). This interviewee appears to be stating that given that the Government tend to act in the public interest, it is, therefore, possible to trust them to manage insurers' use of genetic test results appropriately. However, before endowing the Government with such trust, it is necessary for an individual to acknowledge whether or not they believe that the Government acts in the public interest. Once again, this links trust with public interest.

However, five individuals interviewed disagree that the Government cares for the public, and instead could not be "one hundred per cent sure that they would have our best interests at heart", believing that "the Government would act in political interest rather than in public interest" (see Appendix D.1, Table D-10, Subject 4). Such political self interest is related to the concept that "the Government is pretty much a business, so...they're always looking after their own rather than us" and "they work in their own interests" (see Appendix D.1, Table D-10, Subjects 4 and 2). Feelings of a lack of public interest on the Government's part may be based on an "impression from the overall treatment of the health sector.... that they're (not) putting enough focus on the health sector as a whole and... that this is a part of that" (see Appendix D.1, Table D-10, Subject 1). Backing this theory up, another individual recalls listening to a "radio show this morning and they" (a politician) "completely disregarded a petition, ...the health of children, the health of people... so, definitely and they took no regard for other peoples interests" (see Appendix D.1, Table D-10, Subject 4). The previous history of the Government's action in what is seen as in or not in the public's interest, particularly in areas of health, seem to have a bearing on society's opinion regarding the Government's management of insurers' use of genetic results. According to one individual, it is necessary for "the public keep a very wary and watchful eye over the Government and "just hope that most of the time" the political interests of the Government and the public are "one in the same thing" (see Appendix D.1, Table D-10, Subjects 10 and 6). There is a fear "that the Government would be highly influenced by insurance companies, perhaps a little bit more than by the public" and "that insurance companies would probably be a better interest group than the general public" (see Appendix D.1, Table D-10, Subject 6). This leads to the matter of whether or not the Government listens to concerns of the public and the issue of integrity that shall be examined in Section 6.8.4.

When surveyed, five of those interviewed believed that the Government listens to concerns raised by the public about their management of insurers' use of genetic test results (Subjects 3, 4, 6, 9 and 10), three individuals disagreed (Subjects 1, 2 and 7), with the remainder being unsure about their agreement with this statement (see Appendix D.1, Table D-10). When asked about their agreement with the following statement, that the Government listens to what ordinary people think about the Government's management of insurers' use of genetic test results, three of those interviewed agreed (Subjects 3, 6 and 9) and four disagreed (Subjects 1, 4, 7 and 10). Those who did not indicate agreement or disagreement responded that they did not know if the Government listened to ordinary people's concerns. It can be noted from these results that people saw the public and ordinary people as two different bodies to which the Government responded differently. Looking at this more closely, one interviewee states that "I took the ordinary people to mean a small group and the public to be a large majority", believing that the Government would not take into account what a small group of people thinks but "if there's a large body of people, then… they'd take it into account" (see Appendix D.1, Table D-10, Subject 4). Therefore, the public may be seen as a majority, while the ordinary people may be seen as a minority, whose concerns are treated differently by the Government. This compares with what was earlier mentioned about the fairness of Government's decisions, in that the Government makes fair decisions if the subject matter is "widely known" while for "smaller issues" or minority group concerns, they don't. This may account for the lack of comparability between responses for the above two statements.

Upon interview, eight individuals (Subjects 2, 3, 4, 5, 6, 7, 8 and 9) agree that the Government listens to the public, while none disagree (see Appendix D.1, Table D-10). Only one person mentions the Government listening to ordinary people (Subject 4) and disagrees that this statement is true (see Appendix D.1, Table D-10). This change of opinion from the questionnaire may be indicative of the level of attention paid to this matter after the questionnaire was completed.

The majority of those interviewed, therefore, "believe that the current Irish Government is fairly responsive to public opinion", "simply because we vote them in and we won't vote them in the next time if they don't listen" (see Appendix D.1, Table D-10, Subjects 6 and 3). However, the Government listening to public concerns is dependent upon whether or not "there's big enough public support", that is, if "there's going to be a large majority of people saying this, so if it's a group, they're not going to be seen as if they just completely threw it out the window..." (see Appendix D.1, Table D-10, Subjects 9 and 4). There is an attitude that if an issue catches the "attention" of many people, the Government "have to sit up and listen" for "if they don't listen to them it would create political damage to the Government" (see Appendix D.1, Table D-10, Subjects 2 and 6). Another opinion to add to this is that the Government "listen to concerns but, if it doesn't really suit them, they're not going to do much about it, they'll say they'll take it on board and do the usual, they'll look into it, and sometimes they do things about it but if it doesn't suit them then a lot of the time they don't" (see Appendix D.1, Table D-10, Subject 7). To ensure that the Government reacts to public concerns would "be a matter of somebody pushing the matter on them and having to maybe set up a committee or something" (see Appendix D.1, Table D-10, Subject 8). However, in the long run the Government has no choice but to hear public concerns for if the public "don't like what they do" then the public "won't vote" for that Government at the next election. Therefore, people trust that the Government will eventually take into account public concerns, however, the voices of ordinary people, taken to be small groups, are seldom heard.

6.8.6 Fairness

From the results of the questionnaire, two individuals (Subjects 1 and 10) disagreed that the way the Government makes decisions about the management of insurers' use of genetic test results is fair, two interviewees (Subjects 3 and 6) agreed with this statement while the remaining persons were unsure (see Appendix D.1, Table D-10). This compares with five people (Subjects 3, 5, 7, 8 and 9) who agree and two individuals (Subjects 1 and 4) who disagree when interviewed that the Government makes fair decisions (see Appendix D.1, Table D-10). Once again, the change of opinion on this matter between the time of the questionnaire and the interview may be related to an increased interest in the subject after the questionnaire was completed and the second information sheet provided (see Appendix B-4). Interviewees, once alerted to this issue, may have been more open to information regarding this particular matter.

The consensus of those who agree with the fairness of Government decision-making is that "you're never going to keep everybody happy, …there'll be some decisions that" one may "agree with, there'll be some that you know are for the good of society, but you might not necessarily agree with them, but…overall that they do make good decisions" (see Appendix D.1, Table D-10, Subject 9). One individual who disagrees that the Government makes fair decisions explains that some decisions may "not always" be fair, as the Government are slow to react to or, indeed, do not react to, issues which concern minorities (see Appendix D.1, Table D-10, Subject 4). An interviewee who agrees that the Government makes fair decisions also recognises the ability of the Government to make bad decisions, stating that "mostly I do agree with them but… sometimes I do kind of feel that that the things (decisions) they make, the things that they do…are very incorrect and… that they do have a lot to answer for but… they can only do as best as they can" (see Appendix D.1, Table D-10, Subject 8).

In general, attitudes concerning the fairness of Government's decision-making with regard to insurers' use of genetic test results may once again be based on the Government's decisions on previous issues. Taking this a little further, one individual states that "it seems that they really are trying to stay on top of everything, that's the way it seems at the moment, there's a lot of issues that have been kind of in their faces, and when it's brought to public attention, and once it's aired like that, yeah, they deal with the situation" (see Appendix D.1, Table D-10, Subject 5). For some people it is difficult to "think of a fairer system" given that "it is fairly democratic", although "there's a lot of corruption, bribes and things like that, you can see that with all the tribunals but, you know there's no better way of doing it…they will be…on average more fair than not", indeed, decision-making is "slow, but fair" (see Appendix D.1, Table D-10, Subjects 7 and 3). This 'slowness' to make decisions may be related to the attitude that the Government "have a lot on their plates" so "they don't always have time to" deal with all of the issues affecting society, however, at the same time it is recognised that "the local councillors and the local T.D.'s are" there "to look after the local people", in other words, the minority groups or the small concerns (see Appendix D.1, Table D-10, Subject 4). A second individual who disagrees that the Government makes fair decisions about insurers' use of genetic test results, base their opinion on the fact that "the way they make decisions on issues like this is a little unclear to me and that indicates to me that it is unfair, the fact that decisions are made and are not obvious to the public" (see Appendix D.1, Table D-10, Subject 1). This links two factors of trust - openness and fairness -together, perhaps if the Government were more open about issues being examined and the decisions being made by local and national Government, people may be more open to the belief that the Government is making fair decisions.

6.8.7 Openness

Looking in more detail at the responses of individuals to queries regarding factors related to their general trust in the Government, the most highly discussed topics in order of popularity and sequence of questioning are relevant information, competency, listening to public concerns, public interest and fairness of decision-making. The results of the questionnaire suggest that five of those interviewed (see Appendix D.1, Table D-6, Subjects 1, 5, 6, 9 and 10) disagreed that the Government provides all relevant information about insurers' use of genetic test results, while one person agreed that they did (see Appendix D.1, Table D-6, Subject 3). Of those who discussed the provision of relevant information to the public, six noted that they had not received any information about insurers using genetic test results (see Appendix D.1, Table D-10, Subjects 1, 3, 4, 5, 6 and 9). Seven individuals (Subjects 1, 2, 3, 4, 5, 6 and 8) stated "the information should be out there and be available" as "it would be good to highlight it to people who are applying for life insurance" in that it "is an advantage to know" so that people "can actually participate in decision-making" (see Appendix D.1, Table D-10, Subjects 3, 1, 3 and 4).

Reasoning for such openness is dictated by the belief that the "public put them (the Government) where they are and they're supposed to do a job for the public" (see Appendix D.1, Table D-10, Subject 5). Given that "it's definitely in the public interest to make decisions, the public should get information to make the decisions on this, rather than just the

Government" (see Appendix D.1, Table D-10, Subject 4). "The Government has a duty to the people who elected it and they have to give proper information", as deciding if you "want your results to be given to a company" is "a very personal thing" and in order "to make a choice" it should "be informed" whereby you would "get your fact sheets, get your for and against and why you would vote yes or no" so that one can be adequately "informed before making a decision" (see Appendix D.1, Table D-10, Subjects 3 and 8). The fact that few people received any information about insurers using genetic test results from the Government and, were not asked their opinion on this matter, may influence individuals' trust in the Government to manage this situation appropriately.

Those interviewed had a variety of reasons, both positive and negative, as to why they had not received information from the Government regarding this issue. According to one individual, the Government may not "have had information about it themselves to provide" (see Appendix D.1, Table D-10, Subject 4). Or indeed, it may be the case that "they want to decide what they're going to do first" with regard to the best means of managing insurers' use of genetic test results before circulating information on this matter to the public (see Appendix D.1, Table D-10, Subject 4). Others interviewed explain that since they "never sought that information" and it may be "publicly available", indeed, "you have to weigh out how much interest there is for one thing… if somebody is working in the field, they'll already know or, go out of their way to find out, about it, (while) the general man on the street, or the person on the street… will probably get most of their information from news and from media" (Subjects 1 and 3). For "maybe it's been… televised, or… maybe, it's been on the radio" it's just that they have not been perceptive of this topic or searching for information regarding it (see Appendix D.1, Table D-10, Subject 1 and 5). From the above comments, it is clear that the Government is not blamed for the lack of information regarding this topic and, therefore, has not influenced such individuals' trust in the Government to manage insurers' use of genetic test results appropriately.

However, others are not so neutral with regard to their views as to why they have not received any information regarding this issue, commenting that such a lack of information is a "bit suspicious" (see Appendix D.1, Table D-10, Subject 5 – this Subject had appeared neutral at first with regard to this matter, however, further on in conversation he adopted this less neutral position). Another interviewee states that given such "an awful lot of public ignorance on this issue" it may be that "the

Government is perfectly happy to allow the public to remain in that ignorant position, as opposed to distorting facts, they are just not providing the facts" (See Appendix D.1, Table D-10, Subject 6). This indicates a negative view of the Government, linking a lack of credibility on the part of the Government with a deficiency in openness on this matter.

The comments supplied by interviewees do not only shed light on individuals' trust in the Government but also on the type of information that people would like to see disseminated. "The Irish public should be demanding to know at what level" the use of genetic test results by life insurance companies "currently stands, how much, what access do insurance companies at the moment have to this information and are they currently even using it" and "what it means, or what the consequences are" (see Appendix D.1, Table D-10, Subjects 6 and 8). "The…information documents have to show both sides of the argument…the…Government departments have to reveal all the information like that in a non-emotional…informative fashion" (see Appendix D.1, Table D-10, Subject 3). Currently, there is no such information package available from the Government; indeed, it is questionable if one was introduced if it would affect people's views on this matter and their trust opinions further.

6.9 Trust in the Insurance Industry

When interviewees were asked if they trusted the insurance industry in general, the majority of replies were skewed toward the negative, in that eight individuals were of the sentiment that they "wouldn't have a whole pile of trust in" insurance companies, indeed, "not just in life insurance, in any insurance industry…" (Appendix D.1, Table D-11, Subjects 9 and 4). Although, two individuals did state that they did trust the insurance industry, with one affirming that they "trust them from the point of view that I can see no other system that I would trust more than the system that is already there" (see Appendix D.1, Table D-11, Subjects 6 and 3). According to this individual, therefore, it is not so much the system of life insurance that is under question, as the insurance companies that provide the insurance and society's trust in them to manage their use of genetic test results as the public feels is appropriate.

A variety of responses were provided as to why people did not trust the insurance industry, however, the majority tend to orientate around the size and profit making nature of insurance companies (see Appendix D.1, Table D-11, Subject 2, 5 and 6). The same sentiments are recalled by another interviewee who previously worked in the insurance industry: "having worked in it and on the other side of it, no, the individual has, caveat emptor, let the buyer beware, you really have to be aware what you want, when you want it, how you want it and then see what a particular insurance company or insurance companies would offer you in relation to any of the insurance types, you know you have to be very careful, they are big conglomerates dealing in a massive market, global risks, and in Ireland we're down to a handful, a half dozen of really big insurance companies, you really have to be very careful, no I don't trust them" (see Appendix D.1, Table D-11, Subject 10). The idea of *caveat emptor* suggests that insurance companies may try to swindle or cheat individuals if they are not careful, and one must research the companies' offers very carefully in order to ensure that such a situation whereby one is cheated does not occur. This perception of insurers as "sneaky" appears to mirror other comments, such as "they're always trying to do you out of" your money and "you are signing up to something that is extremely important in your life… so I probably wouldn't trust them a hundred percent" (see Appendix D.1, Table D-11, Subjects 5, 9 and 8). Therefore, before asking interviewees the questions designed to measure the individual trust factors, the majority of interviewees are starting off with an opinion that they do not trust the insurance industry.

It is interesting to note that the measure of one's trust in the insurance industry is twice mentioned in relation to the need for an outside regulator. According to one individual "it's not that I would mistrust them, it's just that I think that there should be a controlling body, independent of…" the insurance industry (see Appendix D.1, Table D-11, Subject 3). Another individual states that "I probably wouldn't trust them a hundred percent, especially if they're not being regulated or don't have to answer to anybody" (see Appendix D.1, Table D-11, Subject 8). This suggests that people may be more willing to place trust in the insurance industry if they were made aware of an independent third party controlling and regulating their activities. This may reduce the lack of trust reflected by the majority of those interviewed, however, this, of course, is not certain. At the time when the interviews were conducted, the IFSRA had taken over the responsibility of ensuring that life insurance companies followed the Code of Practice regarding insurers' use of genetic test results. Responses such as that above, illustrate a lack of knowledge regarding the role of the IFSRA as an independent regulator of the insurance industry. According to one interviewee, "if I found out that insurance companies were using genetic testing to determine peoples premiums, I wouldn't trust…" the insurance industry, perhaps suggesting that trust in the insurance industry is also related to the situation or the circumstances whereby one is measuring trust (see Appendix D.1, Table D-11, Subject 6). However, considering the requirement and the desire for life insurance products in Ireland, one may have no choice but to trust the insurance industry in so far as society is willing to purchase their products.

6.9.1 Competence

When asked whether or not the insurance industry is doing a good job, is competent enough and has the skills to manage their use of genetic test results, the majority of those interviewed agreed in the positive (see Appendix D.1, Table D-7). When interviewed, the responses did not change dramatically, only a couple of interviewees felt that the insurance industry was not competent, while some were "not sure" as to their responses because they "didn't have enough knowledge" about "how much genetic testing they (insurers) have available to them and how they use it" (see Appendix D.1, Table D-11, Subjects 4 and 1). Reasons for not believing in the competence of the insurance industry centre around the need for a regulator, with two individuals stating that they didn't believe the insurance industry to be competent enough, instead "an independent third party should be involved to regulate their use" of genetic test results. "You have to regulate it really" as "all they're worried about is their profits, so… (they're) going to do everything they can to get those figures, so they can not insure those people or charge them a lot of money" (see Appendix D.1, Table D-11, Subject 1 and 7).

For those who felt that the insurance industry was doing a good job, competent enough or skilled enough, reasons varied as to why they believed so. One person stated that "if they're following the constraints laid down by the UK, then they're doing their best. I think they're not abusing the system". In other words, by complying with UK restraints, Irish insurers are proving their competence to manage their own use of genetic test results (see Appendix D.1, Table D-11, Subject 3). Others mentioned the necessity of insurers to achieve a good public image as reason to believe that they can manage their own use of genetic test results, for "if the insurance industry did try and manipulate the situation, they'd know that there'd be a very high chance of creating public scandal and that that pressure will create a certain level of competency or a certain level of caution" (see Appendix D.1, Table D-11, Subject 6). However, one individual points out that just because one might deem the insurance industry competent enough, one may still not "trust them, because the information they get, they will use to their own best interests, so rather than competence, it's how they would deal with this information" (see Appendix D.1, Table D-11, Subject 10). This perhaps refers to the trust factor 'care as being more important than 'competence' in the question on trust with regard to insurers managing their use of genetic test results appropriately.

6.9.2 Credibility

From the results of the questionnaire, five interviewees agree that the insurance industry could distort facts in its favour regarding their management of insurers' use of genetic test results (Subjects 2, 3, 4, 6 and 10), one interviewee disagrees that this is the case (Subject 9), while the remaining are unsure. All of the interviewees questioned on this matter agreed that the insurance industry "could distort facts in its own favour" (see Appendix D.1, Table D-11, Subject 6). One individual states that he doesn't mind if information regarding insurers' use of genetic test results comes from the Government or the insurance industry (see Appendix D.1, Table D-11, Subject 3). Although, this may not be an opinion shared by all or, indeed, the best case scenario given peoples' opinions regarding the insurance industry's propensity to distort facts (see Appendix D.1, Table D-11, Subject 3). It is perceived that the insurance industry "can always accentuate the positive, and, somehow forget some of the negative" in that they can put their "slant on it" (see Appendix D.1, Table D-11, Subject 3). Although, according to another individual, the insurance industry could "possibly distort" information, however, there would be no "gross distortion" (see Appendix D.1, Table D-11, Subject 9). Indeed, there may not be 'distortion' at all, as rather than distorting "facts to a high degree…both from the Government and the insurance point of view… they might more hold back on facts" (see Appendix D.1, Table D-11, Subject 6). And given the "public ignorance in this issue… the insurance industry and the Government are perfectly happy to allow the public to remain in that ignorant position. As opposed to distorting facts they are just not providing the facts" (see Appendix D.1, Table D-11, Subject 6).

When looking at the reasoning behind such opinions, the explanations provided reflect upon the general perceptions of the insurance industry. According to one interviewee, his opinion that the insurance industry could distort facts in its own favour regarding this situation "comes back to my overall impression of the insurance industry, as an industry which is very much self centred, almost verging on greedy and... an attitude like that in any industry would certainly lead them to that type of behaviour" (see Appendix D.1, Table D-11, Subject 1). While another believes that "again they're companies and companies often do" distort facts (see Appendix D.1, Table D-11, Subject 7). Another individual mentions how car insurers can distort statistics in order to charge high premiums to young drivers, an attitude that appears to influence individuals' opinions of the insurance industry in general (see Appendix D.1, Table D-11, Subject 7). In addition, the lack of transparency regarding the workings of life insurance is an issue in people's belief that the insurance industry could distort facts, in that "they could say they're giving you the facts, but you can't go to someone else about this, to check if it's true or not" (Appendix D.1, Table D-11, Subject 5).

In order to prevent the insurance industry from distorting facts regarding their use of genetic test results it is deemed necessary that the Government control the industry for "if they don't have any Government... overlooking what they do and they don't have to answer to anyone... they could distort the information"(see Appendix D.1, Table D-11, Subject 8). It is also considered crucial to protect oneself, in that it's up to the individual "taking out a policy...to read the small writing in between" as insurers will "distort certain information to bring themselves out trumps" (see Appendix D.1, Table D-11, Subjects 1 and 4). It is believed important to note that "insurance companies will do anything to reduce their risk, or get out of paying large settlements, they can get a large amount of legal help and settle on the steps of the court. If it's in court the only people benefiting are the legal side, as the costs are put down to the consumer" (see Appendix D.1, Table D-11, Subject 10). Upon analysis of the above remarks, it is possible to state that individuals feel that there is a need for protection against life insurance companies distorting information and taking advantage of consumers. Such an attitude may well feed into perceptions of who to trust regarding the management of insurers using genetic test results and, indeed, how this issue should be managed.

6.9.3 Reliability

Responses to the questionnaire illustrate that four of the individuals interviewed agree that the insurance industry changes policies about their use of genetic test results without good reasons (Subjects 2, 5, 9 and 10), while two people disagreed (Subjects 3 and 6), and the remaining individuals were unsure (see Appendix D.1, Table D-10). Looking at responses from the interviews, two people agreed (Subjects 5 and 9), while two remained unsure since completing the questionnaire, although one thought that the insurance industry might change policies based on her knowledge of insurers dealings with young drivers (Subjects 4 and 7). This perception of insurers is reflected in the responses provided by an interviewee who agrees that the insurance industry "could change its policies" also detailing insurers treatment of young male drivers as a reason for such a belief (see Appendix D.1, Table D-11, Subject 9). The other individual who agreed based his response on the perception that if the insurance industry is writing the policies "then they can change them" (see Appendix D.1, Table D-11, Subject 5). Such attitudes regarding insurers ability to change polices regarding life insurance practices can be said to be related to perceptions of the dealings of general insurers.

6.9.4 Integrity

Of those interviewed, the questionnaires reveal that only one person disagreed that the insurance industry is too influenced by life insurance companies (Subject 9), three people agreed (Subjects 2, 3 and 10) while the remaining individuals were unsure (see Appendix D.1, Table D-7). Comparing this to responses in the interviews, the two individuals who commented on this matter seemed to think that the insurance industry could be influenced by insurers. With one individual believing that there may be "oversights…in certain areas" "in several years" leading to the industry being influenced by the requests of life insurers, in their management of the use of genetic test results, in underwriting life policies (see Appendix D.1, Table D-11, Subject 3). With regard to the nature of this influence, one of the interviewees argued that the insurance industry could be influenced by life insurance companies in the direction of allowing them access to all genetic information (see Appendix D.1, Table D-11, Subject 6).

6.9.5 Care

Not one individual interviewed felt that the insurance industry is acting in the public interest with regard to their management of life insurance companies' use of genetic test results. Although one interviewee does state that insurers are "doing a job… and…. people who want insurance have to go somewhere for it, so I suppose they are working in the public interest" (see Appendix D.1, Table D-11, Subject 2). Questioned further with regard to insurers use of genetic test results in the interest of the public, the same individual replies "they wouldn't be looking" (for genetic tests results) "for your benefit in anyway" (see Appendix D.1, Table D-11, Subject 2), in other words, such an activity is not in the interests of the public. The many responses provided in interviews indicating that insurers' using genetic test results is not in "the best interests of the public" correlates with responses provided in the questionnaire, which, also indicated disagreement with the statement designed to measure this factor of trust (see Appendix D.1, Table D-11, Subject 1 and Table D-7).

Reasoning provided for such a negative reaction focus on the profit-making interests of insurance companies. A feeling that "insurers would inevitably use it for their own financial gain first, and the welfare of the insured second" is predominant in most responses (see Appendix D.1, Table D-11, Subject 1). This point is reiterated by another individual, who states that we "wouldn't believe that they're going to look after us before they look after themselves" given that "they're not going to let themselves go bankrupt or anything just to look after us" (see Appendix D.1, Table D-11, Subject 4). One interviewee asserts that "insurance companies act with their own interest…in their minds" and will try to "get away with" activities "where they are going to make more money, perhaps, without people realising that they're doing things that they shouldn't be" (see Appendix D.1, Table D-11, Subject 4).

This belief that insurers lack an interest in the security of the public and have a direct interest in profit making, appears to be a great concern for those who indicate a lack of trust in the insurance industry: "I've never really trusted companies that are out to make a profit, they're always going to do whatever suits their margins, so I wouldn't trust them that way" and "I wouldn't necessarily trust them…because they are vested interest, bottom-line profit, bottom-line shareholder, they are going to deal with the information in a way that I suppose secures their best interest and not necessarily the clients" (see Appendix D.1, Table D-11, Subjects 7 and 10). Indeed, the need to "maximise their profits" and "line their own pockets" is the reason provided by some individuals to vindicate a lack of trust in insurance companies to take their "considerations, or considerations of the Irish people in general" into account, believing that one could "never trust a company that's only worried about profit (see Appendix D.1, Table D-11, Subjects 1, 2 and 7). To protect public interest from the business requirements of insurers, it is once again suggested that "consumers have an independent body" to represent them and prevent insurers from taking "opportunities to abuse the situation" or "misuse" genetic information (see Appendix D.1, Table D-11, Subject 3)

Judging from the responses to the questionnaires completed by the interviewees, they were divided equally regarding agreement and disagreement that insurers listen to concerns raised by the public about insurers' management of their use of genetic test results (see Appendix D.1, Table D-10). Three subjects changed their opinion from one of disagreement with the statement, to agreement (Subjects 1, 7 and 10). When interviewed, the majority felt that the insurance industry "at the moment…would be listening to public opinion closely", with some believing "that they will act on the information they gain" (see Appendix D.1, Table D-11, Subject 6). Others believe that such actions will only take place if doing so "doesn't take too much from their profits" (see Appendix D.1, Table D-11, Subject 1).

According to one interviewee, if insurers don't listen "it's going to be very well publicised, it's going to be in the papers, it's going to be in the media, mass media, so, I don't think they'd want their names brought through with negative information… about them". So from fear of the suggestion of a negative image, insurers will listen to the public (see Appendix D.1, Table D-11, Subject 4). Indeed, listening to the public is a necessity of self-preservation, as one individual suggests: "they've got to listen, they are really relying on people at the end of the day to kind of keep them where they are, so, if they're anyway a good company, a good organisation, of course, they're going to listen if their members are interested or worried about something and see how they can sort it out…" (see Appendix D.1, Table D-11, Subject 5). However, as another person suggests "if they were going to go a certain direction, and there was an objection, a major outcry… they would listen, and then they would assess how strong that outcry is and determine which way to go. Would they go in the original direction or would they change their viewpoint? It depends on the level of expert, the level of complaints, or the level of agreement, and they would assess it on that basis" (see Appendix D.1, Table D-11, Subject 3). Therefore, insurers may choose whether or not to react to public concerns, in that "they'd listen, but it mightn't affect them too much unless there's a huge problem and then they'll have to sort it out, but if it's a small one then they probably won't" (see Appendix D.1, Table D-11, Subject 7). Therefore, their reaction and management of the underwriting of genetic test results depends upon the level of public concerns.

Two individuals (Subjects 8 and 10) indicate the need for an independent regulator, such as the Government, to ensure that insurers listen to public concerns, for without a regulator, if insurers "don't want to listen…they really don't have to", for "if they aren't being regulated… they don't have to answer to anyone, they can do their own thing" (see Appendix D.1, Table D-11, Subject 8). Insurers will only listen "when there's enough people out there getting onto Government and getting onto the opposite side of the lobby group to the industry, the Government would say hold on, because people are on the ball, they're interested, and watchful, they're listening and they're pulling the Government to book, or indeed enough of them are getting out there and saying we're watching you or we're not happy with this or no we're not going to pay anymore" (see Appendix D.1, Table D-11, Subject 10). However, once it is "aired… publicly" insurers would "have to listen" as the concerns are "brought before the Dáil" by "somebody or other in the Government" (see Appendix D.1, Table D-11, Subject 1). However, the public need to "keep the Government up to their job and watching the life insurance companies" in order to ensure that public concerns are heard and reacted to accordingly (see Appendix D.1, Table D-11, Subject 10). There is a need for the public to be united in their concern, to have these concerns voiced appropriately, and for an independent regulator to be involved to ensure that the insurance industry care for the public, their consumers, appropriately.

6.9.6 Fairness

When asked in the questionnaire whether the insurance industry makes fair decisions about their management of insurers' use of genetic test results to decide if they will insure an applicant, half of the individuals interviewed deemed that such decisions were unfair (Subjects 1, 3, 5, 6 and 10), the remaining individuals were unsure as to their opinion (see Appendix D.1, Table D-10). When this question arose again during interview, two people felt that it was unfair (Subjects 1 and 7), one person felt that he had "no reason not to believe it" (Subject 3), while two were uncertain (Subjects 4 and 6) (see Appendix D.1, Table D-11).

One interviewee who is unsure as to whether the insurance industry is making fair decisions, explains that she is "not too sure, because I don't know what way they would make decisions regarding genetic testing" (see Appendix D.1, Table D-11, Subject 4). However, based upon her dealings with car insurance and the difficulties faced by "young drivers trying to get insurance" she believes decisions made by insurance companies to be unfair. Therefore, in some cases, dealings with insurance companies outside of life insurance products may influence individuals' perception of the life insurance industry. Another individual who is unsure, clarifies his lack of response as due to the fact that he is currently "unaware of how much access insurance companies have to genetic information, so I don't know whether it's appropriate or not" (see Appendix D.1, Table D-11, Subject 6). The fairness of their decisions could not be judged due to a lack of knowledge regarding the situation. Decision-making is more difficult when there is incomplete knowledge.

People who disagreed with the fairness of the insurance industry's decisions explain that they don't think such decisions are fair "because you're discriminating against people for something that's not their fault. If you're born with a family, genetic history of disease, it's not your fault" (see Appendix D.1, Table D-11 Subject 7). Such an answer illustrates that some individuals may not agree with insurers using genetic test results based upon the assumption that if the insurance industry were to manage this situation, they would ensure insurers have the right to use genetic test results in their underwriting practices. Those who think that insurers are not making fair decisions on this matter "don't think they should be doing it in the first place" (see Appendix D.1, Table D-11, Subject 1). The person, who agrees that insurers will make fair decisions, recognises that "they will make decisions, looking at their future growth…and reduce their risk". However, it is demonstrated earlier that it is this knowledge that causes some of those interviewed to not trust the insurance industry, fearing that insurers look after their own interests before those of the public (see Section 6.9.5).

6.9.7 Openness

From the results of the questionnaire, not one of those individuals interviewed agreed that the insurance industry provides all relevant information about their use of genetic test results to the public, although four individuals were unsure if they did or not (see Appendix D.1, Table D-10, Subjects 4, 5, 7 and 8). When interviewed, one individual presumes "that they would have to provide information" (see Table D-11, Subject 8). While two interviewees (Subjects 1 and 5) disagreed that the insurance industry has provided enough information on this matter, one individual based his answer on the fact that he hadn't "seen any or heard anything about it" (see Appendix D.1, Table D-11, Subject 5). Two individuals (Subjects 4 and 9) were unsure, both had not "received information from insurance companies or heard or seen anything about it", although one of them assumes that the insurance industry have the information laid out "in some sort of book somewhere" and believes that the reason she never heard about this matter is that insurers "might not just have passed it out very adequately to the public" (see Appendix D.1, Table D-11, Subject 4). The same respondent, giving benefit of the doubt to the insurance industry, states that "...it's probably not well enough known if it's going to be undertaken at this stage...maybe that's why they haven't given information about it", however, "...if it's a big query and the life insurance companies and the government are thinking about it, well then I wouldn't consider it very fair at all to not bring it to the peoples attention". It can be said, therefore, that although people had not received information on insurers' use of genetic test results, they feel that such information should be provided, with some believing that this may currently be the case, it's just a matter of researching the issue.

6.10 Risk Acceptance

From the results of the questionnaire analysis, seven individuals who were subsequently interviewed disagreed that the life insurance industry should be allowed to use genetic test results to decide if they will insure an applicant (see Appendix D.1, Table D-16, Subjects 1, 5, 6, 7, 8, 9 and 10). Two individuals agreed with this arrangement (see Table D-16, Subjects 2 and 3), and one person was unsure (see Appendix D.1, Table D-16, Subject 4). This compares with eight of the ten individuals who, upon further examination in an interview, disagreed that the life insurance industry should be allowed to use genetic test results to decide if they will insure an applicant (Subjects 1, 2, 4, 5, 6, 7, 8, and 10), while two interviewees agreed (see Appendix D.1, Table D-16, Subjects 3 and 9).

Firstly, looking at the majority of interviewees who disagreed, some could "see arguments in favour of having them use genetic test results... but looking at it from the perspective of someone, as a client of an insurance company", they preferred that insurers "didn't" have access to such information (see Appendix D.1, Table D-16, Subjects 1, 4, 5, 8 and 10). Some people mentioned that it was "unfair" that insurers should be allowed to use genetic test results to underwrite individuals (see Appendix D.1, Table D-16, Subjects 2, 4). Hence, "if there was a public vote over... allowing life insurance companies to use genetic testing or not" most of those interviewed "would probably vote not, that they would not, that they should not be allowed to use it" (see Appendix D.1, Table D-16, Subject 4).

Some people believe that "the risks are too high" in that they "don't see huge benefits for the person or society", leading to an attitude that "the benefits don't outweigh the risks of it ..." (see Appendix D.1, Table D-16, Subject 8). It is deemed that "because companies are by their nature... commercial organisations, they'll use it (genetic test results) to their own best advantage, they'll not necessarily use it benignly, they'll not necessarily use it ethically, morally or anything else because insurance companies are financial entities, they're based on shareholders, and in most cases even the most avid of mutuals, such as Standard Life, is going to demutualise shortly so there's going to be no mutual companies which would return the benefits to their clients, kind of venture holders, mutual company members, they're all plc's, bottom line share price, profit, that sort of way. An awful lot of ethics, morals, and everything else gets lost in that. And therefore, access to this information will only ever be used, on the basis of what's good for the company and not necessarily what's good for the client" (see Appendix D.1, Table D-16, Subject 10). Another interviewee questions the necessity of insurers to access genetic test results, given that "the basis of insurance is to share out risks amongst a large number of people and I believe that by following that insurers should be able to minimise the risk to them from genetic diseases which people pick up". Therefore, "in the short run...it should not be used" and "in the long run, it would be optional" (see Appendix D.1, Table D-16, Subject 6). Given such reasoning, the majority of those interviewed would prefer if insurers were not allowed to use genetic test results in their underwriting of applicants.

Moving on to those who agree that insurers should be allowed to use genetic test results, there exists the argument that if an "individual knows the result, while the insurance company doesn't and is being asked to insure the person" the insurer "has a right to know" (see Appendix D.1, Table D-16, Subject 3). In addition another individual states, that "if the person isn't forced to take" genetic tests "then... there's no reason not to" use such results in underwriting (see Appendix D.1, Table D-16, Subject 9). By insurers using genetic test results, it ensures that underwriters provide "a fair assessment of the risk" (see Appendix D.1, Table D-16, Subject 3). However, along with such an air of 'readiness' to allow insurers to use genetic test results, there is also an emphasis upon limitations. One individual asserts that if he "had to take out life insurance then, fair enough if you have to pay a little extra", but people should not be excessively penalised, given that there's enough people in the world scheme of things to absorb any..." extra costs incurred by insurers by individuals knowing results of genetic tests (see Appendix D.1, Table D-16, Subject 9). The same interviewee suggests that a monetary excess should apply for applicants, so that one can take out life insurance to cover "a mortgage plus fifty or a hundred thousand" euros without incurring a loading on their premium for the results of a genetic test. The monetary amount on the mortgage should also be related to the region whereby the house is located, as some counties and areas are more expensive than others (see Appendix D.1, Table D-16, Subject 9). So, although "we definitely should be pushing out the frontiers…they should be controlled" (see Appendix D.1, Table D-16, Subject 3). There should also "be more of an over-riding authority on the insurers" (see Appendix D.1, Table D-16, Subject 3). Controls on insurers using genetic test results are examined in further detail in the next section on risk management (see Section 6.11).

It may be likely that an individuals' medical background may influence their perceptions of the risks related to insurers using genetic test results and their acceptance of such risks. A total of five interviewees mentioned their personal family history, that is, their personal medical background. From the information provided it was possible to conclude that four individuals had medical backgrounds whereby they could have a premium loading by insurers (see Appendix D.1, Table D-22, Subjects 1, 4, 5 and 8). However, none of those interviewed are likely to be denied life insurance based upon their family history. Of those who may be charged higher premiums because of their personal and/or family history, one individual had a form of heamachromatosis (Subject 1). While another interviewee is predisposed to ovarian cancer and high blood pressure (Subject 4). A third individual has a family history of diabetes (Subject 5). The remaining individual had skin cancer and her mother died of cancer, while her father's side of the family has a history of heart problems (Subject 8). All of the above interviewees belong to the Representative Case. Others also reported medical conditions, for example a low thyroid level is a reported medical defect by one individual (Subject 6). A history of osteoporosis affects the female members of another individual's family (Subject 10). However, neither medical issue would indicate a need to increase the premiums charged for life insurance. It is interesting to note that all those individuals who may have a genetically inherited disease that could possibly affect their life insurance premium, prefer if the insurance companies should not be allowed to use genetic test results. This may indicate that when an individual is prone to genetic difficulties they are more likely to perceive more risks related to the use of such information.

A finding that supports this theory illustrates how the only two individuals that accepted insurers using genetic test results, do not report any 'serious' health problems or family history of disease (Subjects 3 – Extreme Case and 9 – Extremely Varied Case). Therefore, one may conclude that proximity to the risks may be a factor that influences risk acceptance.

Upon further examination of the issue of the provision of genetic information to a life insurance company, it appears that those individuals that constitute the representative case believe that not only should insurers not be allowed to use genetic test results, they should also not be allowed to use family history information when underwriting applicants. Instead these individuals indicate that family history information should only be used by the medical profession and not by commercial enterprises (see Appendix D.1, Table D-22, Subjects 1, 2, 4, 5, 7 and 10). One of the interviewees that make up the extremely varied case is also of the same opinion (Subject 6). Although, there is an interviewee from the Representative Case, who states that there is little difference between revealing genetic test results and family history information (see Appendix D.1, Table D-22, Subject 10). This person is the only individual within this group that is of this opinion. He believes family history and genetic test results to be "one and the same", in that "your genetic history is your family history" (see Appendix D.1, Table D-22, Subject 10). It is interesting to note that all of the above individuals, bar one, reported having a family history of disease (Subject 7), once again illustrating that proximity to the risks related to revealing any form of genetic information may be conceived as a factor that influences the acceptance of the use of genetic test results, or indeed, family history information by life insurance companies.

Although, one interviewee from the representative case, who has a history of genetic disease, believes that the provision of family history information and genetic test results to a life insurance company are "two different things" (see Appendix D.1, Table D-22, Subject 8). This statement is based upon a belief that family history information reveals less probability of a disease occurring in a person than genetic tests, hence, insurers should be allowed to use family history information. This attitude takes away slightly from the above insistence that proximity to the risks may influence risk acceptance. However, it also reveals that individuals with a genetic history of disease may not wish for insurers to obtain a more accurate account of their genetic predisposition to disease. This once again illustrates how having a family history of disease can generate a particular attitude toward the acceptance of the use of genetic test results by the life insurance industry. The following section examines interviewees' opinion of risk management of this situation.

6.11 Analysis of Risk Management

Risk management is directly linked to risk acceptance. It delineates the preferred circumstances and the conditions whereby an individual will accept risk. In this case, the interviewees were asked who they would like to see managing insurers' use of genetic test results, that is, which authority individuals would prefer to see controlling the risks related to insurers using genetic test results, and who should be involved in the management of such perceived risks. It is considered that "over a period of time there could be misuse" by insurers of genetic test results", therefore, "it's better if consumers have an independent body" to oversee this situation (see Appendix D.1, Table D-19, Subject 3). This need for an independent "police system" or "watchdog" leads to the call for "a specific group to deal with" insurers' use of genetic test results (see Appendix D.1, Table D-19, Subjects 1, 3, 5, 6, 7 and 8). One interviewee states that "there needs to be controls" and the insurance industry should not "be controlling it themselves" (see Appendix D.1, Table D-19, Subject 3). The Government "should really be on top of the situation" and not the insurance industry, as "the Government is the law of the land rather than the insurance companies" (see Appendix D.1, Table D-19, Subjects 5 and 2). Although one individual believes the insurance industry capable of managing their own use of genetic test results in their underwriting practices (Subject 9), eight out of ten interviewees mention the Government itself or, a "Government appointed agency" should be set up, as the authority that "should certainly legislate to provide some restrictions" and to make sure that things are managed better" (Subjects 1, 2, 3, 4, 5, 6, 8 and 10, Subject 1).

Some case study interviewees discussed whether or not they believed themselves to have the means to hold the Government and the insurance industry accountable for their decisions regarding insurers' use of genetic test results. Interviewees held a different view of their means to hold the insurance industry accountable as compared to the Government. Only three individuals felt that they had the means to hold the insurance industry accountable for their decisions "to some degree" (see Appendix D.1, Table D-21, Subjects 6, 5 and 8). Four interviewees stated the opposite (Subjects 1, 3, 9 and 10), with one asserting that unlike controlling the Government, there is no voting system. And unless one can afford to take a life insurance company to court, it is unlikely that they will be held accountable for their actions (see Appendix D.1, Table D-21, Subject 10). In addition, given that insurers have more money than the average individual to afford all of the court fees, it is the insurer that will win the case at hand (see Appendix D.1, Table D-21, Subject 10). Therefore, it is very difficult for an individual to have the means to hold the insurance industry accountable for their decisions (see Appendix D.1, Table D-21, Subject 10).

With regards to the Government, five people (see Appendix D.1, Table D-21, Subjects 3, 6, 8, 9 and 10) stated that they felt that they could hold the Government accountable due to the system of general election, although this means of control is considered "far removed" (see Appendix D.1, Table D-21, Subjects 10, 9 and 3). Two individuals state that they do not have the means to hold the Government accountable for its decisions, but did not provide an explanation for their responses (see Appendix D.1, Table D-21, Subjects 1 and 5). There do not appear to be commonalities of thinking regarding control between cases. All of the individuals that indicated that they could or could not hold the insurance industry accountable did not trust the insurance industry to manage its use of genetic test results. It is interesting to note, however, one of the three interviewees that felt they could hold the insurance industry and the Government accountable for their actions, viewed no risks related to insurers using genetic test results (Subject 6). The two individuals that accept insurers using genetic test results (Subjects 3 and 9) indicate that they do not feel that they can hold the insurance industry accountable for its decisions but they can hold the Government accountable for their decisions. Perhaps this indicates that they feel more comfortable regarding insurers using genetic test results due to being able to control the Government's decisions regarding this matter to some extent.

If the Government were to appoint a body, such an organisation would need to be "given the appropriate powers to hold the Government accountable, I guess that's one of the key things for me, is that in some ways, such as this, a third party can be more powerful than Government itself, in that it can hold Government accountable as well as the insurance industry, but be representative of the people" (see Appendix D.1, Table D-19, Subject 1). Another interviewee also suggests having a "consulting group who are not...politically driven" (see Appendix D.1, Table D-19, Subject 8). In creating such a body, the Government would "have to appoint who they would think would be best…to manage it" (see Appendix D.1, Table D-19, Subject 2). One interviewee recommends that the group consist of "a mixture of… experts in the field and people who have experience of regulation as well" (see Appendix D.1, Table D-19, Subject 7). Thus the body should include "private but also…public sector people" (see Appendix D.1, Table D-19, Subject 8). Other individuals state that "the best practical solution would be to have an independent body made up of members of the public along with experts", believing it important to include "laypeople on the committee expressing general concerns" as "sometimes experts don't always see the wood for the trees and ordinary people, without such focused outlooks, may have different opinions that are very important" (see Appendix D.1, Table D-19, Subjects 7 and 1).

In addition, it is suggested that "all findings from such a meeting should be made public and should be open to the public" (see Appendix D.1, Table D-19, Subject 6). One individual sums up the entire management of insurers' use of genetic test results, asserting that "the politicians make the law, they decide, they get the best possible advice, right, now you can go to ten experts and get ten different focuses, or directions on that advice, at the end of the day, they decipher that, the civil servants do it, not the politicians, and the civil servants, they are, not necessarily scientific, but expert in their own right, interpret that, advise the ministers, the ministers draft the laws and the Government passes them". Such a "system is there to work properly" and should be able to manage the risks related to this situation (see Appendix D.1, Table D-19, Subject 3).

6.11.1 Public Participation

Two interviewees question if the Government "would employ a group of people… to go to that kind of trouble and expense of employing… a committee or a task force to overview…genetic testing (see Appendix D.1, Table D-19, Subjects 8 and 10). It is deemed necessary for the public to "get on to the Government ministers, get on to the Government media… making their voice, making their point, saying no" as "the less we do, it's not going to be dealt with accurately because the vested interest will deal with it in their own way, and then the Government tends to be incompetent, lacking in information, too busy doing other things, there's not a specific group to deal with this sort of way and it just gets lost in the maelstrom" (see Appendix D.1, Table D-19, Subject 10). The majority of interviewees mentioned the need for public involvement in the decisions made regarding insurers' use of genetic test results (see Appendix D.1, Table D-19). According to one individual, the Government "have to react in some way, but like that, it wouldn't be off their own bat... it'd be a matter of somebody pushing the matter on them and having to maybe set up a committee or something" (see Appendix D.1, Table D-19, Subject 8). In addition, the public should have a say in the decisions made regarding insurers' use of genetic test results "because it affects everybody that needs insurance, it's a major deal, and all these decisions are being made without most people knowing about them" (see Appendix D.1, Table D-19, Subject 7). "It's definitely in the public interest to make decisions", however, in order to do so, "the public should get information to make the decisions on this", "rather than just the Government making all of the decisions for us" (see Appendix D.1, Table D-19, Subject 4).

Although some interviewees feel that many people would be interested in providing their opinions, another individual asserts that she "can't imagine...there being a referendum on whether this information should be given out or not", instead, "it will be very much between the Government and the insurance companies". The interviewee concludes that "people wouldn't have a say" and even if she had, she is unsure how she "could influence a decision like that" (see Appendix D.1, Table D-19, Subject 8). Another individual agrees with this, stating that "a nationwide referendum... for every decision that had to be made wouldn't be a practical solution." However, it is also noted that "there are lots of ways to poll public opinion" via "text message voting, websites, (or) gather opinions from radio listeners", although it is noted that such methods of obtaining opinions are "open to a certain level of bias" but given this drawback it is still necessary "that public opinion should be factored in" (see Appendix D.1, Table D-19, Subject 1).

When interviewees were asked whether or not they believed that they had a say regarding the content of decisions made, or to be made, on insurers using genetic test results, five people felt that they did not have a say (see Appendix D.1, Table D-19, Subject 1, 2, 6, 7 and 9), while seven people stated that they would like to have their opinion counted, as "everybody has the right to have their say" (see Appendix D.1, Table D-19, Subject 2 and Subjects 1, 2, 3, 4, 6, 8 and 9). Out of those who stated that they would like to have their opinion heard by the governing authority on this matter, two individuals stated that they would not contribute as they wouldn't "be able to articulate" their "opinions strongly enough to work in that kind of role", or "wouldn't know enough about what I'd be trying to say about this" (see Appendix D.1, Table D-19, Subjects 1 and 5). Two interviewees stated that in order to express their concerns regarding insurers using genetic test results, they would "write a letter to somebody in the Government", to "T.D.'s and get them to pass it on" (see Appendix D.1, Table D-19, Subjects 7 and 9).

One individual states that "we can form our own opinions and make our own decisions, but… individually we won't be making decisions pertaining to industry. ….it is necessary that Government makes those decisions, and they do it by the best sources available, the best scientific sources available. The politician doesn't make the decision, we just see it as the minister makes the decision and drafts the law and the Government decide on that too. But it is on solid scientific advice". "If you don't like the decision your Government makes or, the local politician as part of your Government makes, then you change direction the next time round, you punish them for a lack of decision, or a lack of doing what you want them to do. So… you have a say, but it is well removed…only by putting the Government into place" (see Appendix D.1, Table D-14, Subject 3). Therefore, the decision may not be made by the public; however, the decision can be directed in some format by choice of Government ministers and can be changed through democratic voting of different Government ministers.

Therefore, one can conclude that, based upon analysis of the interviewees comments, the majority of them would prefer the Government to be in charge of the management of the risks related to insurers using genetic test results. Although a Government appointed body that is independent of political desires and incorporates public and expert opinion is considered preferable.

6.12 Summary

This chapter further examined the nature of the relationship between knowledge, trust, risk perception, benefit perception and risk acceptance. It is concluded that it may be the case that unless individuals work in the insurance industry or a related body, the vast majority of individuals are not aware, nor have they been informed, of insurers' use of genetic test results, a factor which may influence the statistical finding that there is incomplete knowledge of insurers' use of genetic test results in Irish society. The risks as perceived by those interviewed are also examined and divided into common recurring themes or factors that manifest themselves as risks in the minds of those interviewed. The factors examined relate to the privacy of genetic test results, the denial of insurance applications, the charging of high or unaffordable premiums, unfair discrimination, and finally, the selection of only low risk individuals.

The outcome from the analysis indicates that the nature of the privacy of individuals and genetic test results needs to be explored in relation to whether or not such data is private property to be shared with others only at the individual's discretion, or, indeed, whether commercial organisations should have the right to avail of such information. In addition, concerns regarding the workings of life insurance and the lack of transparency regarding underwriting methods and calculations needs also to be addressed. It is recognised that these concerns may be amplified by the 'probability' nature of genetic test results with a view that such data is far too limited and unreliable to be used as an evidence basis from which to make underwriting decisions. It is concluded that the risks foreseen as relating to insurers using genetic test results may also be acerbated by the perceived need for life insurance, concerns regarding the present and future welfare of society, the fear of lax privacy laws covering access to applicants' information provided to life insurance companies, and a belief that controls are lacking on insurers methods of asking for the results of genetic tests. All of these factors seem to connect and feed into individuals' perceptions of the risks related to insurers using genetic test results. If these concerns were addressed appropriately, it may be that the risks perceived can be reduced.

Individuals are also questioned regarding their perceived benefits of insurers using genetic test results, and in doing so a highly negative view of insurance companies and, indeed, their willingness to use genetic test results in underwriting practice is revealed. Only two individuals agreed that there would be a benefit for society, their families and themselves should insurers use genetic test results, that of a fairer premium being charged to applicants, although most interviewees note the financial benefits for insurers. As during discussions on risk perception, it is possible to note concern with regard to how insurance companies use information in their underwriting practices, developing from a lack of transparency regarding this issue, which leads to a negative view of life insurance companies. Examining the nature of the relationship between risk perception and benefit perception it is found that the majority (8/10) of such relationships are inverse with more inverse relationships identified during interview than in the survey. An inverse relation refers to the phenomenon that when risk perceptions are greater, benefit perceptions are smaller and vice-versa.

When interviewees are asked regarding their trust in the Government and in the insurance industry, three remarked that they did trust the Government, while only two stated that they trusted the insurance industry. This rather general trust question is then broken down to dimensions of trust, as in the survey. To examine whether such dimensions are, indeed, representative of trust, eight interviewees were asked to define trust. The definitions provided confirm that the concepts of care, integrity, reliability, competence, fairness, credibility and openness, all of which are examined in the survey and the interview, are indeed, dimensions of trust. In addition, trust as a positive concept is iterated intimating the lack of negative connotations attached to a trusting relationship.

Findings regarding trust in the Government indicate that individuals would like to receive information about insurers using genetic test results from the Government in order that they may have the ability to make decisions regarding this situation, thereby, linking knowledge with the ability to make one's own decision. Currently, there is no such information package available from the Government and it is questionable if one was introduced, if it would affect peoples' views on this matter and their trust opinions. However, the analysis indicates that a call for information must come from a majority, as the Government seldom hears small groups, or voices of ordinary people. Besides this, many individuals believe that the size, wealth and the economic value of the insurance industry in Ireland may influence Governmental management of insurers' use of genetic test results and, as a result, public opinion may not be heard on this issue. Interviewees were very much divided as to their view of the competency of the Government to manage this situation, however, it is evident that receipt of information regarding the Government's lack of competence in areas other than insurance management, influences society's views of trust in the Government to manage the use of genetic test results by the life insurance companies. In addition, previous history of the Government's action in what is seen as in, or not in, the public's interest (particularly in areas of health) seem to have a bearing on society's opinion regarding the Government's management of insurers' use of genetic results. Findings from the interviews also illustrate that attitudes with regard to the fairness of Government's decision-making and integrity may also be based upon the Government's decisions and actions on previous issues. Such findings seem to indicate that the history of the Government's decisions and actions play a large role in people's decisions of trust.

With regard to the insurance industry, when interviewees were asked if they trusted the insurance industry in general, the majority of replies were skewed toward the negative, with most indicating that the size and profit-making nature of insurance companies as factors which cause a lack of trust. In addition, dealings with insurance companies outside of life insurance products, influences individuals' trust in the life insurance industry. Findings also indicate that the trust factor 'care' is more significant than 'competence' in the question on trust with regard to insurers managing their use of genetic test results appropriately. It is likely, judging from the comments of interviewees, that people may be more willing to place trust in the insurance industry if an independent third party was involved in controlling and regulating insurers' activities.

Only two individuals believe that life insurance companies should be allowed to use genetic test results in their underwriting processes, while others believe the risks outweigh the benefits, with insurance companies using the results of such tests for their own financial good rather than for the betterment of society.

Finally, risk management perspectives are discussed which allow for further insight into the conditions whereby individuals agree to tolerate or to accept the risks perceived by insurers using genetic test results. There are several calls for an independent authority to manage the risks related to insurers using genetic test results, with some requesting a Government appointed agency. Indeed, this was the case at the time the interviews took place, with the Irish Financial Services Regulatory Authority (IFSRA) overseeing the matter, however, it appears to be the case that interviewees were unaware of this situation. In addition, there is a call for increased public participation in decision-making regarding insurers use of genetic test results.

The reasoning behind the qualitative analysis was to further explore any association between trust, risk perception, benefit perception and risk acceptance, and to understand the reasoning behind individual's responses to the questioning posed in the questionnaire and the factors that may have had a role to play in people's responses to the questions relating to knowledge, social trust, risk and benefit perception and risk acceptance. Indeed, it was not possible to directly examine this relationship based on the findings from the semi-structured interviews. The reason for the lack of ability to do so, could be related to the questions asked, whereby some individuals were asked all of the eight questions related to general trust and the three questions related to sceptical trust, while others were only asked a selection of such questions, depending upon their ability and desire to respond to the interview questions asked. For instance, some individuals were asked questions which they did not answer as they followed their train of thought which the interviewer followed. In addition, time restraints meant the interviewer was unable to pursue all of the questions as dictated by the questionnaire in an effort to keep the interview duration under the recommended one hour (see Section 4.7.3). In other cases, interviewees appeared to get frustrated when a question was repeatedly asked and gave an answer that did not coordinate with the question provided. So as not to vex interviewees, an attempt was made to ask questions only once if at all possible.

This chapter has outlined some of the factors and conditions that generate the results from the statistical analysis. In addition, the preferred circumstances and the conditions whereby an individual will accept the perceived risks related to insurers using genetic test results are discussed. The following chapter shall pursue the line of enquiry as to the factors and conditions that produce the results from the study throughout the social reality surrounding acceptance of the risks related to insurers using genetic test results (Layder, 1993). In addition, the statistical findings from the study shall be examined and compared, in light of the qualitative findings, to further understand the risk acceptance pattern delineated in the present study. It will be demonstrated how the questions posed in Chapter 1 regarding the risk acceptance pattern of insurers' use of genetic test results have been answered.

CHAPTER SEVEN

DISCUSSION OF KEY FINDINGS

7.1 Introduction

The main objective of this research was to investigate the role that incomplete knowledge, social trust, risk perceptions and benefit perceptions have in influencing society's acceptance of the risks perceived as relating to insurers' use of genetic test results in the Republic of Ireland. In addition, it was intended to identify the factors and conditions that impact upon this risk acceptance pattern. Findings from the semi-structured interviews demonstrated the influential significance of a variety of factors and conditions on knowledge, trust, risk perceptions, benefit perceptions and risk acceptance. The surveys illustrated a weak correlation between knowledge, trust, risk perceptions, benefit perceptions and risk acceptance. In addition, regression analysis indicated a low association between these factors. Some socio-demographic factors were demonstrated as having a slight impact upon knowledge, trust, risk perceptions, benefit perceptions and risk acceptance. The aim of this chapter is to explore the findings from the semi-structured interviews and the surveys and to compare and contrast their results in order to develop a further understanding of the relationship between knowledge, trust, risk perceptions, benefit perceptions and risk acceptance and the factors that influence this relationship.

Given the background and objectives of this study, it was decided to divide the findings into three sections. The first section shall examine the implications that the findings have for research in the 'risk' field. The following section shall explore the general implications for the insurance industry in Ireland, while the third section shall offer a criticism of the procedure and means whereby the findings from the present study were generated. Given the detailed and in-depth nature of the findings from the qualitative analysis in comparison to the rather broad and surface-based results of the survey, it was decided to allow the findings from the semi-structured interviews to guide as to which of the quantitative findings deserves most emphasis and to demonstrate the extent to which these findings corroborate what the in-depth discussions propose.

Based upon the case analysis in Chapter 6, it is clear that the case categorisation is different between the interview and subsequent analysis (see Appendix E.1). The four cases were initially chosen according to the responses of individuals to the survey, however, upon interview these responses changed (see Section 7.4 for reasons). Based upon the survey analysis, there was one representative case, one extremely varied case, one critical case and one extreme case. There were three individuals representative of each case, bar one - the extreme case corresponded with only one respondent (see Appendix E.1). Subsequent to interview analysis, four subjects changed cases. Three of those interviewees (Subjects 2, 7 and 8) previously belonged to the critical case, however, after interview analysis they appeared to satisfy the characteristics of the representative case. While the fourth interviewee moved from being classified as belonging to the extremely varied case to what constituted a representative case (Subject 1). This rearrangement of interviewees resulted in seven subjects corresponding to the characteristics of the representative case (Subjects 1, 2, 4, 5, 7, 8 and 10), two subjects indicative of the extremely varied case (Subjects 6 and 9) and one subject representative of the extreme case (Subject 3). There were no subjects left to represent the critical case, whereby a particular response appeared that should not have typically done so. Instead, there were extremely varied cases that were very different from each other with regard to two or more of the dimensions of importance for the study. All the representative cases remained representative, two of the extremely varied cases stayed as such, and the one subject identified as belonging to the extreme case remained constant in his views. A criticism of this alteration of cases is provided in Section 7.4, however, it should be noted that in order to expose the factors and conditions guiding peoples' attitudes and perceptions, it was necessary to determine what was common and what was particular about each case (Stake, 1998:90). If this comparison was based upon incorrect case alignment, the findings of the study could be considered invalid (see Section 7.4). Hence, case restructuration appeared to be a legitimate step in analysing the findings from the in-depth interviews. The following section shall commence this analysis with an exploration of the implications that the findings from the qualitative and quantitative analyses have for the field of 'risk' research.

7.2 Implications of the Findings for Risk Research

The main objective of this research was to investigate the role that incomplete knowledge, social trust, risk perceptions and benefit perceptions have in influencing society's acceptance of the risks perceived as relating to insurers' use of genetic test results in the Republic of Ireland. In doing so, it was intended to identify whether a relationship existed between these factors as demonstrated by previous research and to identify the factors and conditions that affect this risk acceptance pattern (Siegrist, 2000; Siegrist and Cvetkovich, 2000 and Poortinga and Pidgeon; 2005). Looking at the attitudes and evaluations of respondents, the majority of individuals sampled had incomplete knowledge of the use of genetic test results; high risk perceptions; low benefit perceptions; low trust in the government and the insurance industry; and a lack of willingness to accept insurers' use of genetic test results to underwrite applicants for life insurance. When exploring the factors and conditions that influence knowledge, trust in the Government and the insurance industry, risk perceptions, benefit perceptions and risk acceptance, it was possible to identify a link between each of these elements, except for knowledge. This finding suggested that there may indeed exist a relationship between social trust, risk perceptions, benefit perceptions and risk acceptance. In other words, social trust, risk perceptions and benefit perceptions may play a role, to a greater or lesser extent, in influencing acceptance of the risks related to insurers using genetic test results.

A summary of the factors and conditions found to influence knowledge, social trust, risk perceptions, benefit perceptions and risk acceptance across the dimensions of social reality are illustrated in Table 7-1. Discussion of the main findings of the research can be found in Section 7.2.1. To take a holistic view of the factors and conditions that were found to affect knowledge, trust, risk perceptions, benefit perceptions and risk acceptance, it was decided to use Layder's (1993:71) research map as a guide, given that an explanation for risk acceptance occurs throughout the dimensions described by Layder (1993:71). In Section 4.2.4, Layder's (1993) dimensions of social reality were briefly discussed. It was outlined that there are no clear boundaries between the dimensions. Instead they influence each other similar to the strands of a spider's web (Greenford, 2001). Section 7.2.1 shall examine how risk acceptance may be influenced at each of these dimensions. Firstly, the *'self' dimension* shall be examined. This does not indicate the relative importance of this dimension compared to the others, but emphasises this layer's predominance within the primary research of the present study. The primary research conducted was intended to collect data regarding the *'self' layer* of social reality. As the individual makes decisions regarding risk acceptance within a *'situated activity'* within a *'setting'* and a *'context'*, it was intended to investigate each dimension in this order. However, it should be noted that there that there may be factors and conditions other than those noted in the present study that influenced the findings. It is impossible to note all influences upon evaluations of knowledge, social trust, risk perceptions, benefit perceptions and risk acceptance, however, the present study has made an attempt to acknowledge as many as possible.

Table 7-1 Factors and Conditions Affecting Knowledge, Social Trust, Risk Perceptions, Benefit Perceptions and Risk Acceptance and their Relationship

	Context	Setting	Situated Activity	Self
Knowledge				Employment in the insurance industry
				Familiarity with genetic testing
Risk & Benefit Perceptions	Dominant culture	Utmost Good Faith	Privacy	Trust in risk managers
				Lack of underwriting transparency
		Underwriting transparency: no risk management	Vulnerability	Nature of genetic tests
				Need for life insurance
				Undesired impact on future generations
				Involuntary exposure to risk
				Life cycle factors

				Familiarity with the risk
				Control over predispositions to genetic disease
				Financial effects
				Fairness of the distribution of risks and benefits
				Gender
				Economic Status
				Proximity to the risk; personal & family medical history,
Trust Evaluations	Dominant culture	Underwriting transparency: no risk management	General trust	Care
		Data Protection Act	Integrity	Fairness
			Vulnerability	Competence
				Reliability
				Integrity
				Credibility
				Openness
				Characteristics of the individuals that make up the Government
				Independent expertise and regulation
				Successful risk management history
				Size of the group voicing concerns

				Size, wealth and economic value of the insurance industry
				Lack of underwriting transparency
				Dealings with general insurers
				Access to comprehensive and trustworthy information
				Incomplete knowledge
				Being part of a power elite
Risk Acceptance	Dominant culture	Need for risk management of insurers' use of genetic test results		Personal and family medical history, proximity to the risk
				Risk defined equity and fairness
				Economic status
				Gender
				Opinion of fairness regarding discrimination
				Ability to control risk management decisions and accountability
				Commercial nature of insurers
				Social trust
				Risk perceptions

				Benefit perceptions
				Nature of genetic tests
				Perceived need for life insurance
				Life cycle factors

7.2.1 Main Findings

Looking at the list of factors and conditions affecting knowledge, social trust, risk perceptions, benefit perceptions and risk acceptance (see Table 7-1), it is possible to note the entangled nature of the factors and conditions listed and the dimensions of social reality. For example, lack of underwriting transparency appears to influence risk perceptions, benefit perceptions and trust evaluations throughout several dimensions. Such interrelations not only identify the interconnectivity of Layder's (1993) dimensions of social reality, but also demonstrate an association between trust, risk perceptions and benefit perceptions. The factors and conditions listed in Table 7-1 illustrate some congruence with what Renn and Rothmann (2000) describe as the psychological factors that influence risk perception (see Figure 3-2). In addition to the factor *trust in risk management*, affecting risk perception, it is also noted that *involuntary exposure to the risk, life cycle factors, familiarity with the risk, ability to personally control the degree of risk, undesired impact on future generations, impression of fair distribution of benefit and risk, reliability of information sources,* and *clarity of information on risk,* all affect risk perception (Renn and Rothmann, 2000). However, in the present study, each of these elements are noted as factors that affect knowledge, social trust, risk perceptions, benefit perceptions and risk acceptance in the study, rather than risk perception alone. This further illustrates the interconnectivity between these factors, despite the findings from the survey analysis.

Findings from the survey analysis indicate that there may not be a direct association between social trust, risk perceptions, benefit perceptions and risk acceptance. This observed non-actualisation of effects may have derived from countervailing tendencies, or indeed, demonstrate the misrecognition of underlying causal mechanisms of risk acceptance. The more in-depth qualitative findings however, reveal that the underlying factors and conditions of trust evaluations, risk perceptions, benefit perceptions and risk acceptance serve as a link between these factors suggesting that there is a causal association between social trust, risk perceptions, benefit perceptions and risk acceptance in the present study. This causal tendency should not be interpreted as a universal 'rule', rather the findings from the present study illustrate a rather local mechanism or tendency. Given that it is the factors and conditions underlying trust, risk perceptions, benefit perceptions and risk acceptance that serve to enforce the association between these factors, this demonstrates that the scenario and area of research very much influences the nature of this relationship. Section 8.2 shall further examine the nature of the causal association between social trust, risk perceptions, benefit perceptions and risk acceptance based upon the findings of the present study.

Although the findings from the study indicate an association between social trust, risk perceptions, benefit perceptions and risk acceptance, the strength of this relationship is unknown, and it may be that this relationship is slight and indirect. The following sections shall explore the factors and conditions listed in Table 7-1 so that their existence as variables that influence knowledge, trust, risk perceptions, benefit perceptions and risk acceptance is understood and explained. In addition, the contribution that these findings present to the body of knowledge surrounding risk perception research is identified.

7.2.2 Self

The impact of 'setting', 'context' and 'situated activity' upon the sample population is similar among all members of that population, however, the acceptance of risks differ between members of society. The explanation for such difference occurs mostly at the centre of the web, where the historical background, attitudes and perceptions of the individual influence their acceptance of insurers using genetic test results. The dimension 'self' is representative of Irish society, or more particularly, those individuals that participated in the survey and the interviews. The 'self' is the biographical experience of these individuals and reflects their understanding and reactions to the 'situated activity' and the 'setting' in which they are operating. Therefore, an examination of the 'self' requires an analysis of the historical background, attitudes and perceptions of these individuals. The following section shall consider the biographical background of the sample population and the influence that peoples' personal history has upon the study's findings. In addition, individuals' perceptions regarding the risks and benefits of insurers using genetic test results and their perception of the Government and the insurance industry are considered.

7.2.2.1 Knowledge

The overall qualitative and quantitative results from the present study suggest that incomplete knowledge does not directly influence social trust, risk perceptions or risk acceptance. Only one interviewee, who previously worked in the insurance industry, was aware of insurers' interest in gaining access to genetic test results (see Table 6-2, Subject 10). Although this individual had knowledge of insurers using genetic test results, he satisfied all other criteria deemed to qualify him as a representative case. This particular finding raises the question as to whether or not knowledge has a direct effect upon one's judgement of social trust, risks and benefits. Analyses of results from the survey were unable to indicate whether or not a relationship existed between these factors. It was impossible to compare correlations between lower and higher self-assessed knowledge and any other variable, due to the fact that there were no individuals who deemed themselves knowledgeable on this subject.

The survey revealed that 70 per cent of the sample population felt that they had no knowledge about the use of genetic test results by life insurance companies, while 81 per cent indicated a lack of any knowledge of current policies guiding life insurance companies' use of genetic test results (see Section 5.5.1). Originally, this lack of knowledge was attributed to a lack of familiarity with genetic testing. However, given the findings from the interviews, whereby eight out of the ten interviewees were aware of what genetic testing is and knew some of the diseases that the tests are designed to diagnose or determine predisposition toward, this appears to be a questionable explanation (see Section 6.3). However, it was also noted that all interviewees received information sheets that provided information on genetic testing and insurers use of genetic test results after the survey was completed, therefore this finding may be slightly deceptive (see Appendix B-4). Indeed, interviewees indicated that the surveys were the first time that they had ever heard of life insurers' intention to use genetic test results (see Section 6.3). It is, therefore, not possible to overrule that familiarity with genetic testing and insurers' use of this information influences knowledge. The only factor conclusively identified as impacting knowledge in the present study is employment in the insurance industry (see Section 6.3). Given the limitations of the degree of knowledge of the sample population in the current study, it would be interesting to note in future studies the effect that complete knowledge has on risk acceptance and if familiarity with genetic testing has a direct effect upon one's knowledge of this matter.

7.2.2.2 Risk Perceptions

According to the results of the survey, 63.7 per cent of the sample population agreed that life insurance companies using genetic test results to decide if they will insure an applicant expose society and their families to many risks (see Section 5.5.2). When interviewees were asked questions regarding their perceived risks, responses indicated concerns regarding the privacy of genetic test results, the denial of insurance applications, the charging of high or unaffordable premiums, unfair discrimination and the selection of only low risk individuals by life insurance companies. Genetic tests were not considered as equivalent to other medical tests. These perceptions of the risks related to insurers using genetic test results appear to be acerbated by several factors:

- Lack of transparency regarding underwriting calculations
- Lack of control regarding predisposition to a genetic disease
- Fairness of the distribution of risks and benefits
- Trust in risk managers
- The nature of genetic tests
- A perceived need for life insurance
- Concerns regarding an undesired impact on future generations
- Familiarity with the risk
- Life cycle factors
- Involuntary exposure to the risk

The majority of these factors have been demonstrated to influence perceptions of risk in previous research (Renn and Rothmann, 2000). The following paragraphs shall explore each of these factors and their impact upon risk perceptions in more detail.

7.2.2.2.1 Lack of Transparency

In Sections 6.4 and 6.5, it was briefly discussed how some individuals view the practice of underwriting in a negative light. As far as applicants are concerned, the underwriting deliberations and calculations performed by the life insurance company are carried out behind a closed door. It is not possible to examine underwriters' calculations of premium or decisions regarding insurability in order to understand the overall premium quoted or the decision made. Such a lack of transparency surrounding the workings of life insurance companies may lead to suspicion that life insurance companies' premium calculations are not based upon a common method and means of computation, hence, policyholders may be inappropriately charged. Ability to purchase insurance and the premium charged often varies between life insurance companies. This indicates that the system is not universal, and given the closed-door nature of underwriting, this factor may lead to a view that the life insurance underwriting system is unfair. Given that the ability to underwrite underpins the system of insurance, such a perception may influence the negative view of insurers using genetic test results and hence the risks related to such usage.

7.2.2.2.2 Control over Predisposition to a Genetic Disease

It was considered by at least half of those interviewed that a risk should not be imposed upon an individual based upon a factor that is beyond their control (see Subjects 2, 4, 6, 7 and 10, Table D-4, Appendix D-1). Given that one cannot be blamed for having a genetic predisposition to disease, as it something that is impossible for an individual to control given that one is born with this link to a particular disease. Thus, several respondents found that insurers should not be allowed to discriminate according to this factor (see Section 6.4). It was earlier noted that this form of discrimination is perceived as a risk emanating from life insurers using genetic test results in deciding the premium or insurability of an applicant (see first paragraph, Section 7.2.2.2). One's ability to personally control the degree of risk to which one is exposed was a factor previously noted by Renn and Rothmann (2000) as a variable which influences risk perceptions. In this case, interviewees' perceptions of risk related to insurers using genetic test results may be effected by their belief that it is impossible to personally control one's genetic data and, hence, control the financial risks incurred should insurers demand knowledge of this information.

7.2.2.2.3 Fairness of the Distribution of Risk and Benefits

Section 6.4 notes the questioning of the fairness of the distribution of risks and benefits as reasoning behind interviewees' judgements of risk. Several of those interviewed questioned whether or not it is fair to use genetic test results that reveal adverse factors for one population but not to use those that reveal risk factors for another population. The 'other' population in this case consist of those individuals who have not yet partaken of a genetic test or for who no tests are available and persons who already have life insurance and do not wish to purchase any further products. The result is that certain segments of the population are discriminated against and may be charged high or unaffordable life insurance premiums, while other segments are not exposed to such perceived risks, although they may possess a gene that makes them equally susceptible to disease. This unequal distribution of risks and benefits appears to influence risk perceptions.

It should also be noted that the fairness of the distribution of risk and benefits can be seen as a factor that influences risk perceptions. It was considered unfair to financially burden individuals with a high priced life insurance product given that they are exposed to physical health risks. Indeed, it is those individuals who have discovered their genetic predisposition to disease that need life insurance the most. Given their chances of developing a genetic disease and their misfortune of having this link to disease, it was considered unfair by some to further burden such individuals with more risks, in relation to their ability to purchase life insurance (see Subjects 2, 5 and 10, Table D-4, Appendix D-1). From this perspective, 'unfairness' appears to influence risk perceptions.

7.2.2.2.4 Trust in Risk Managers

Risk perceptions appear to be influenced by the belief that privacy laws are inadequate and insurers may divulge personal information provided to them by applicants (see Section 6.4). This could be interpreted as a lack of trust in the Government to ensure the privacy of genetic test results when passed onto insurers. In addition, there appears to be low trust in the insurance industry to ensure such privacy. In the Republic of Ireland, the Data Protection Act 1988 and 2003 dictate that underwriters are legally bound with the same confidentiality rules as doctors when dealing with medical reports (see Section 2.3.5). However, the interview analysis reveals that such a possibility is of concern to some individuals, perhaps demonstrating feelings of low trust in the Government's management of the risks related to privacy matters (see in particular Subject 2, Section 6.4). Indeed, there are also concerns that if life insurance underwriters are allowed to use genetic test results that an insurer may manipulate such an allowance to their advantage. Once again this presents a rather negative view of life insurers and, indeed, a lack of trust in the insurance industry and it appears that this lack of trust positively influences risk perceptions.

7.2.2.2.5 Nature of Genetic Test Results

Such an anxiety regarding the provision of information to insurers appears to be acerbated by the nature of the data in question. The risks perceived by individuals seem amplified by the nature of genetic tests, in that the results of such tests were considered private property by some or, indeed, the results are too inaccurate a source of medical data to be used by underwriters. As time passes, an increasing number of tests may be considered reliable and accurate enough to be used as underwriting information (GAIC). It is interesting to note that five of the seven individuals that make up the Representative Case mentioned the private nature of genetic test results (see Section 6.4), while none of the other cases discussed this matter. Members of the Representative Case are also the only individuals interviewed that suggest that genetic test results are often fallible regarding future mortality and should, therefore, not be used as evidence in assessing insurability or indeed, the premium to be charged (Subjects 2, 4 and 8). This is an interesting finding since the Representative Case viewed the risks related to insurers using genetic test results as high and did not accept insurers using genetic test results, unlike all of the other cases. This suggests that perceptions regarding the nature of the information provided by genetic tests highly influences risk perceptions and risk acceptance.

7.2.2.2.6 Perceived Need for Life Insurance

Looking at the expectations of individuals, there was a perceived need for life insurance to financially protect one's family and in order to purchase housing (see Section 2.3.1). Interviewees expected to be able to purchase life insurance and would prefer that the cost of doing so is inexpensive. In addition, based upon the interview analysis, the ability to purchase life insurance, as a means of protecting one's family from financial loss and not just as mortgage protection insurance was also considered a necessity (see Section 6.4). There appears, therefore, to be an expectation that given such needs it should be possible for all members of society to purchase affordable life insurance. Although, life insurance may be perceived as a necessity in current Irish society, it is a commercial product designed to make its suppliers a profit and, therefore, it cannot be provided to all individuals or, indeed, supplied to all at an affordable rate. Findings from the present study indicate that this perceived need for life insurance may impact upon risk perception, whereby if this need is not accounted for by insurers, many individuals may be faced with the risk of not being able to financially protect their family or, indeed, purchase a home. Insurers may earn a lack of respect as a result of turning individuals down for life insurance or charging them a premium which is considered over-priced, which may influence the negative view of life insurers illustrated by this study's findings. Theoretically, this negative view of insurers may feed into trust evaluation and, in doing so, serve to reinforce risk perceptions.

Only two individuals in the representative case owned a life insurance product at the time of interview (see Appendix E.1, Table E-2, Subjects 1 and 10). Four interviewees that made up the representative case intended purchasing life insurance in the future, while one did not. The extreme case did not have life insurance, nor did he intend to purchase a life insurance product in the future. It may be the perceived need for insurance that affects the risk acceptance pattern of the Representative Group. The necessity factor is further emphasised by the extremely varied case, whereby judgements of risks and benefits do not seem to influence risk acceptance. However, the individual that did not accept the risks related to insurers using genetic test results, in other words did not wish for insurers to be allowed to use genetic test results, intended to purchase life insurance in the future. In this case, the need for life insurance may not affect risk perceptions to any great extent, but it may have a direct influence upon agreement for insurers to use genetic test results.[1] Findings from the survey analysis also suggest that individuals in the life insurance purchasing category are most against insurers using genetic test results (see Section 5.6.4). While respondents, who already have life insurance or do not intend purchasing it, are the least against insurers using genetic test results. However, further analysis demonstrates that risk perceptions and risk acceptance does not differ between those who own and those who don't own life insurance and those who intend to purchase life insurance in the future and those who have no intention of doing so. Given this discrepancy between findings, there still appears to be a link between a perceived need for life insurance and risk acceptance.

7.2.2.2.7 Concerns regarding Undesired Impact on Future Generations

Analysis in Section 6.4 also illustrates how the risks foreseen from insurers using genetic test results may be acerbated by concerns regarding the present and future welfare of society. It is interesting to note that those who mentioned such concerns were members of the representative case, which is indicative of the average responses of the sample population (Subjects 2, 4 and 10). Renn and Rothmann (2000) identify this phenomenon

[1] This possibility may be slightly refuted by the other interviewee who makes up the extremely varied group, who perceives some risks from insurers using genetic test results, but appears satisfied to allow insurers to use genetic test results in their underwriting practices and, yet still, would like to purchase life insurance in the future.

of concern regarding an undesired impact for future generations as an important psychological influence of risk perception identified by many studies across Austria, Germany, Great Britain, The Netherlands and the U.S.

7.2.2.2.8 Familiarity with the Risk Source

Although the in-depth interviews indicated that the majority of individuals were aware of what genetic testing is, this form of medical test was considered as being different to other medical tests (see Sections 6.3 and 6.4). Objectively, this difference may be based purely upon the possibility that the present-day results from a genetic test may be alternatively interpreted in the future or indicate a link to a different disease (see Section 2.3.3). This assumption is based upon the potential findings of future research. However it could equally be proposed that future developments in medical research could produce equivalent findings from more familiar or widely used medical tests. Nonetheless, it may be that genetic tests are considered as encouraging perceptions of new risks to society from their use by insurers, based upon their relatively new introduction to the range of medical tests previously available rather than the precise nature of the test itself. Although, the nature of genetic testing does appear to influence risk perceptions (see 7.2.2.2.5).

7.2.2.2.9 Life cycle Factors

Findings from the in-depth interviews did not indicate that age influences perceptions of the risks related to insurers using genetic test results, indeed, correlation analyses of survey responses demonstrated that age does not influence self-assessed knowledge, trust evaluations, risk perceptions, benefit perceptions or risk acceptance (see Section 5.6.2). However, the basic descriptive results from the survey appear to indicate that life cycle factors may influence risk perceptions to a certain degree. Findings from the survey demonstrated that the age group most likely to be in the purchasing bracket for life insurance in Ireland (26-56) were those that most strongly disagreed with insurers using genetic test results, perceived the most risks, and the least benefits (see Section 5.6.2). While those who were least likely to apply for life insurance, respondents that had raised their families and paid off their mortgages (over 46 years old) or, were quite young (18-25), were the group with the highest declaration of agreement with insurers using genetic test results and perceived most risks. Correlation analysis revealed that the strength of the relationship between age and life insurance ownership and intention to purchase life insurance was weak (see Section 5.6.4). Nonetheless, there appears to be some argument that life cycle factors influence risk perception and risk acceptance attitudes.

7.2.2.2.10 Involuntary exposure to the risk

The in-depth interviews indicated that involuntary exposure to perceived risks influences overall perceptions and acceptance of the risks related to insurers using genetic test results. It was believed by some individuals that applicants for life insurance could be forced to take genetic tests and hence become involuntarily exposed to higher or unaffordable life insurance premiums (see Subject 10, Section 6.4). It was considered that one has the right not to be exposed to such risks in that one has the right not to know the results of a genetic test (see Subject 5, Section 6.4). Voluntary acceptance of risk is a factor that previous research has demonstrated to influence risk perceptions (Renn and Rothmann, 2000). The involuntary exposure of individuals to psychological anguish and financial loss caused by insurers demanding or pressuring them into taking genetic tests appears to increase the perceptions of risk by some individuals (see Section 6.4 and Table D-23, Appendix D-1).

7.2.2.3 Benefit Perceptions

Only two individuals, out of the ten interviewed, agreed that there would be benefits for society, their families and themselves should insurers use genetic test results (see Section 6.5). This is comparable with a total of 69.8 per cent of respondents to the survey who disagreed that there were many benefits for society or their families if life insurance companies were to use genetic test results (see Section 5.5.2). Benefits mentioned included a fairer and a reduced premium being charged to applicants as a result of a fairer assessment of all applicants' risks. Although only two individuals agreed that there were benefits for society, four more indicated that there may be a financial benefit for some (see Section 6.5). As during discussion on risk perceptions, it was possible to identify concerns related to insurance companies' use of genetic test results in the underwriting practices which appeared to develop from a lack of transparency (see Sections 6.4, 6.5 and 7.2.2.2.1) However, other factors also appeared to influence benefit perceptions, such as financial effects and distribution of benefits and risks. Both influences contain fairness considerations, an aspect known to influence risk perceptions (Renn and Rothmann, 2000). The nature of these factors is discussed in the following paragraphs.

7.2.2.3.1 Financial Effects and Fairness

The findings from the interviews suggest that the financial effect, whereby those individuals without a predisposition toward a genetic disease had a positive financial advantage over those individuals who may have discovered their susceptibility to genetic disease, influences benefit perceptions. In addition, although this benefit is understood by six out of ten of the interviewees, the majority prefer to ignore its importance and do not wish to acknowledge it as a benefit (see Section 6.5). Although the majority of interviewees acknowledged this financial effect, when asked if there were any benefits for society, they replied that there were none (see Section 6.5). Perhaps this indicates that for these individuals, the risks outweigh the benefits to such a great extent, that any betterment for society resulting from insurers using genetic test results appears negligible and not worth noting in particular. It is interesting to note what is considered as 'fair' to these individuals. The insurance industry believe an insurance scheme to equitable or fair it respects all of the actuarial rules. However, one of the respondents indicated that there is a difference between financial benefits and moral benefits, which appears to suggest a difference between the insurers' notion of equity and society's aspirations for fairness (see Subject 3, Section 6.5). It is perhaps this view of ethics and fairness that leaves so few people able to acknowledge any benefits related to insurers using genetic test results. It would be interesting to pursue this line of inquiry in further research.

7.2.2.3.2 Fairness of the Distribution of Benefits and Risks

Although the majority of respondents' viewed few benefits for society from insurers using genetic test results, interviewees were aware that such usage would have a beneficial effect for insurers. Most interviewees considered the distribution of benefits as unfair (see Section 6.5). The insurance company was perceived as the sole benefactor from insurers using genetic test results, however, some interviewees acknowledged that applicants without a predisposition to a genetic disorder may also benefit from insurers usage of this information (see Section 6.5). It appears that interviewees view the distribution of benefits as skewed in favour of the insurance industry and this factor may influence their perceptions of risk related to this situation. Overall, it appears that disagreement with the distribution of benefits influences the majority of individuals to ignore the financial benefits for insurance companies and genetically 'healthy' individuals. The requirement for insurance companies to be risk averse was noted by interviewees, along with their need for accurate information and the equivalent medical information supplied to applicants (see Section

6.5). However, concerns were raised with regard to how insurance companies use information in their underwriting practices, developing from a lack of transparency regarding this issue, leading to a negative view of life insurance companies (see Section 7.2.2.2.1). Such a lack of transparency regarding the workings of underwriters appears to cause suspicion and a heightened sense of risks surrounding insurers accessing this information, perhaps outweighing any perceptions of benefits, thereby influencing an individual's evaluation of the distribution of risks to benefits.

7.2.2.4 Trust in the Government

Findings from the interviews indicate that only three individuals (see all of Section 6.8 and Appendix E.1) had general trust and did not have sceptical trust in the Government. These individuals were in the minority (extreme and extremely varied cases), with the representative case indicating that individuals do not have general trust, have sceptical trust or are unsure regarding their trust in the Irish Government. The majority of respondents to the survey were unsure whether or not to trust the Government on this issue. This may be due to peoples' general lack of knowledge of life insurance companies' use of genetic test results and/or peoples' low or non-existent knowledge of any risk management policies (see Sections 5.5.1 and 6.3). This assumption is confirmed by the findings from the in-depth interviews by interviewees who stated that the surveys were the first time that they had ever heard of life insurers' intention to use genetic test results (see Section 6.3). Other explanations for this finding may be hesitancy in responding to questions regarding trust in the government due to the personal nature of the questions. Respondents to the survey may have felt uncomfortable responding to trust questions in the Government due to fears regarding the privacy of their responses or suspicion regarding future use of the survey findings.

Although the majority of survey responses indicated that respondents were unsure whether or not to trust the Government or the insurance industry on this matter, the remaining individuals were largely of the opinion that the Government and the insurance industry could not be trusted to manage insurers' use of genetic test results. To a large extent, respondents disagreed that the Government is competent, fair, or caring for the public or individuals with regard to managing life insurance companies' use of genetic test results (see Section 5.5.3). Sceptical trust was high, in that respondents agreed that the Government could distort facts in its own favour regarding the above, that it could change policies without good reasons and is too influenced by life insurance companies in its decision-making regarding this matter (see Section 5.5.3). According to the respondents of the survey, the Irish Government lacks credibility, reliability and integrity with regard to managing life insurance companies' use of genetic test results to decide if they will insure an individual (see Section 5.5.3).

Based upon findings from the survey, a decision was made to exclude survey responses regarding 'openness' from those factors designed to measure general trust in the insurance industry and general trust in the government due to certain irregularities (see Sections 4.6.2.3.1 and 4.6.2.3.2). However, the interview findings illustrated how openness could be considered a dimension of trust that is linked to other trust dimensions (see Section 6.7). For example, two factors of trust, openness and fairness seem bound together. Findings demonstrated that if the Government were more open about issues being examined and the decisions being made by local and national Government, people may be more open to the belief that the Government is making fair decisions and can be trusted to do so (see Section 6.8.7). Credibility on the part of the Government also appears to aid in individuals evaluations of openness (see Section 6.8.1). In addition, Section 6.8.2 explored how competency and openness perceptions, linked together, serve to influence trust evaluations. Therefore, openness should be regarded as a general trust evaluation despite quantitative findings. However, the discrepancy with regard to 'openness' and its inclusion as a dimension of trust, should be examined further in future studies.

Several interesting findings regarding the factors that might influence individuals' trust perceptions in the Government were revealed upon interview. The lack of information provided to individuals describing the pros and cons of insurers using genetic test results and an understanding that public opinion has not been gauged regarding insurers using genetic test results appear to influence judgements of the openness of the Government and hence trust evaluations in this authority (see Section 6.8.7). News of the Government's lack of competence in areas other than insurance management also appears to influence society's views of trust in the Government to manage this situation. An emphasis was placed on the need to recruit experts to aid in the research required to formulate decisions regarding this matter. Statements by interviewees seem to reflect an opinion that the Government can be trusted to manage the use of genetic test results by life insurance companies if they employ experts to research the issues at stake (see Section 6.8.2). This finding indicates that competence depends upon expertise and this factor influences trust notions.

Interviewees appear to trust that the Government will eventually take into account public concerns, although factors such as the size of the group voicing concerns was considered to have a significant impact upon the ability to be heard. The voices of ordinary people, taken to be small groups, were perceived as seldom heard (see Section 6.8.5). Given such findings, it is advisable that if the public wish to have their opinion regarding insurers using genetic test results taken into account upon decision-making (which from the population sampled may well be the case, given that 91.7 per cent of respondents to the survey agreed that the Irish public should participate in any decisions made regarding this matter) it may be wise to establish a committee that can articulate the concerns of the public (see Appendix E.2, Table E-11).

The majority of those interviewed believed that the size, wealth and the economic value of the insurance industry in Ireland may influence Governmental management of insurers' use of genetic test results and this, in turn, appears to affect trust perceptions (see Section 6.8.4). In addition, findings from the interviews indicate that one's trust in the Government's management of insurers' use of genetic test results may depend upon the characteristics of the individuals that make up the Government. Attitudes with regard to the fairness of Government's decision-making surrounding insurers' use of genetic test results appear to be based upon the Government's decisions on previous issues. It appears that trust perceptions are heavily influenced by the previous history of the Government's risk management successes and failures (see Section 6.8.2). A belief that the Government may distort facts or withhold information is another factor that affects trust evaluations (see Section 6.8.2). Given that there is an opinion that it may be possible to find the truth if one searches hard enough, easier access to such information may increase trust (see Section 6.8.2).

7.2.2.5 Trust in the Insurance Industry

The qualitative and quantitative findings did not only reveal low trust in the Government but also a lack of trust in the insurance industry to manage insurers' use of genetic test results. Analysis of the interviews indicated that five individuals did not have general trust in the insurance industry, while none of the individuals interviewed demonstrated general trust in the industry (see Section 6.9 and Appendix E.1). Three individuals had sceptical trust in the insurance industry, while one individual did not (see Section 6.9 and Appendix E.1). All other interviewees were unsure whether to trust the insurance industry regarding this matter. The population sampled in the survey trusted the insurance industry less than the Government with regard to the management of life insurance companies' use of genetic test results (see Section 5.5.3 and Section 5.5.4). The majority of respondents to the survey disagreed that the insurance industry is competent, fair, or caring for the public with regard to managing life insurance companies' use of genetic test results (see Section 5.5.4). Sceptical trust was high, with most participants agreeing that the insurance industry could distort facts in its own favour, that it changes policies without good reasons and is too influenced by life insurance companies regarding this situation (see Section 5.5.4). This demonstrates that the greater part of respondents were of the opinion that the insurance industry lacks credibility, reliability and integrity with regard to managing life insurance companies' use of genetic test results. It can be said, therefore, that both the quantitative and qualitative findings from the study illustrate a lack of trust in the insurance industry to manage life insurers' use of genetic test results.

A variety of responses were provided as to why people did not trust the insurance industry, however, the majority tend to orientate around the size and profit making nature of insurance companies (see Section 6.9). Regarding the trust dimension 'openness', although people had not received information on insurers' use of genetic test results, they felt that such information should be provided, with some believing that this information is available. There were some suggestions that the insurance industry may try to swindle or cheat individuals (see Section 6.9). With regard to the dimensions of trust, it appears that the trust factor 'care' is more important than 'competence' with regard to insurers managing their use of genetic test results appropriately. Most individuals believed that the insurance industry could competently manage this situation but may not care to do so in the interest of the public (see Sections 6.9.1 and 6.9.5). None of those interviewed felt that the insurance industry is acting in the public interest with regard to their management of their use of genetic test results. The belief that insurers lack an interest in the security of the public and, instead, have a direct interest in profit making, appears to be a great concern for those who indicate a lack of trust in the insurance industry (see Section 6.9.5). To ensure that insurers are not allowed to lose sight of the public interest, there is a need for the public to be united in their concern, to have these concerns voiced appropriately and for an independent regulator to be involved in order to ensure that the insurance industry care for the public or their consumers appropriately (see Section 6.8.5).

Dealings with insurance companies outside of life insurance products may influence individuals' perceptions of fairness, credibility and reliability of the life insurance industry and, hence, their trust in life insurance companies (see Sections 6.9.2, 6.9.3 and 6.9.6). Findings from the semi-structured interviews illustrate how decision-making, regarding the fairness of insurers using genetic test results, is made more difficult when there is incomplete knowledge (see Section 6.9.6). The lack of transparency regarding the workings of life insurance also appears to be an issue that affects individuals' beliefs regarding whether or not the insurance industry could distort facts or withhold information regarding their use of genetic test results (see Sections 0, 0 and 7.2.2.2.1).

It is interesting to note that interviewees felt that there is a need for protection against life insurance companies distorting information and taking advantage of consumers (see Section 6.9.4). Such an attitude may well feed into perceptions of who to trust regarding the management of insurers using genetic test results and, indeed, how it should be managed. Interviewees demonstrated a belief that the insurance industry could be influenced by life insurance companies in the direction of allowing them access to all genetic information, which could be interpreted as a feeling of low integrity on behalf of the insurance industry and a reason for little trust (see Section 6.9.4). This is noteworthy, as rather than trust indirectly affecting risk acceptance, a decision to not accept insurers' use of genetic test results appears to directly influence integrity perceptions and hence trust evaluations to a certain extent. This finding corresponds somewhat to an explanation of risk acceptance provided by Eiser *et al.* (2002) and Poortinga and Pidgeon (2005). According to the results of these studies, the acceptability of a risk could be the determinant of trust. It may be interesting for future studies to examine this theory in more detail and compare both approaches to risk acceptance.

The negative view of insurers outlined above may lead to the notion of the existence of risk in their practices when it comes to the inclusion of the results of genetic tests in their analysis of applications for insurance and subsequent premium to be charged. This negativity could be interpreted as a lack of trust in the insurance industry, a finding that is illustrated in Section 5.5.4. This study previously noted that trust can be divided into two components, described by Poortinga and Pidgeon (2003) as general trust and sceptical trust (see Section 4.6.2.3). It is possible to broadly describe these components as being related to perceptions of competence and morality. However, the qualitative analysis revealed a negative view not only towards the insurance industry (see Section 6.8 and Section 6.9), but also towards the Government. This finding indicates the presence of a third component or dimension of trust, possibly related to perceptions of a tendency to corruption (in a broad sense). This characteristic of trust may be based upon an individual's beliefs, ideology and emotional reactions to the nature of the insurance industry and the Government. This dimension of trust is similar to the findings of Viklund (2002), who includes items regarding an organisation being part of a power elite in society and working primarily in its own interest within the dimensions of trust.

Looking at this finding more closely, it is an accepted fact that the Irish Government has much power in society, given its status as prime legislator and risk manager. However, the life insurance industry also wields a lot of power in Irish society, given the requirement to own life insurance for house purchase. In addition, during the course of the primary research of the present study, the Irish Insurance Federation had the power to manage the risks related to insurers using genetic test results. According to Lipset and Schneider (1983), people do not trust institutions that are regarded as part of a power establishment. Thus, concentration of power (or the belief that an organization is part of the power elite) is one attribute of possible importance for people's levels of trust in an organization (Viklund, 2002). This aspect of trust was not regarded as a dimension of trust in this study; however, it appears as an element in trust evaluations during the qualitative analysis. In the case of the Government, peoples' attitudes to the characteristics and history of members of the Government were revealed to have an influence upon their trust in this organisation (see Section 6.8.1 and 6.8.2. The misuse of the power provided to Governmental authorities and, media reports of Governmental corruption, negatively affect trust in this institution (see Section 6.8.1, 6.8.2, 6.8.4 and 6.8.5). The insurance industry is looked at with suspicion by some individuals and may be seen as abusing its power to underwrite individuals without any requirement to reveal such calculations (see Sections 6.9, 6.9.2 and 6.9.5).

From the findings discussed above, it is possible to say that trust evaluations in the Government and the insurance industry appear to be composed of several factors that are in turn influenced by a number of further elements. Judgements of competence, care, reliability, fairness, openness, credibility and integrity each feed into evaluations of trust in the Government and the insurance industry regarding their management of insurers' use of genetic test results.[2] There are no commonalities between cases based on their individual responses to the dimensions designed to measure trust (see Appendix D.1, Table D-23 and Table D-24). Therefore, it does not appear that any particular dimension is striking for a specific case. To briefly summarise the above discussion of trust in the Government and the insurance industry, the following elements are found to indirectly affect trust judgements by directly influencing evaluations regarding the dimensions of trust:

- Characteristics of the individuals that make up the Government,
- Independent expertise and regulation,
- Successful risk management history,
- Size of the group voicing concerns,
- Size, wealth and economic value of the insurance industry,
- Incomplete knowledge
- Lack of underwriting transparency,
- Dealings with general insurers,
- Access to comprehensive and trustworthy information
- Being part of a power elite

7.2.2.6 Risk Acceptance

Unacceptable risk, in this scenario is defined as allowing life insurance companies to use genetic test results to decide upon the insurability of applicants and the premium to be charged. Interview findings were similar to those of the survey, indicating that eight of the ten interviewees disagreed with such usage of genetic test results, while two agreed (see Section 6.10). Findings from the survey illustrated that 73.4 per cent of

[2] There appear to be some problems with the wording of the questions designed to measure reliability, therefore, it is difficult to understand the influence that this factor may have upon trust evaluations (see Section 6.8.3). However, qualitative analysis demonstrates that reliability may be a factor that influences trust evaluations (see Sections 6.7 and 6.9.3)

the sample population believes that life insurance companies should not be allowed to use genetic test results to decide if they will insure an applicant (see Section 5.5.6). Of those individuals that were of the opinion that there are more benefits than risks emanating as a result of insurers using genetic test results, 41 per cent believed that insurers should not be allowed to use genetic test results (see Section 5.5.6).

Reasoning for the lack of risk acceptance varied, with some declaring their decision as based upon a measure of the risks and benefits involved (see Section 6.10). Others mentioned the commercial nature of life insurance companies as a reason for their decision, a factor that appears to affect trust perceptions (see Table 7-1). This indicates that trust, risk perceptions and benefit perceptions may influence risk acceptance judgements. The nature of genetic tests, the perceived need for life insurance and life cycle factors, all of which appear to influence risk perceptions, seem to be associated with risk acceptance also (see Sections 7.2.2.2.5, 7.2.2.2.6 and 7.2.2.2.9). Other reasons for risk acceptance appears to be a belief that it is not necessary for insurers to use genetic test results given that they should be able to share out the higher risks with the masses, and to use such results would be unfair (see Section 6.10). This response denotes a sentiment that the present system of insurance is fair and in order to ensure continued fairness of this system, society should absorb the higher risk introduced to the life insurance fund by individuals not providing genetic test results to insurers. Arguments for insurers using genetic test results (volunteered by the extremely varied case and extreme case) revolve around the belief that insurers have the right to know information that may impact on their life insurance fund. In addition, it was mentioned that by providing insurers with this information, everyone is provided with a fairer assessment of their risk and a fairer premium is charged to all (see Section 6.10). This response indicates that what is fair and acceptable is risk-defined equity. The above findings illustrate the diversity among individuals as to what is considered fair and unfair discrimination.

It is interesting to note that although, there were individuals that agreed that insurers should be allowed to use genetic test results, there was still an appeal for an overriding authority to manage this situation (see Section 6.11). This perhaps demonstrates a lack of trust in the insurance industry to manage the risks related to life insurance companies' use of genetic test results, perhaps indirectly linking risk acceptance with trust evaluations. The ability to control the decisions made by the life insurance industry may also be a factor that influences a negative view of insurers and risk acceptance. Only 19.2 per cent of those individuals surveyed agreed that they had some say on the content of risk management decisions and an input into how these decisions are made (see Appendix E.2, Table E-3 and Table E-4). This finding is complemented by the fact that only three interviewees felt that they had the means to hold the insurance industry accountable for their decisions "to some degree", while four interviewees stated the opposite (see Section 6.11). According to the findings from the survey 7 per cent more of the sample population felt that they did not have the means to hold the Irish Government accountable for their decisions than those who were of the opinion that they did. This statistical difference is rather small when one examines attitudes regarding accountability for decision-making and the insurance industry (see Appendix E.2, Table E-7). Approximately half of the population sampled (55.6 per cent) felt that they did not have the means to hold the insurance industry accountable for its decisions regarding life insurance companies' use of genetic test results, a considerable 24.5 per cent more than those who did feel that they had some control in this matter (see Appendix E.2, Table E-8). This lack of control over risk management decisions made and the lack of accountability of the Government and the insurance industry for any decisions that are made may feed into whether or not people are willing to accept the risk management strategies decided upon by both or either parties.

Reasons provided for difficulties regarding having the means to hold the insurance industry accountable included the lack of a voting system (unlike the Government) and the large financial backing of insurance companies (see Section 6.11). With regards to the Government, five people stated that they felt that they could hold the Government accountable due to the democratic system associated with the Irish Government, whereby the Irish public can vote for a political party which best represents their values. However, this means of decision-making control was considered "far removed", in that one may not always agree with all of the decisions made by an elected party and there may be little action that one can take to reverse such decisions (see Section 6.10). Two individuals stated that they did not have the means to hold the Government accountable for its decisions (see Section 6.10). These findings demonstrate that those individuals who believe that they have some power to control the Government's decisions regarding this matter are more accepting of insurers using genetic test results than those who feel that they have no power. Given this finding, it appears that there may be some relationship between perceived ability to control risk management decisions and risk acceptance.

Finally, it may be likely that an individuals' medical background may influence their perceptions of the risks related to insurers using genetic test results and their acceptance of such risks (see Section 6.10). Findings from the case study analysis indicate that when an individual is prone to genetic difficulties they are more likely to perceive more risks related to the use of such information and to not accept insurers using genetic test results (see Section 6.10). This finding suggests that proximity to the risks may be a factor that influences risk perceptions and risk acceptance.

7.2.2.7 Gender

Both gender and economic status are considered to be significantly related to risk and trust judgements (Satterfield *et al.,* 2004; Slovic, 2000; Siegrist, 2000; Gustafson, 1998; Davidson and Freudenburg, 1997; Barke *et al.* 1997; Slovic *et al.,* 1997; Gutteling and Weigman, 1993) (see Section 37.2). In particular, previous research has indicated that women and economically disadvantaged persons have a higher risk perception than men and economically advantaged individuals. Explanations as to the reasoning behind such differences tend toward a focus on issues such as differences in technical knowledge, the influence of social and biological 'roles', and power (see Section 3.7.2).

Upon reorganisation of the cases subsequent to interview analysis, there were only three female interviewees, all of who satisfied the characteristics of the Representative case. All three females perceived high risks and either some or no benefits regarding insurers' use of genetic test results. In addition, they did not accept the risks related to such usage, in that they did not agree that insurers should be allowed to use genetic test results in their underwriting processes. There were no women that characterised an alternate risk acceptance pattern, and the only individuals to view few risks and accept the risks related to insurers using genetic test results were male. Therefore, based upon the case study findings, it is not possible to refute the claim that women perceive higher risks and are less likely to accept risks than their male counterpart.

The statistical findings from the survey also illustrate that the female population are more likely to perceive risks related to insurers using genetic test results than their male counterpart, although it has to be noted that the difference in risk perception and gender was minimal (see Section 5.6.1). Risk acceptance was only slightly higher for males than females, with the statistical findings suggesting relatively little difference in risk acceptance between the two sexes. Both the results of the survey and the interview analysis indicate that overall trust in the Government and the insurance industry does not differ greatly between genders. Findings from the in-depth interviews demonstrate that individuals have some knowledge of genetic testing, but previous to the survey, knowledge of insurers' use of genetic test results was unlikely. It is also interesting to note that differences in knowledge regarding insurers using genetic test results in their underwriting practices does not appear to be influenced by gender (see Appendix E-2, Table E-1). Instead, the amount of knowledge regarding this matter appears to be influenced by occupation (see Section 6.3). Results from the survey also indicate that knowledge does not appear to differ between genders (see Appendix E-2, Table E-1). These findings also go against previous findings that determine that differences in technical knowledge between gender influence risk judgements (Slovic, 2000).

Given the findings from the in-depth interviews and the surveys it is possible to conclude that gender may be a factor that influences risk perception and risk acceptance, however, social trust and knowledge may not be a reason for gender difference. Therefore, there may be some other factor not reflected in the present study that exerts an influence upon risk perceptions and acceptance judgements. Perhaps these factors include intangible power differentials or are related to one's socio-biological role (Bord and O'Connor, 1997:832 and Finucane *et al.,* 2000:19-172). It would be interesting for future studies to explore the reasoning behind the influence that gender appears to play upon risk perceptions. However, it should be noted from the present study, that due to the relatively small sample size and, in the case of the survey, a disproportionate number of males to females, it may be that if the study is repeated with a different sample, gender may affect these factors differently. Or indeed, future studies may demonstrate that gender does not influence trust, risk perceptions and/or risk acceptance.

7.2.2.8 Economic Status

Turning to economic status as an explanation for risk perceptions, the extreme case supports Siegrist's (2000) theory, whereby the interviewee viewed no risks, some benefits, trusted the Government and the insurance industry, and accepted the risks related to insurers using genetic test results. The case comprised of an individual who wields a high amount of power and status in their job and has the highest economic position occupied by all ten interviewees. Such an observation appears to satisfy Finucane *et al.'s* (2000:19-172) theory that those with more power and greatest socio-economic advantage are less likely to see the world as dangerous as others. This is as a result of power differentials dependent upon status and authority.

With regard to the Representative Case, all of the individuals interviewed worked in positions that were generally lower paid than the Extreme Case. The only two individuals interviewed without a third level degree were part of the representative case. One of the two individuals was a Special Needs Assistant (Subject 2), while the other was a self-employed artist (Subject 5). The remaining interviewees who were characteristic of the Representative Case had third level qualifications and were working in jobs related to their degrees. The nature of the extreme case in comparison with the representative case appears to validate Slovic's (2000) conviction that economically advantaged individuals view less risk in society because they create, manage, control, and benefit more from many of the modern technologies and activities. It is likely that the interviewee representative of the extreme case may feel that he has more influence over the use of such technologies and decision-making processes regarding the management of related risks. Hence, this individual appears to have higher trust in those managing the risks. Therefore, according to these results from the qualitative analysis, the influence of status and power appears to influence the risk acceptance pattern related to insurers using genetic test results. It affects perceptions of the risks and benefits related to insurers using genetic test results and trust judgements in the insurance industry and the Government and, in doing so, perhaps, indirectly influences risk acceptance.

The two individuals representative of the extremely varied case had general trust in the Government but did not have general trust in the insurance industry. The student, although he perceived no risks related to insurers using genetic test results, did not accept insurers using genetic test results. The second individual was a college graduate of five years who works as an electronic engineer. He perceived some risks and no benefits for society from insurers using genetic test results. Yet, he accepted the risks from such a practice. It is questionable why this is the case, given the representative risk acceptance pattern, whereby risk perceptions do not seem to influence risk acceptance. The perception of risks and benefits in both such cases does not appear to have a direct effect upon risk acceptance. Perhaps it is the case that economic status influences risk acceptance in a more direct manner than risk perception as discussed above. Both individuals had general trust in the Government, no sceptical trust in the Government or the insurance industry, but did not, or were not, sure if they had general trust in the insurance industry. The extremely varied case may be indicative of a more direct influence of trust, status and power upon the risk acceptance pattern than as discussed above.

In support of such a finding, judging from the survey results, economic status had less influence upon risk perception, benefit perception and trust that it had upon risk acceptance (see Section 5.6.3). Initial results from the survey suggested that those individuals who were most vulnerable with regard to being able to afford life insurance (unemployed) were those most likely to disagree with genetic stipulations on access to insurance products (see Section 5.6.3). And those who were least likely to apply for life insurance, for example, respondents who had retired, were the group with the highest declaration of agreement with insurers using genetic test results. However, correlation analysis revealed that the strength of the relationship between economic status and life insurance ownership and intention to purchase life insurance was weak (see Section 5.6.4). Tests for the magnitude of the relationship between occupational status and knowledge, trust, risk perceptions, benefit perceptions and risk acceptance demonstrated that it was unlikely that a relationship exists between these factors (see Section 5.6.3).

Despite several findings to the opposite of Siegrist's (2000) theory that the position of one's economic status influences one's judgements of risk, there appears to be more evidence between the in-depth interviews and the surveys to suggest that those with more power and greatest socio-economic advantage are less likely to see the world as dangerous as others. In addition, there appears to be an indication that economic status influences risk acceptance to a greater degree than it impacts on social trust, risk perceptions and benefit perceptions. Perhaps this indicates that an individual who may be considered relatively powerful, based upon their status and authority, bypasses decisions regarding trust, risk and benefit and accepts the decisions of whom they consider as other authoritative individuals. This relationship between economic status and risk acceptance may exist due to a belief that fellow authoritative individuals will act in a manner similar or in a manner deemed appropriate by individuals with an equally high and powerful status. It may be interesting for future studies to explore this theory in greater detail, as the findings might shed more light on the reasoning behind risk acceptance attitudes.

7.2.2.9 Age

Based upon the qualitative responses of interviewees', age does not appear to influence knowledge, trust judgements, risk perceptions, benefit perceptions or risk acceptance (see Table E-2, Appendix E-1). Examining whether or not age influences individuals' allocation of resources to alleviate a future problem, five out of the ten individuals interviewed mentioned that they would pay extra on top of their premium to ensure that individuals did not have to reveal the results of genetic tests to life insurance companies (see Appendix D.1, Table D-20, Subject 1, 6, 7, 9 and 10). Although the sample frame was small, it appears that the youngest member of the sample (Subject 6, aged 18-25) was the most willing to pay a "substantial amount" of 15-20 per cent extra on top of an existing premium to ensure that the perceived risks related to insurers using genetic test results do not occur. The eldest interviewee (Subject 10, aged 56-65) was willing to pay 10 per cent extra on top of their premium, but would not consider going higher than this amount. The other three individuals, aged between 26 and 35, were willing to pay 5-10 per cent in addition to their premium to create an intervention against future potential threats to applicants' ability to purchase affordable life insurance products.

Findings from the survey demonstrate that the age group most likely to be in the purchasing bracket for life insurance in Ireland (26-56) were those who most strongly disagreed with insurers using genetic test results, perceived the most risks, and the least benefits (see Section 5.6.2). While those who were least likely to apply for life insurance, respondents that had raised their families and paid off their mortgages (over 46 years old) or, were quite young (18-25), were the group with the highest declaration of agreement with insurers using genetic test results. However, correlation analysis revealed that the strength of the relationship between age and life insurance ownership and intention to purchase life insurance was so weak that it may be considered as non-existent. In addition, further analysis demonstrated age does not influence self-assessed knowledge, trust evaluations, risk perceptions, benefit perceptions or risk acceptance.

The findings from the present study demonstrate a contradiction to claims by Siegrist *et al.* (2005) and Dosman *et al.* (2001) that age is positively correlated with risk perception. In addition, age does not appear to influence the extent to which individuals are willing to allocate present resources to alleviate a future problem, despite a report stating the opposite (Nielsen *et al.* 2003:39). Therefore, it is not possible to deduce an association between these items from the findings of the present study.

7.2.2.10 Life Insurance Ownership

The concept that some individuals will be excluded from life insurance ownership depending upon the results of a genetic test proved distasteful for many interviewees (see Section 6.4). Only two individuals in the representative case currently owned a life insurance product (see Appendix E.1, Table E-2, Subjects 1 and 10). Four interviewees that made up the representative case intended purchasing life insurance in the future, while one did not. The extreme case did not have life insurance, nor did he intend to purchase a life insurance product in the future. It may be the perceived need for insurance that affects the risk acceptance pattern of the Representative Group. The necessity factor is further emphasised by the extremely varied case, whereby judgements of risks and benefits did not seem to influence risk acceptance. However, the individual that did not accept the risks related to insurers using genetic test results, in other words did not wish for insurers to be allowed to use genetic test results, intended to purchase life insurance in the future. In this case, the need for life insurance may not affect risk perceptions to any great extent, but it may have a direct influence upon agreement for insurers to use genetic test results.[3]

Initial responses from the survey indicated that it is those individuals in the life insurance purchasing category that are most opposed to insurers using genetic test results, whilst it is those respondents who already have life insurance or, do not intend purchasing it, that feel the least opposed to insurers using genetic test results (see Section 5.6.4). However, further statistical analyses conducted in Chapter 5 suggested that life insurance ownership and future intention to purchase life insurance play a minimal role in determining evaluations regarding knowledge, trust, risk perception, benefit perception and risk acceptance (see Section 5.6.4). Judging from

[3] This possibility may be slightly refuted by the other interviewee who makes up the extremely varied group, who perceives some risks from insurers using genetic test results, but appears satisfied to allow insurers to use genetic test results in their underwriting practices and, yet still, would like to purchase life insurance in the future.

the responses to the in-depth interviews and surveys, ownership of life insurance and the perceived need for life insurance appears to influence risk and risk acceptance evaluations. There is a thin boundary between ownership of life insurance and perceived need for life insurance, and this may account for the influence that these factors have upon risk assessments.

7.2.2.11 Inverse Relationship between Risk Perception and Benefit Perception

The inverse relationship between perceived risks and perceived benefits found in previous studies (Siegrist & Cvetkovich, 2000; Alhakami & Slovic, 1994; Frewer *et al.,* 1998; Gregory & Mendelsohn, 1993) was replicated in the present study. Findings from the interview and survey analyses revealed predominantly inverse relationships between risk perceptions and benefit perceptions, with more inverse relationships identified during interview (see Section 5.3.3 and Section 6.6). Individuals' who perceived the use of genetic test results by life insurance companies as risky, ascribed less benefits than people who viewed such usage as not risky. According to Frewer *et al.,* (1998), perceived risks can be reduced by emphasising the benefits associated with a technology. However, results of the survey suggest that this may not be the case, given that a partial correlation controlling for the general trust dimension and the sceptical trust dimension in the Government and the insurance industry reduced the association between perceived-risk means and perceived-benefit means (see Section 5.3.3). These findings are in line with other studies in which the trust on gene technology was investigated (Siegrist and Cvetkovich, 2000; Siegrist, 2000 and Siegrist, 1999). This finding further emphasises the interrelationship between trust, risk perception and benefit perception.

Section 7.2.2 has explored the main findings from the present study that can be related to the 'self' layer of social reality. It was demonstrated that findings that can be associated with this layer of social reality appear to demonstrate a relationship between social trust, risk perceptions, benefit perceptions and risk acceptance with regard to insurers using genetic test results. The following section intends to further develop an understanding of this association by examining the findings from the study that appear to exist within the 'situated activity' layer of reality.

7.2.3 Situated Activity

'Situated activity' concentrates on the interaction between the public, the insurance industry and the government. It centres on the dynamics of face-to-face interaction (Layder, 1993:80). Rather than focusing on the individual, as in the 'self' dimension, the 'situated activity' is concerned with the activities of several individuals undertaking a common endeavour. The situated activity, in this research, occurs at the point where individuals and representatives of life insurance companies interact in order to process an application for life insurance. The way in which individuals perceive insurers and life insurance products, and the perceptions of life insurance companies of applicants, influences the relationship and interaction between the two parties. In this way the self level is not only influenced by the situated activity, but it also has some bearing upon relations at the 'situated activity' dimension of social reality. The following paragraphs shall examine some aspects of the 'situated activity' in order to provide an understanding of its effect upon the 'setting' and 'self' dimensions, thereby, influencing the risk acceptance decisions of individuals.

When an individual approaches a life insurance company for insurance, they are usually told of the products provided by the company and the suitability of these products to the individual's needs is discussed. This involves the applicant revealing information regarding his personal situation and his perceived needs for life insurance to cater for a particular scenario. Therefore, the issue of privacy arises upon initial interaction between the applicant and the life insurance company representative. When a decision is made regarding which product(s) is of interest to the applicant, an application form for the particular product(s) is completed, which requires further divulgence of the applicant's personal information. This information, which is considered private and protected by privacy laws, is allowed for usage by the underwriter to process the application and to determine the selection and classification of the candidate for insurance. This leaves the individual in a rather exposed situation, relying upon the discretion and appropriate underwriting of the insurer.

On the other hand, the underwriter is vulnerable with respect to the candour and honesty of the applicant. Sections 6.9, 7.2.2.2, 7.2.2.3, 7.2.2.5 describe how, at the 'self' dimension, individuals take a rather negative view of life insurance companies. A great concern in Ireland, with regard to negative attitudes towards the insurance industry, is an outlook that it is okay to fraud insurers (IIF Press Release, 2003). The problems encountered by insurers in relation to fraudulent claims were briefly examined in Section 2.2, and the similarity between the effect that fraud has on the life insurance industry and an inability for insurers to request the results of genetic tests was explored.

It can be observed from the above analysis of the situated activity, how this dimension can influence the self level, where it appears that both parties have to trust each other to achieve a satisfactory outcome. This outcome can be viewed in two ways. Firstly it may be defined as the purchase of a life insurance product, suited and appropriately priced for the applicant, where information provided by the applicant is private and secure. Secondly, the outcome may be described as all relevant information that may influence the selection and classification of the applicant being revealed to the insurer so that the policy is fittingly underwritten. Therefore, each party involved in the contract must rely on the integrity of the other party and/or manage the risks related to such outcomes.

It is important to note at this point how the introduction of genetic testing affects both parties. This new form of information influences the 'situated activity' in that it provides a new factor to determine the risk status of an applicant. The survey and interview results obtained as part of the present study indicate that the majority of individuals do not wish for life insurers to use genetic test results in their underwriting processes (see Sections 5.5.6, 6.10 and 7.2.2.6). In addition, the results from the survey illustrate that those individuals who intend to purchase insurance in the future are more likely to disagree that insurers should be allowed to use genetic test results compared to those who have no intention of doing so (see Section 5.6.4).[4] On the other hand, insurers require revelation of genetic test results from an applicant to calculate the insurability of an applicant and to decide upon the correct premium to charge. The influence that genetic test results have upon the 'situated activity' further emphasises the strain upon the trust relationship between both parties and the requirements for risk management to protect the interests of the applicant and the insurer. This situation, although occurring at the 'situated activity' level, cannot be explained solely at this dimension, as it is influenced by 'setting' and 'context', which will be described in the following sections (7.2.4 and 7.2.5).

---

[4] Although, there does not appear to be a correlation between intention to purchase life insurance and risk acceptance

7.2.4 Setting

The 'setting' is the dimension of reality where the situated activities occur. In this study, it is possible to observe the life insurance system, intertwined with the risk management system, at the 'setting' dimension of social reality. The life insurance industry has its own risk management system while, the Government manages the risks faced by applicants for life insurance. Governmental risk management strategies tend to override those of the insurance industry, where deemed necessary to protect society. Hence, anybody entering the 'setting' is confronted by a set of rules that they must abide by, although the rules that prevail in the 'setting' are not considered static and may change over time. For example, the factors involved in underwriting may change (e.g. introduction of genetic testing) resulting in the need for a change in risk management strategies from both the insurance industry and the Government.

The 'situated activity' explored several issues that interplay at this dimension and it is briefly discussed how these affect the 'setting' dimension. It is explored how the insurance industry is vulnerable to non-disclosure and misleading disclosure on the part of the applicant. This aspect of the 'situated activity' influences the life insurance industry to create a risk management system at the 'setting'. One of the basic risk management strategies that insurers enshrine is the principle of utmost good faith (lat.: uberrimae fidae). This principle describes how applications must be completed with absolute and perfect candour or openness and honesty, in the absence of any concealment or deception (see Section 2.2). Any breach of this duty by the applicant may entitle the insurer to repudiate liability to any subsequent claim. This principle, in effect, protects the insurer by managing the risks related to non-disclosure or misleading disclosure on the part of the applicant. Therefore, for the insurance industry, the outcome of the exchange of information and insurance products is, to a certain extent, protected.

However, recently, many lines of insurance business, including life insurance, are facing a decline in earnings, reserve deficiencies, rising loss costs and other insurance expenses, as well as pricing difficulties (Viaene and Dedene, 2004:313). Cost control is now claimed to be of vital importance for the industry's financial attractiveness and future viability. In particular, the issue of fraud control, where the system of utmost good faith breaks down, has gradually been gaining momentum as a means of keeping down insurance costs (see Section 2.2). Management of the risks related to fraud cannot be managed alone by the insurance industry. In the

Republic of Ireland the Government has also had a role to play in controlling insurance fraud.[5] However, life insurers' precautions against fraudulent claims may also affect the processing of individual claims. Claims may take longer to settle as insurers investigate any possible fraudulent activity on the part of the insured. This 'treatment with suspicion' of claims may lead to further reasoning for negative attitudes towards life insurance companies, affecting activity not only at the 'situated activity' dimensions, but also perceptions and attitudes at the 'self' layer of social reality. The relationship between trust in the insurance industry and fraudulent activity is an area that may be of interest for future study. In particular, it will be interesting to note if widespread use of genetic test results coupled with legislation denying insurers' access to such information shall have a similar effect upon the life insurance industry as fraudulent claims today (see Section 2.2). It is questionable in this scenario whether or not the Government will back the insurance industry, as it is doing so in the case of fraud, or if it will continue to back society, believing the risks for society to demand control above the risk facing the insurance industry.

It can be seen from the above discussion, that the insurance industry and the Irish Government have enacted risk management strategies to protect the outcome of the exchange between the insurance industry and the applicant, in the insurance industry's favour. With regard to the outcome of this interaction for the applicant, the Irish Government enacted privacy laws, namely the Data Protection Act, 1988 and 2003, which are designed to manage risks faced by the applicant from the insurer in relation to any information disclosure regarding the applicant's policy details and personal information to a third party (see Section 2.3.5). However, the vulnerability of the situation for the applicant is not entirely overcome. This is due to the lack of transparency of the insurer's underwriting system. Given that the applicant is not provided with details regarding the calculations made by the underwriter regarding their selection and classification methods and calculations, the applicant remains unsure as to the appropriateness of his premium or the insurers' decision regarding their insurability. Since different life insurance companies charge a variety of prices for similar products, it is expected that some doubt should remain on the part of the applicant (IFSRA:2004). This doubt in the

[5] In 2003, the Tánaiste, Mary Hearney, launched the insurance industry's anti-fraud awareness campaign. This initiative was followed, in 2004, by the IIF and the Garda Síochána launching guidelines for the reporting of suspected fraudulent insurance claims.

relationship between the applicant and the insurer may affect perceptions of the life insurance industry at the self level. Indeed, the interviews conducted as part of the present study, demonstrate that a lack of transparency affects trust in the insurance industry and perceptions of the risks related to allowing insurers access to genetic test results. This illustrates the interconnections that exist between the 'setting', 'situated activity' and the 'self' dimensions of social reality.

With the advent of reliable and accurate genetic testing, the risk management strategies, designed to protect the insurer from information asymmetry and the applicant from violations of privacy, are under question. These principles and regulations do not appear to protect both parties form the ethical dilemma introduced by genetic testing. There may even be a requirement for the rules, which have remained static in this 'setting' for several decades, to change. The risk management of genetic test results by the life insurance industry and the Government is examined in Section 2.5. Attitudes towards risk management of this issue by the insurance industry and the Irish Government are rather negative and culminate in a lack of trust in the two authorities (see Sections 5.5.3, 5.5.4, 6.8, 6.9, 7.2.2.4 and 7.2.2.5). Such negative attitudes towards the insurance industry and the Government may impinge upon the effectiveness of their management of this risk situation, as distrusted authorities have to continually explain and justify their actions, which can diminish their ability to effectively manage risk situations (Kramer, 1999:585). This inability to react accordingly to negative situations may deepen feelings of low trust and suspicion in authorities, producing a somewhat cyclical effect between self, 'situated activity' and 'setting'. Without the trust of the public behind the authority(s) governing such risks, the authority's power to manage this situation decreases. In order to develop successful risk management strategies to adequately control the perceived risks related to insurers using genetic test results, it is necessary for the insurance industry and the Government to cultivate the trust of the public. This, once again, demonstrates how the 'setting' is interwoven, influences and supports the other layers of social reality.

However, the influence that a lack of trust has on the 'situated activity' and 'setting' is not all undesirable; indeed, there are some positive connotations for the 'setting' in this case. Although a lack of trust in the insurance industry and the Government can be portrayed in largely negative terms, it is important to note that a certain amount of weariness and suspicion may constitute appropriate and even highly adaptive stances toward institutions (Kramer, 1999:590). Vigilance and wariness about institutions, some have argued, constitute essential components of adaptable and healthy organisations and societies (Barber 1983, March & Olsen 1994, Shapiro 1987). A lack of trust in, and between, the insurance industry, the Government and members of society, ensure that each constituent is kept in check for misdemeanours that affect the overall welfare of that society. From this perspective, a lack of trust may constitute an important form of social well-being. Therefore, it can be said that a low level of trust may be an integral aspect of a healthy society. Indeed, this situation of a lack of trust may provide a resolution for the risks perceived regarding insurers using genetic test results. For without such attitudes, it may be the case that these risks will materialise due to a lack of a call for risk management. And the severity of the consequences may be irreversible. From the above discussion, it is possible to note how the 'setting' is sustained by the activities occurring at the 'situated activity' and self levels and vice versa. The following section shall demonstrate how such activities occur within the overall 'context' (Layder, 1993:89).

7.2.5 Context

The 'context' lies at the outer edges of the web and directs attention to macro social forms, such as the national and dominant culture. The dominant culture influences the attitudes and evaluations of individuals and shapes their perceptions of risk and risk acceptance decisions. This dimension encompasses all other levels of social reality and effects and is influenced by each dimension. The dominant culture that exists in the Republic of Ireland may influence individuals' risk acceptance decisions at the 'self' dimension, and in doing so, may affect the 'situated activity' and the 'setting', as the issue regarding insurers using genetic test results progresses. However, given the limitations and criticisms of previous work on cultural theory, it appears to be impossible to test for the national culture of Ireland that distinguishes its society's perceptions of risk and trust attitudes from other countries (see Section 3.7.1), although it is not denied that culture appears to influence risk perceptions to some extent. However, it is beyond the scope of the present study to define what this culture is and therefore it is not possible to explore this dimension and its effect upon risk perceptions and trust attitudes.

7.2.6 Conclusions on the Implications of the findings for Risk Research

The findings from the present study illustrate the uncertainties surrounding the risks and the management of insurers' use of genetic test results. The risks described by participants of the present study are related to the privacy of genetic test results, the denial of insurance application, the charging of high or unaffordable premiums, unfair discrimination and the selection of only low risk individuals. On the other side of the equation lie the risks faced by the insurance industry which mostly pertain to the risk of adverse selection. Each of these risks is innate to the industrialised and self-advancing existence of modern society. However, unlike the risks described by Beck (1992), which tend to be associated with environmental concerns, the present study emphasises how these new risks are related to developments in technology and how they affect society on a global level. There is an air of uncertainty surrounding the risks facing society and the insurance industry as a result of developments in the science of genetics, a field of knowledge that aimed to make life more predictable. The pace and growth of genetic testing and the extent of its uses are unknown, a factor that contributes to our uncertainty regarding its consequences. Given the unpredictable nature of genetic technology, it appears that the world, rather than being within our control, an aim of science and technology, has developed into an erratic and dislocated world. Giddens (2000) terms this overview of modern society as a 'runaway world', where the only thing we can be sure of is that we cannot be certain (Beck *et al.*, 1994:59). The present study appears to corroborate this theoretical view of modern society and to push out the confines of Beck's (1992) risk society from a German and environmental context. The risk society can be extended to include first and second world industrial countries, where the consequences of recent technological innovations are being seen increasingly as potentially harmful and as having a serious negative impact upon society. As originally hypothesised by Beck (1992; 1999), the risk society is evolving beyond its beginnings and in line with industrial, technological and social developments.

The analysis of the difficulties surrounding the management of risks related to insurers using genetic test results, illustrates the problems and complications associated with controlling risks in a 'runaway world' (see Sections 2.5 and 2.6). According to Luhmann (1979), we cannot be in control of everything, nor is it always possible to make informed choices. However, trust can act as a mechanism to reduce such problematic social complexity. By trusting others, institutions or systems, trust acts as a positive mechanism that helps individuals simplify their lives (Luhmann, 1979). According to Beck (1994), living in a risk society generates worries and distrust. However, in doing so, individuals are stimulated to be more active, reflexive and critical (Beck 1994). Luhmann's (1979) and Beck's (1994) theories of the role of trust in modern society differ, in that Luhmann (1979) focuses on the necessity of reducing complexity by individuals choosing to trust or distrust, while Beck (1994) focuses on the necessity of reducing risk through reflexivity which is activated by distrust. Giddens (1990, 1991) explains that individuals avoid the paralysing contemplation of risks by adopting a generalised trust in expertise. Although, the theoretical reasoning behind the role of trust in late modernity differs between Luhmann (1979), Beck (1992 and 1994) and Giddens (1990 and 1991), all three theorists indicate that trust, in some format, is thought to reduce or minimise the dangers to which particular types of activity are subject.

Section 3.1 describes how there exists little research examining the dynamics of the relationship between incomplete knowledge, social trust, risk perception, benefit perception and risk acceptance apart from some theoretical work by Giddens (1990) and Earle and Cvetkovich (1995) and some empirical work by Siegrist (2000), Siegrist and Cvetkovich (2000) and Poortinga and Pidgeon (2005). Instead, the focus of academic endeavours has centred upon the relationship between just two or three of these factors across a broad range of issues and subjects. Siegrist (2000) and Siegrist and Cvetkovich (2000) used structural equation modelling procedures to test the plausibility of the postulated causal model (see Figure 3-2). Assessment of the model fit indicated a well-fitted model. The present study adopted this model, and examined the associations between incomplete knowledge, social trust, risk perceptions and benefit perceptions. It was revealed that within the subject and societal boundaries of the present study, there is an influential relationship between these factors, apart from knowledge. This interrelationship appears to be influenced and enforced by a variety of factors and conditions particularly related to the social reality surrounding this study (see Section 7.2). The findings from the present study, therefore, support the above theorists' view of the nature of this relationship to some extent, however, it is recognised that it is the factors and conditions present within the social reality surrounding this relationship, and the nature of the risks examined, that demarcate this association, rather than a universal causal theory relating to trust and risk evaluations. To apply psychometric results across all cultural environments could prove invalid, as each culture may have unique perceptions on the risk of insurers using genetic test results.

Findings from the survey and semi-structured interviews indicate that general trust in the Government and the insurance industry is low, while sceptical trust is high, with participants' trusting the insurance industry less than the Government. Given such low trust in the ability of both authorities to manage this situation, it is not surprising that risk perception is high, if understood from the perspective of the above risk theorists. The majority of respondents do not accept the risks related to insurers using genetic test results and would prefer if life insurance companies were not allowed to use genetic test results to decide upon the insurability of applicants and the premium to be charged. If risk theory to date is followed, this lack of risk acceptance is to be expected, and if social trust were present then perhaps this situation may not exist.

By loosely following Giddens' (1990, 1991) risk theory, it could be expected that if trust in the authorities managing this situation was promoted, risk perceptions would be reduced and the associated risks may become acceptable. The findings from the present study indicate that a relationship does exist between these social trust, risk perceptions, benefit perceptions and risk acceptance, whereby it was possible to identify a link between each of these elements and the factors and conditions that underlie their evaluations. Therefore, risk theory suggesting that trust and risk perceptions are related and that trust may lower perceptions of the risks related to insurers using genetic test results is not falsified by this study. Although, it should be noted that the findings illustrate that this relationship may not be direct and that there are multiple factors and conditions that influence the nature of this relationship.

It can be said, therefore, that analysis of the risks related to insurers using genetic test results extend Beck's (1992) risk society theory outside of the original German and environmental context to include risks emanating from the use of new technologies. In addition, the findings from this study suggest that the theorised relationship between trust and risk perception as defined by Giddens (1990 and 1991) may exist in the present study.

This study concludes that social perception of risk is multi-dimensional and based on several characteristics. Indeed, the findings from the present study confirm the psychometric theory that society judges risk not on facts, but by what they *think* and how they *feel* about the activity or technology (Slovic *et al.,* 2002). Psychometric research characterises risk into a number of discrete items (see Renn and Rothmann, 2000 for overview). The qualitative findings from this study correspond with several of the psychometric influences on risk perception, including familiarity with the risk source, voluntary acceptance of the risk, ability to personally control the degree of risk, undesired impact on future generations, impression of fair distribution of benefit and risk and trust in risk managers (see Section 7.2.2). In doing so, this study validates the several factors identified by Slovic (1992) as influencing risk perceptions.

This study also confirms that the major dimensions of trust identified by Poortinga and Pidgeon (2003) as being important in determining trust are indeed identifiable as defining trust judgements. It is concluded that Irish society's trust in the Government and the insurance industry to manage insurers' use of genetic test results was based upon several dimensions of trust, including competence, reliability, credibility, integrity (vested interest), care, fairness, and openness (see Sections 4.6.2.3, 6.8, 6.9, 7.2.2.4 and 7.2.2.5). Based on this conclusion, it is recommendable that future studies of trust should use the measurements of this combination of trust characteristics to distinguish a measurement of trust, given the difficulties in identifying one meaning of trust (see Section 3.4). Amalgamated, these observable and measurable dimensions serve to provide a particular view of overall trust, meaning that it is possible to measure social trust without applying a strict, universal definition of trust, as Siegrist (2000) and Siegrist and Cvetkovich (2000) attempted, despite the social construction of trust. The use of various trust characteristics to gain an understanding of trust sheds light on reasons as to why people trust or do not trust organisations and this proves useful when examining trust evaluations.

In addition, the quantitative analysis verified that trust can be divided into two components based upon the dimensions of trust examined, described by Poortinga and Pidgeon (2003) as general trust and sceptical trust (see Section 4.6.2.3). However, the qualitative analysis reveals signs of a further dimension of trust, possibly related to a negative view of the Government and the insurance industry in a broad sense, defined as 'being part of a power elite' (see Section 7.2.2.5). This dimension of trust is similar to the findings of Viklund (2002), who includes the item regarding an organisation being part of a power elite in society within the dimensions of trust (see Section 7.2.2.5). This characteristic of trust may be based upon an individual's beliefs, ideology and emotional reactions to the nature and power of industry and the Government. It appears that some interviewees assume that society has less power with regard to influencing Governmental decision-making than industry. In this regard, both the Government and industry are seen as possessing more power than society when it comes to making decisions that affect industry and society. It may be valuable for future research exploring social trust to examine and explore being part of a power elite using both quantitative and qualitative methods to obtain a fuller perspective of the role of this item in determining trust evaluations and its associations with general or sceptical trust.

Finally, findings from the study demonstrate the value of using both qualitative and quantitative methods in examining the nature of the relationship surrounding knowledge, social trust, risk perception, benefit perception and risk acceptance. A combination of methods forms alternate views and attacks research questions from different methodological perspectives, hence, the flaws of one method are often overcome by the strengths of the other, as was the intention of the present study. However, as discussed in Section 4.4.1, mixing methods in order to obtain a richer picture may lead to contradictions and confusions, particularly if it is hard to reconcile discrepant sources of data that suggest different conclusions. It is possible to note that this issue can be observed in the present study, whereby results from the qualitative and quantitative analysis appear to reveal different conclusions regarding the same set of objects. In the case of the present study, qualitative methods, in the form of in-depth interviews, provided a clearer view of the relationship between these factors, whereas the numerical boundaries of the quantitative analysis hid those conditions which serve to link social trust, risk perceptions, benefit perceptions and risk acceptance together.

The differences in findings between the quantitative and qualitative analysis suggest an anomaly between the two approaches. However, different methods show up different aspects of reality. Quantitative analysis is rather ‘surface based’, in that it does not allow the researcher to explore beyond the simple single phrase responses of participants to the survey. Therefore, analysis of these responses leads to conclusions that cannot possibly explain the underlying reasons for the responses. Qualitative analysis, on the other hand, allows the researcher to dig underneath the simple responses of the survey in order to explore the factors and conditions that lie behind such simple responses. One cannot say that findings observed by the two methods are inconsistent, merely the methods are investigating separate dimensions of the reality surrounding this study. The social reality surrounding acceptance of the risks related to insurers using genetic test results is considerably more complex than quantitative data collection techniques are capable of identifying or demonstrating (Easterby-Smith *et al.*, 2002:41-42). Therefore, the quantitative analysis was only able to reveal that there is no direct association between incomplete knowledge, social trust, risk perception, benefit perception and risk acceptance. However, it was possible for the case study to go beyond these findings and to enter the more complex reality surrounding these factors. Analysis of the results of the qualitative analysis demonstrated that it is the factors and conditions that lie behind social trust, risk perception, benefit perception and risk acceptance evaluations that link each of these factors with each other. Therefore, it is advisable that future research in the risk field should adopt qualitative analysis as the leading methodology in which to identify the nature of the relationship between knowledge, social trust, risk perception, benefit perception and risk acceptance.

7.3 Implication for the Insurance Industry in Ireland

This study has demonstrated that the public's reactions to the risks perceived as relating to insurers using genetic test results can be attributed to sensitivity to technical, social, and psychological qualities. The identification and understanding of the factors and the role they play in influencing risk perceptions and risk acceptance is of great importance to the insurance industry and several implications for the future of the insurance industry in Ireland can be noted based upon the findings of the present study. If the findings of the entire study demonstrated that there is no relationship between trust, risk perceptions, benefit perceptions and risk acceptance, it would not be necessary for either the insurance industry or the Government to take the findings of this study into account in its future decisions regarding the management of the risks related to insurers using genetic test results. As disjointed variables, the findings merely demonstrate that there is low trust in the insurance industry, the risks perceived as relating to insurers using genetic test results are high, and people do not wish to accept these risks. Without a connection between these variables there appears to be no need to take action to improve trust or risk perceptions, as there is no advantage in doing so.

However, the findings conclude that there is an association between these factors and therefore the findings should be considered as important, and should not be ignored by either party. If trust levels in the Government and the insurance industry are low across society as a whole and a relationship exists between trust, risk perceptions and risk acceptance as defined by previous researchers, it is likely that the risks perceived as relating to insurers using genetic test results will not be accepted by society. Hence, there will be pressure placed upon the Government and the insurance industry to create policies forbidding usage of genetic test results in underwriting applicants. Given this relationship, if trust levels are promoted in these risk managers, it may be likely that such usage of genetic test results will be accepted by society. This may be a necessary contingent for the insurance industry given what can be considered rapid developments in genetic research and the constant stream of discoveries regarding genetic links to disease. If an increasing number of genetic test results become available in the Irish market, and the uptake of these tests becomes higher, it may be necessary for insurers to request the results of genetic tests in order to prevent the consequences of adverse selection.

If trust in the insurance industry is low, the risks related to insurance companies using genetic test results may not be accepted and there will be pressure applied to the insurance industry from the public to not use genetic test results. However, if trust in the Government is high, society may trust the Government to manage these risks appropriately and if the Government decides to allow insurers access to genetic test results, this policy may be accepted by society. However, if trust is low in the Government, the public will not accept these policies. This could lead to a situation of conflict, and perhaps even instability in the political and, indeed, in the insurance arena. If adverse selection does become a problem for insurers, it is likely that they will move out of an area where they are not making a profit. This creates a situation whereby the Government may be expected to provide such 'necessary' life insurance products to society, and, of course, this may not be financially feasible. Therefore one is left with a situation where nobody has access to life insurance and its benefits.

The findings from the interviews and the questionnaire suggest that the Government and the insurance industry should be working harder to build the trust of society. Perhaps, the insurance industry could improve the transparency of its underwriting procedures and calculations. For this concealment of information appears only to breed distrust among applicants for insurance (see Section 7.2.2.5). The insurance industry should also attempt to promote the benefits of providing genetic test results to insurers to deflate the risks perceived (see Section 7.2.2.11). The Government should encourage the trust of the population it is designed to protect through an emphasis of its risk management successes as it appears from the findings that previous risk management history influences individuals trust and, hence, risk acceptance judgements (see Section 7.2.2.4). Unfortunately, it has been demonstrated that failures tend to carry more weight than successes; however, this may be a problem that can be ameliorated with the publicised inflation of success rates (Siegrist and Cvetkovich, 2001 and White *et al.,* 2003). Given that the public would also like to be involved in the decision-making process regarding this situation, it is also advisable that in the promotion of trust, that a citizen panel be instigated that allows members of the public to voice their opinion on this matter and where open debate can take place on the risks and benefits of insurers using genetic test results (see Section 7.2.2.4). This feeling of involvement and openness may contribute to trust and hence acceptance of future risk management decisions made regarding this matter.

Participants involved in this study know relatively little about the use of genetic test results by the life insurance industry, yet they perceive risks as relatively high and benefits as low from such usage for society and, in particular, for their families (see Section 5.5). Should a small negative incident regarding this matter occur and get reported in the media, this attention could dramatically affect public perception of genetic testing, the life insurance industry and indeed the Government's ability to manage this situation. In addition, given the lack of trust in the Government and the insurance industry to manage this situation, people may misperceive the risks and be swayed by exaggerated claims of those opposing the use of genetic test results by insurers (see Section 5.5). The public perception of risk surrounding such usage, coupled with low trust of the risk managers, may increase or amplify public concerns about this technology, fuel further controversies regarding this issue, and reduce the possibility of managing the associated risks successfully (Kasperson *et al.,* 1992; Slovic, 1993).

It would appear from initial assessment of the Disability Act, 2005 that the regulations surrounding insurers obtaining or assessing the genetic test results of an applicant for life insurance are rather stringent. Therefore, it may be believed that the possibility of any negative situations occurring relating to insurers access to genetic test results is negligible given such tight controls. However, this approach to managing insurers' use of genetic test results has been criticised (see Section 2.5.3). Perhaps most notable, is the 'exceptionalistic' view of genetic test results, where genetics is considered unique among medical information because it reveals sensitive and detailed information about someone's future health risks. (Murray, 1997) (see Section 2.5.3).

This 'exceptionalistic' view of genetic test results is identifiable in the findings of the present study. As indicated by several respondents, genetic information can be regarded as extremely personal property (see Section 6.4), perhaps deriving from an opinion that a person's genetic code contains, and reveals, the essence of their individuality. Findings from the study also illustrate a belief that insurers should not be allowed to access genetic test results on the grounds that one's genetic status is beyond individual control (see Section 6.4). In connection with this conviction is a belief that people should not be negatively affected by insurers for a condition which they are not responsible for (see Section 6.4). These findings represent an 'exceptionalistic' view of genetics, an outlook that is mirrored by the Disability Act, 2005. The problems introduced by an 'exceptionalistic' view of genetic test results were discussed in Section 2.5.3, and shall not be reiterated in this paragraph. However, it suffices to state here that the Government by adopting the Disability Act, 2005 as an approach to manage the risks related to insurers using genetic test results may merely reintroduce the risks it is designed to prevent in another guise.

For example, recent research into behavioural genetics suggests that genetic factors may contribute to conditions such as alcoholism and nicotine addiction (Enoch and Goldman, 2001), both of which are considered as lifestyle choices by insurers and are penalised accordingly. However, given the genetic predisposition to these conditions, diseases caused by what has long been considered individual choice, may be genetically related and, hence, should be protected as such (Lemmens, 2000). At present, such genetic associations can be considered controversial and require further investigation, however, they serve to expound the difficulties encountered in distinguishing the relationship between genetic and other determinants of health. In addition, it serves to illustrate the problems that society may encounter should those whose health status is considered genetically related, receive greater protection against insurance discrimination than those who are unable to invoke a genetic predisposition for their health status. Therefore, in an effort to thwart risks such as unfair discrimination, by preventing insurers' access to genetic test results, the Government has simply opened the door to further discriminatory claims, which cannot be counteracted without dismissing the requirements of insurance practice.

Such issues bring into question the future effectiveness and appropriateness of the Disability Act, 2005. In addition, given the nature of the relationship between trust and risk perception, the Disability Act, 2005 may only serve to decrease trust in the Government's management capabilities to control genetically related risks and the insurance industry's fairness in underwriting procedures. In doing so, there may be an increase in the public's perceptions of the risks related to denying insurers rights to access genetic test results. Therefore, the Disability Act, 2005, given the association between trust, risk perceptions and risk acceptance, may not be a long-standing feature of the legislation surrounding insurers' use of genetic test results, and the means of managing the risks related to insurers using genetic test results may be undecided for some time to come.

Given the above risks relating to the Disability Act, 2005, and those described in Section 2.5.3, in addition with the association identified between trust, risk perceptions and risk acceptance, it may be wise for the Government to hold regular meetings regarding the progress and popularity of genetic testing, the effect of the Disability Act, 2005 upon insurers' earnings and changes regarding the nature of the risks related to insurers using genetic test results. However, rather than just involving experts from the Government and the insurance industry in such discussions, as has been the case so far, it would be of immense value to include members from society in these discussions also. Findings from the survey analysis indicate that 91.7 per cent of the population sampled agreed that the Irish public should participate in any decisions made regarding this matter (see Section 5.5). Qualitative analysis reveals that the majority of interviewees mentioned the need for public involvement in the decisions made regarding insurers' use of genetic test results (see Section 6.11.1). This study has also demonstrated that there are value-laden issues underlying the multiple dimensions of public perceptions of the risks related to insurers using genetic test results, trust in the risk managers and acceptance of the risks and risk management strategies. In order for risk management to be effective, these factors need to be considered in risk-policy decisions (Slovic, 1996). It is suggested that the Government should establish a committee that can articulate the concerns of the public regarding the Disability Act, 2005 and any changes to be made to the Act (see Appendix E.2, Table E-11). The creation of a forum for public debate should be the next step to attempt successful management of the perceived risks related to insurers using genetic test results.

Although, this study has contributed to our knowledge regarding trust and risk perceptions and has implications for the future of the insurance industry in the Republic of Ireland, the study did suffer from several limitations that became apparent during the progress of the research. These criticisms of the study are discussed in the following section.

7.4 Criticism

Upon completion of the quantitative and qualitative analyses, it was possible to identify several criticisms of the present study. Although these issues are not fully capable of remedy, it is important to note their presence nonetheless, so that future researchers in this area may benefit from a further understanding of the present study. One of the criticisms that may be considered least possible to resolve stands in relation to the questions administered as part of the quantitative analysis. Although four pre-tests of the questionnaire were administered to pilot groups to readjust and improve the questions and design of the survey, the questions were originally devised from a review of the relevant literature (see Section 4.6.4). However, in adopting this strategy to design the questions, it is possible to note a number of omissions that give rise to ambiguities in the analysis of responses from the survey. It is possible to view such omissions in light of the results from the qualitative analysis. Guba and Lincoln (1994:105-117) refer to this problem as 'context stripping', and indicate that this issue is an implicit problem associated with quantitative research. The survey is designed to focus on only a selection of variables, however, in doing so, it strips from consideration, through appropriate controls and randomisation, other variables that might, if allowed to exert their effects, greatly influence findings.

The qualitative analysis revealed the presence of a third component or dimension of trust, possibly related to perceptions of a tendency to corruption (see Section 7.2..2.5 and 7.2.6). Section 7.2.2.5 explained how this dimension of trust appears similar to the findings of Viklund (2002), who includes items regarding an organisation being part of a power elite in society and working primarily in its own interest as a factor that influences trust decisions. This aspect of trust was not regarded as a dimension of trust in the present study, however it appears as an element in trust evaluations during the qualitative analysis. By conducting the interviews following completion of the survey, it was not possible to explore this dimension of trust further. This omission of what might be a further dimension of trust from the questions that were asked as part of the survey can be interpreted somewhat as a loss of information that could have been sought had the interviews been completed prior to the surveys. By omitting this particular factor from the trust evaluation section in the survey, the findings from the survey regarding trust can be considered slightly ambiguous, whereby the concept of trust may not be defined as closely to the opinion of the population surveyed, as it ought to be.

From the findings of the interviews it was possible to note that the perception of risks appears to be acerbated by a lack of transparency regarding underwriting calculations, the nature of genetic tests, a perceived need for life insurance, concerns regarding an undesired impact on future generations, lack of control regarding predisposition to a genetic disease, fairness of the distribution of risks and benefits, familiarity with the risk, life cycle factors and involuntary exposure to the risk (see Section 7.2.2.2). While factors such as lack of underwriting transparency, financial effects, fairness and distributions of benefits can be observed to influence benefit perceptions (see Section 7.2.2.5). These findings are highly interesting and if it had been possible to question these factors and their relationship with risk, benefit, trust and risk acceptance as part of the survey, the findings produced may have considerably added to our knowledge of risk perception and its associations with these factors. However, as the in-depth interviews were not completed prior to the survey, these findings could not be tested.

In answering questions regarding causality, Section 4.2.3.1, notes that much depends on how the objects are conceptualised, for example, what is meant by risk and what is included in this understanding of risk (Sayer, 2000:17). It is important to carefully define the concepts of risk acceptance, risk, trust and knowledge in order to present an accurate portrayal of what is meant by each concept and how they relate to each other. The survey quantitatively measured each of these concepts as defined by previous literature and research in the area. However, findings from the qualitative analysis provide additional information pertaining to our understanding of these concepts and their relationship with each other. For example, Sections 3.3, 3.4, 3.5 and 3.6 demonstrate how several studies indicate that social trust is negatively correlated with perceived risk and positively correlated with perceived benefit, while risk acceptance is directly influenced by society's perceptions of risks and benefits whereby high risk perceptions and low benefit perceptions negatively correlate with risk acceptance. The conclusion assumed by such research is that trust indirectly influences risk acceptance in such a manner that where there is no trust between a population and a risk management authority, perceived risks are not considered acceptable. It is with this theory in mind that the survey was designed and conducted. However, the interview analysis, to a certain extent, presents a challenge to this particular theory. The findings from the interviews demonstrate that although some individuals may not trust the insurance industry and perceive greater risks than benefits as a consequence of insurers using genetic test results, they would still be willing to accept these risks given a perceived need to purchase life insurance for personal or employment reasons (see Section 7.2.5.3). Therefore, in some cases, although a lack of trust and a view of higher risks than benefits is presented as part of the equation, the need for life insurance acts as a driver for risk acceptance, overriding trust, risk and benefit perceptions as factors that influence risk acceptance. This finding indicates that trust may only be a necessary part of the risk perception/risk acceptance equation when there are no other reasons for accepting the risks. Had the interviews been conducted prior to the surveys in order to prepare the conceptual groundwork for the survey alongside the use of previous literature, it may have been possible to obtain a clearer picture of the relationship between trust and risk acceptance by examining the influence that other factors, such as the perceived need for life insurance, have upon this equation.

In-depth interviews conducted prior to the survey may have also aided in defining the concept of knowledge and instructing the way in which knowledge was measured as part of the survey. 'Knowledge' in the survey was measured using three items, each measuring a separate aspect of knowledge of genetic testing and its use by life insurance companies (see Section 4.6.2). The first question related to how much an individual knows about genetic testing while the second and third questions were associated with how much an individuals knows about insurers using genetic test results and the policies guiding insurers use of genetic test results. In addition to this section to account for an individual's knowledge, a further section of the survey related to whether or not the respondent believed that there are many risks or benefits for their family and for society from insurers using genetic test results. Although, this section was designed to illicit information regarding perception of risks and benefits, it also further tested the respondent's knowledge of risks and benefits and the use of genetic test results by life insurance companies.

The interview analysis demonstrated that although some people know very little about genetics and only a little more about actuarial practice in regard to this matter, they still believed that they knew the social costs and benefits of taking genetic testing into account in insuring lives (see Sections 6.3, 6.4 and 6.5). Therefore, the interview analysis demonstrated that the items designed to measure knowledge did not necessarily map onto each other and are perhaps not the best indicators of knowledge as adopted by the survey. This finding further promotes the criticism that in-depth interviews, such as those conducted as part of the qualitative analysis, should have been completed before the survey to improve the accuracy of the basic concepts measured in the survey. This finding also indicates that the survey results may be rather ambiguous given the omission of interviews as part of the conceptual groundwork prior to the survey.

Further ambiguity as a result of not completing in-depth interviews prior to the quantitative analysis is demonstrated when examining the perceptions of risk as defined by interviewees. It is discussed in some detail in Section 2.4 how 'risk' does not have transparent meaning and how a single definition of 'risk' does not exist. Instead, risk is defined differently depending on one's background and personal views. The present study demonstrated how risk perceptions of the public differ from the insurance industry's perceptions of risk relating to their use of genetic test results (see Sections 2.2 and 2.3). The insurance view of risk is tied up between a view of 'risk' and 'fairness' whereby the scientific principle of risk-defined equity ensures that the premium paid by each insured life is adequate to cover the risk that an individual introduces to the life insurance fund. The term 'equity' implies that the system is one that is based upon fairness, whereby policyholders with the same or similar expected risk of loss are charged the same. The risks for the insurance industry in the context of this study present themselves in the form of the principle of risk-defined equity being broken and insurers have to supply life insurance products to individuals at a premium lower than that required. The interview analysis demonstrates that the public have an alternative, yet equally valid, view of 'risk' and 'fairness' which are also bound together. Issues of blame enter into the definition of risk and fairness by the public. Section 6.4 outlines how some interviewees believe that given the blameless nature of genetic disease, it would be 'unfair' of insurers to make individuals exposed to genetic disease pay a higher premium than those who are not exposed. The view that no one is to blame for having an increased likelihood of suffering from a genetic disease influences interviewees to believe that insurers should not be allowed to use this information. The insurance industry's view of 'fairness' differs, yet the consequence of a violation of fairness as interpreted by each group is deemed as 'risky' and unacceptable. It can be seen from the above discussion that although the same words are used to describe the good and bad of the same situation, the two groups are at cross-purposes. Both groups are speaking the same language, yet their definition or interpretation of the words spoken is distinctly different. It may have been useful to develop an understanding of risk and fairness from in-depth interviews as part of the groundwork behind developing the concepts questioned as part of the survey. Had this been done, a more thorough quantitative study could have been completed whereby, for example, the relationship between risk and fairness could have been tested and the issue of blame expanded within the same context.

The nature of the majority of the above criticisms might well have been avoided with improvements within the conceptual groundwork for the survey. In-depth interviews conducted prior to the survey may have decreased the ambiguity of the findings from the survey by further developing the concepts beyond what was sourced from previous research in the area, thus making the questions asked more suitable to obtain the information required by the present study and reducing the number of omissions. In doing so, it may have proved more likely that the results from the survey reflected respondents' opinions and perceptions as accurately as possible.

How the concepts are indicated and measured in the survey generates another criticism of the present study. For example, the analysis of the questionnaire depends highly upon respondents' self-assessments of knowledge in order to quantify individuals' knowledge of insurers using genetic test results. However, it is important to note that perception of knowledge about insurers using genetic test results may not be related to actual knowledge. Indeed, research in cognitive psychology suggests that a discrepancy between objective and perceived knowledge is common (Rock *et al.*, 2005; Park *et al.*, 1998; Radecki and Jaccard, 1995; Glenberg *et al.*, 1982; Gentner and Collins, 1981). By adopting self-assessed perceived knowledge as a means to identify and indicate knowledge, the present study does not apply an objective means of measuring individuals' state of knowledge. In contrast to assessing actual knowledge, the respondent judges their own state of knowledge according to their own interpretation of what knowledge is and how much they believe themselves to know about a certain subject. Given the multiplicity of our understanding of knowledge, what one individual considers as knowing a lot, may be what another person believes to be as knowing very little. This is unlike a situation whereby the researcher objectively assesses the respondent's knowledge to specific questions to determine an individual's actual knowledge. If actual knowledge were being assessed, knowledge would be defined as "correct" answers as judged by the researcher to factual questions. Therefore, it is possible to conclude that the responses provided to the survey indicating knowledge should not be translated as actual knowledge. Although perceived knowledge is not the same as how much people actually know, though what they think they know may influence their behaviour considerably.

In addition to the above criticism regarding the means of measuring knowledge in the survey, it is important to reflect upon the object(s) of knowledge being measured. The first section of the survey examined perceived knowledge of genetic testing and of insurers using genetic test results, while the second section investigated individuals' perceived knowledge of the risks and benefits of insurers using genetic test results for society and for the families. Perceived knowledge may differ between objects of knowledge, for example an individual may know a lot about genetic testing but may feel that they know very little about insurers using genetic test results. Indeed, statistical analysis from the survey suggested that the three knowledge items were not measuring the same object, although two appear to do so (see Section 4.6.2.1). Within the present study, these two aspects of knowledge measuring perceived knowledge of genetic test results by insurers were often conflated to provide an overall view of knowledge. This can be criticised given the variation of objects of knowledge tested, in that they were measuring separate and distinct objects and should not have been conflated to provide one overall view of knowledge. However, the statistical analysis demonstrates that given the responses to the survey, it appears to have been possible to merge both items and to interpret both as measuring a similar aspect of knowledge.

The second section of the survey examined perceived knowledge of the risks and benefits associated with insurers using genetic test results were divided by sections and pages within the questionnaire itself and they were analysed separately to the first section. The risks/benefits for society and for family were also separately measured and statistically explored as it was considered important to take into account the framing of the risk in relation to perceived personal or general risk (see Section 4.6.2).

However, to obtain an understanding of overall risk perception and overall benefit perception the associated responses (i.e. for society and for family) were sometimes merged during statistical analysis and this may be criticised from the above point of view (i.e. separate objects of knowledge). However, the statistical analysis demonstrated that both sets of perceptions appeared to be measuring the same object, thereby allowing for such conflation at times (see Section 4.6.2.2).

Other criticisms of the present study relate to the nature of the tests used to analyse the data from the questionnaire. Parametric tests were used although several assumptions underlying them were marginally violated (see Section 4.6.4). However, given the assumptions it could be argued that non-parametric tests would have been a more appropriate form of analysis.[6] Non-parametric assumptions include the requirement for the observations to be independent and for the variables under study to have underlying continuity (Hebel, 2002). Both assumptions were satisfied by the present study and no violations for non-parametric tests were observed. Non-parametric tests are considered distribution-free methods in that they do not rely on any underlying mathematical distribution. Non-parametric tests treat samples made up of observations from several different populations and can treat data that are inherently in ranks, as well as data whose seemingly numerical scores have the strength in ranks. It is possible for non-parametric tests to treat data that are classificatory and tend to be easier to apply than parametric tests (Hebel, 2002). A further advantage of using non-parametric tests relates to the probability statements obtained, in that they are exact probabilities regardless of the shape of the population distribution from which the random sample was drawn (Hebel, 2002). In addition, if sample sizes as small as N=6 are used there is no alternative to using a non-parametric test (Hebel, 2002).

[6] The nonparametric equivalent to MANOVA is the Kruskal-Wallis analysis of ranks and the Median test. Nonparametric equivalents to the standard correlation coefficient are Spearman R, Kendall Tau, and coefficient Gamma. If the two variables of interest are categorical in nature (e.g., "passed" vs. "failed" by "male" vs. "female") appropriate nonparametric statistics for testing the relationship between the two variables are the Chi-square test, the Phi coefficient, and the Fisher exact test. A simultaneous test for relationships between multiple cases is available: Kendall coefficient of concordance. This test is often used for expressing inter-rater agreement among independent judges who are rating (ranking) the same stimuli. Instead of examining the mean of the ordinal data, it is possible to compute the geometric mean. Non-parametrics and Distributions will compute a wide variety of measures of location (mean, median, mode, etc.) and dispersion (*variance*, average deviation, quartile range, etc.)

However, non-parametric tests have been criticised as losing precision or being wasteful of data in that they usually result in loss of efficiency in the ability to detect a false hypothesis (Hebel, 2002). In this case, they are believed to be less powerful than parametric tests. Statistical power relates to the probability of rejecting the null hypothesis when it is in fact false and should be rejected. The power of parametric tests can be calculated from formula, tables, and graphs based on their underlying distribution. While the power of non-parametric tests is less straightforward, it can be calculated using Monte Carlo simulation methods (Mumby, 2002). Given that non-parametric statistics may be considered as less statistically powerful or sensitive than their parametric counterparts, it is important to use a parametric test if at all possible if one wishes to detect very small effects as is the case with the present study (for example, high risk acceptance versus low risk acceptance) (Statsoft, 2003). According to Hebel (2002), one is open to a false sense of security when using non-parametric tests. A non-parametric test protects against some violations of assumptions and not others. Non-parametric alternatives to parametric tests may not rely on the normality assumption, but it is necessary to ensure that one still meets the equal variances and independence assumptions. The data from the questionnaire in the present study satisfied these assumptions.

According to Vickers (2005) many sociological and biological researchers priase the superiority of non-parametric methods for non-normal distributions given that data from these fields rarely follow a normal distribution. Micceri (1989) compares the normal curve to the unicorn and other improbable creatures, given the unlikelihood of demonstrating normality within such data sets. This has led to implicit recommendations that non-parametric techniques should be considered the method of choice in sociological and biological studies (Bridge and Sawilowsky, 1999). It has been argued, however, that these prior investigations leading to such declarations can be considered as flawed (Vickers, 2005). Bridge and Sawilowsky (1999) compared parametric and non-parametric statistics and outlined how the magnitude of the differences between the tests was trivial. According to Vickers (2005), there is no simple and obvious manner in which non-parametric methods becomes superior once the distribution of data shifts away from normal. It is true that under normality parametric methods are trivially more efficient, but for non-normal data, the relative power of parametric and non-parametric statistics varies from distribution to distribution (Vickers, 2005). It has also been noted above that non-parametric methods are most appropriate when the sample sizes are small. However, when the data set is large (e.g., $n > 100$), as is the case in the present study, it often makes little sense to use non-parametric statistics (Vickers, 2005) (see also Section 4.6.4.3 "central limit theorem"). As outlined in Section 4.6.4, the choice of statistical approach and the statistical tests used for analysing the data from the questionnaire were corroborated by a statistician from the Statistical Consulting Unit. In addition, all conclusions drawn from the analyses were reviewed and agreed with by the Executive Director of the Statistical Consulting Unit in the University of Limerick. Given the above arguments and those contained in Section 4.6.4, parametric tests may still be considered as the more appropriate form of analysis for the present study. However, it would be interesting to note the difference in the findings should equivalent non-parametric tests be used to analyse the same set of data.

It is possible to argue that logistic regression may have been more suitable to use than the tests performed in the present study (see Section 4.6.4). However, none of the variables were measured in binary form, instead they were treated as interval data. The majority of previous quantitative research with similar objectives to the present study did not use binary variables but used Likert type scales (Siegrist, 2000; Siegrist and Cvetkovich, 2000 and Poortinga and Pidgeon; 2005). To compare the present study to the findings from recent research, it was necessary to adopt a similar format to the scales of measurement. Given this treatment of the variables, it can be said that logistic regression was not suitable for the study of the association between the predictors and the dependent variable in the present study. Instead, correlation analysis and ordinary regression were used to interpret the nature of the association between knowledge, trust, risk perceptions, benefit perceptions and risk acceptance.

Based upon the case analysis in Chapter 6, it is clear that the case categorisation was different between the interview and subsequent analysis (see Appendix E.1). The four cases were initially chosen according to the responses of individuals to the survey, however, upon interview these responses changed. Reasons for such changes vary. Judging from the interview responses, it is likely that changes could be linked to the misreading of questions in the survey, or indeed, the changing of opinion upon receiving and being more receptive of information regarding life insurers using genetic test results after the survey was completed (see Section 6.3). The questionnaires were completed in accordance with the respondents' knowledge at the time of the survey. If the questionnaires were re-administered, the above factors could cause respondents to change their answers, similar to the scenario whereby survey respondents were asked more detailed questions upon interview. It is questionable whether or not it was valid to reorganise the four cases based upon their interview responses, given that there was not the same opportunity to reorganise the survey responses that formulate conclusions to the study. However, if the interviewees were no longer representative of the case they occupy, it would have been invalid to typify their responses to their initial underlying case, since the responses of some individuals during interview indicated an association with an alternative case structure. Conclusions drawn regarding the factors and conditions that affect risk acceptance, based upon incorrect case alignment, would lead to an inaccurate portrayal of the social reality surrounding risk acceptance. Therefore, it was deemed more valuable for this study to base the conclusions upon findings from the survey and analysis of the cases formed subsequent to the interview process to obtain a more accurate understanding of the prevailing factors and conditions that underlie risk acceptance.

7.5 Summary

The intention of this chapter was to explore the findings from the semi-structured interviews and the surveys. It was intended to compare and contrast their results to develop an understanding as to whether there exists a relationship between knowledge, social trust, risk perceptions, benefit perceptions and risk acceptance within the frame of Irish society's perceptions of life insurers underwriting genetic test results of applicants. In addition, it was intended to examine the factors and conditions that appear to have an impact upon knowledge, trust, risk perceptions, benefit perceptions and risk acceptance. The chapter concludes that there is a relationship between these objects within the present study, apart from knowledge, however, this association may be considered as somewhat indirect, and the direction of the relationship is sometimes a little unclear. To a large extent, it is the factors and conditions that occur throughout the layers of reality that appear to support and demarcate the nature of this relationship. In exploring the factors and conditions that underlie this association, it was possible to reconstruct the theoretical understandings of social trust, risk perceptions and risk acceptance and to provide an account of the explanatory mechanisms surrounding the association between these factors. In several cases, new explanatory schemes were provided for the appearance of an association between social trust, risk perceptions, benefit perceptions and/or risk acceptance. Hence, this study has added to previous research on the relationship between these factors and, in doing so, has restructured views of causality between social trust, risk perceptions, benefit perceptions and risk acceptance.

Section 4.2.3.1 outlined how Sayer (2000:16) recommends answering three questions in order to clearly identify causal responsibility in the complex open system inherent in the observations of any social study. These questions demanded knowledge of the preconditions of the objects under study, an understanding of the conditions of existence of the relationship between the objects and finally, an explanation as to what it is about these objects that enable them to do certain things. The present study has inexplicitly responded to all three questions and in doing so has clarified the causal nature of the relationship between social trust, risk perceptions, benefit perceptions and risk acceptance in the present study. Sections 7.2.2, 7.2.3, 7.2.4, 7.2.5, 7.2.6 explained the factors and conditions underlying the existence of knowledge, social trust, risk perceptions, benefit perceptions and risk acceptance and the relationship between these objects, thereby responding to the first two questions regarding the existence of these objects and the nature of their relationship with each other. Sections 7.2.6 and 7.3 responded to the third question explaining how the objects explored in the present study can transform the insurance system. It was concluded that there appears to be a causal pattern between social trust, risk perceptions, benefit perceptions and risk acceptance, however, as noted above, the direction of this causal path was not clearly distinguished and the directness of the causal pattern remains questionable.

The following chapter shall provide a conclusion as to the achievement of the objectives of the present study. In addition, the limitations of the present study shall be outlined and areas for future research, based upon the findings of the present study, shall be suggested.

CHAPTER EIGHT

CONCLUSIONS

8.1 Introduction

The present study has demonstrated that an understanding of society's acceptance of the risks related to insurers using genetic test results must be based upon consideration of the macro and micro levels of social reality. This study aimed to explore the factors and conditions that influence knowledge, social trust, risk perceptions, benefit perceptions and risk acceptance, and the relationship between these factors. These factors and conditions can only be fully understood by taking into account how the various strands of the web of social reality in which society's perceptions are entangled, interact and serve to influence such attitudes and decisions. The previous chapters identified a variety of factors and conditions that appear to influence knowledge, social trust, risk perceptions, benefit perceptions and risk acceptance within the self, situated activity, 'setting' and 'context' dimensions. It was demonstrated how a number of factors (the dominant culture, the insurance system, the risk management system, the nature of genetic tests and the interaction between the public, the insurance industry, and the government) influence society's perceptions of the risks related to insurers using genetic test results and opinions regarding the acceptance of perceived risks. The preceding chapter explored the findings of the present study and formulated several conclusions about the relationship between social trust, risk perception and risk acceptance and the factors and conditions affecting these perceptions and evaluations. The intention of the present chapter is to amalgamate these conclusions and to present them in a coherent and distinct manner. Moreover, achievement of the study's objectives will be discussed, followed by an outline of the limitations of the research and suggested avenues for further research in this area.

8.2 Achievement of Objectives

The main objective of this research was to investigate the role that incomplete knowledge, social trust, risk perceptions and benefit perceptions have in influencing society's acceptance of the risks perceived as relating to insurers' use of genetic test results in the Republic of Ireland. In addition, it was intended to identify the factors and conditions that influence this risk acceptance pattern. Looking at the attitudes and evaluations of respondents the majority of individuals sampled had incomplete knowledge of the use of genetic test results, high risk perceptions, low benefit perceptions, low trust in the government and the insurance industry and a lack of willingness to accept insurers' use of genetic test results to underwrite applicants for life insurance. The main risks foreseen related to the privacy of genetic test results, the denial of insurance applications, the charging of high or unaffordable premiums, unfair discrimination and the selection of only low risk individuals by life insurance companies. Benefits mentioned included a fairer and a reduced premium being charged to applicants as a result of a fairer assessment of all applicants' risks.

Regarding the main objective of the present study, the findings are two fold, with answers to the research question varying between findings from the quantitative analysis and the qualitative analysis. Firstly, initial analysis of percentage agreements and disagreements with the survey questions revealed that there was incomplete knowledge, low trust, high risk perceptions, low benefit perceptions and low risk acceptance. This indicated that there may be a relationship between these factors corresponding to that dictated by previous research (Siegrist, 2000; Siegrist and Cvetkovich, 2000 and Poortinga and Pidgeon; 2005) (see Sections 3.2, 3.3, 3.5 and 3.6). Further statistical analysis of the survey revealed that the magnitude of the relationship between knowledge, social trust, risk perceptions, benefit perceptions and risk acceptance was small and the likelihood of a correlation between these factors was minimal (see Section 5.2). Although the influence that the level of knowledge had upon the association between risk perception and trust and benefit perception and trust could not be examined, it did not appear that trust was a significant determinant in judgements of risks or benefits in the absence of knowledge (see Section 5.3). Social trust did not influence perceptions of risks or benefits (see Section 5.3). Indeed, risk perceptions and benefit perceptions were found to have a minimal influence upon risk acceptance (see Section 5.3). Hence, it could be concluded from the results of the quantitative analysis that knowledge, social trust, risk perceptions and benefit perceptions did not play a significant role in influencing risk acceptance of insurers' use of genetic test results. However, given that the margin of error at the 95 per cent confidence interval with a sample population of 180 individuals is approximately 7.3 per cent (American Research Group), the margin of error in the present study provides a less precise confidence interval for the results than a study that uses a larger sample population.

The qualitative analysis displayed a somewhat alternative set of results to the quantitative findings. An association between social trust, risk perceptions, benefit perceptions and risk acceptance was demonstrated. However, findings regarding knowledge demonstrated that, similar to the survey results, knowledge does not appear to enter into this causal pattern of events. The factors and conditions that were found to influence trust in the Government and the insurance industry, risk perceptions, benefit perceptions and risk acceptance appeared to reinforce the link between each of these elements (see Sections 7.2, 7.2.1, 7.2.2, 7.2.3, 7.2.4, 7.2.5 and 7.2.6). These findings suggest that there exists a relationship between social trust, risk perceptions, benefit perceptions and risk acceptance. In other words, social trust, risk perceptions and benefit perceptions may play a role, to a greater or lesser extent, in influencing acceptance of the risks related to insurers using genetic test results.

A summary of the factors and conditions found to influence knowledge, social trust, risk perception, benefit perception and risk acceptance across the dimensions of social reality were illustrated in Table 7-1. Table 7-1 and Sections 7.2.2, 7.2.3, 7.2.4 and 7.2.5 demonstrated the entangled nature of the factors and conditions listed and the dimensions of social reality. It was noted that there may be other factors and conditions that influenced the findings from the present study, although not identified throughout the course of this research. The factors and conditions illustrated in Table 7-1 were considered to interplay and exert an influence upon each other to a greater or lesser extent. Therefore, some factors and conditions may have had a more powerful effect upon other elements, and upon knowledge, social trust, risk perceptions, benefit perceptions and risk acceptance evaluations, than others. However, it was the dynamics and overall interaction of each of these factors and conditions that delivered the findings from the present study.

Drawing conclusions from this analysis may be considered as rather difficult, given the diverse set of findings between the quantitative analysis and the qualitative analysis. Section 4.4.1 noted that using mixed methods in the study of the same phenomenon could lead to contradictions and confusions. However, multi-method research has proved valuable in the present study in its ability to show up different aspects of reality surrounding social trust, risk perceptions and risk acceptance of insurers using genetic test results (see Section 7.2.5). It was not possible for the survey to identify the complex reality surrounding the nature of the relationship between trust, risk perceptions, benefit perceptions and risk acceptance. The survey focused on a narrow selection of variables, and in doing so stripped from consideration other variables that were demonstrated to exert their influence upon social trust, risk perceptions, benefit perceptions and risk acceptance in the qualitative analysis (see Section 7.4). However, findings from the survey indicated that a direct relationship between knowledge, social trust, risk perceptions, benefit perceptions and risk acceptance did not appear to be present in the study. The survey allowed for an extensive view of knowledge, social trust, risk perceptions, benefit perceptions and risk acceptance across a large number of individuals. In doing so, the findings from the survey provided a good overview of percentage agreement with insurers using genetic test results, the level of trust in the insurance industry and the Government, the level of risks and benefits viewed by society as relating to insurers using genetic test results and the extent of knowledge regarding this matter. Due to the relatively small number of individuals that occupy the cases, it would have been impossible, to any great extent, to generalise from their responses to a broader group. Although the sample obtained for the survey can be considered as a convenience sample, rather than one chosen for representativeness, efforts were made to obtain a cross-section of gender, age and income across the population (see Section 4.6.3). In doing so, it was possible to suggest that the responses of the sample obtained could be generalised beyond the group selected to a broader population and perhaps narrowly indicate Irish society's view of trust, risk and benefits from insurers using genetic test results in their underwriting processes.

The findings from the qualitative study displayed a degree of similarity with the quantitative findings, in that they appeared to demonstrate that the relationship between social trust, risk perceptions, benefit perceptions and risk acceptance was largely indirect. The qualitative analysis, however, also allowed for an explanation of the underlying reasons for such responses and, in doing so identified and demonstrated that there existed factors and conditions that linked social trust, risk perceptions, benefit perceptions and risk acceptance. Hence, rather than a direct causal pattern existing between these factors, both methods confirmed an indirect connection between these factors. By combining quantitative and qualitative analyses within the same study, it was possible to investigate different dimensions of the reality surrounding risk perceptions and to obtain a more holistic view of the relationship between social trust risk perceptions, benefit perceptions and risk acceptance.

The association between social trust, risk perceptions, benefit perceptions and risk acceptance may be understood as demonstrating a causal pattern of risk acceptance. According to Sayer (2000:21) causal powers of objects are generalisable to other contexts as they are necessary features of these objects (see Table 4.3). However, the present study's findings demonstrate that an indirect causal relationship exists between social trust, risk perceptions, benefit perceptions and risk acceptance, that, to a large extent, is made possible and reinforced by the factors and conditions underlying each of these objects. Therefore, the causal pattern identified in the present study may be demonstrated to operate only within a very narrow set of circumstances, where the same or similar factors and conditions are operating behind social trust, risk perceptions, benefit perceptions and risk acceptance. Therefore, based upon the findings from the present study, the causal risk acceptance pattern demonstrated may not be considered as extendible in a universal fashion. Rather the findings indicate a more conservative view of the generative mechanism of risk acceptance in that they are representative of narrow local tendencies.

The findings from the present study have contributed to knowledge surrounding the field of risk research (see Section 7.2). This study allows, to a certain extent, a restructured theoretical understanding of social trust, risk perceptions and risk acceptance by providing an account of the explanatory mechanisms surrounding the association between these factors within the context of insurers using genetic test results. In several cases, new explanatory schemes were provided for the appearance of an association between social trust, risk perceptions, benefit perceptions and/or risk acceptance. Hence, this study has added to previous research regarding the relationship between these factors and, in doing so, has restructured views of causality between social trust, risk perceptions, benefit perceptions and risk acceptance. In addition the present study has contributed to previous research with regard to the understanding of trust in relation to the components and dimensions of social trust (see Sections

7.2.2.4, 7.2.2.5 and 7.2.5). The study also confirmed the psychometric paradigm of risk perception and concludes that risk perception is multi-dimensional and based on several characteristics (Slovic *et al.,* 2002), (see Section 7.2.5). Finally, the findings illustrated the uncertainties surrounding the management of insurers' use of genetic test results and has served to extend Beck's (1992) risk society from a German and environmental context to include first and second world countries and new innovative technologies (see Section 7.2.5).

The implications that this study has for the insurance industry were discussed in Section 7.3. Overall, there appears to be a need for the Government and the insurance industry to promote the trust of the public to ensure their stability and ability to care for the public and to continue providing life insurance within a profitable market. It is demonstrated that there is an 'exceptionalistic' view of the risks related to insurers using genetic test results; hence the Disability Act 2005 may reintroduce the risks it is designed to prevent (see Section 2.5.3, 6.4 and 7.3). The risks relating to the Disability Act 2005 as described in Section 2.5.3 and the risks identified within the findings from the present study indicate that the Government should hold regular meetings regarding the progress and popularity of genetic testing, the impact of the Disability Act 2005 upon insurers' earnings, and changes regarding the nature of the risks related to insurers using genetic test results (see Section 7.3). However, rather than just involving experts from the Government and the insurance industry in such discussions, as has been the case so far, it is considered that it would be of immense value to include members from society in these discussions also (see Section 7.3).

In conclusion, the objectives that the study set out to achieve have been realised. In achieving these objectives, the findings from this study have made a contribution to knowledge in the immediate discipline of trust and risk perception research. Moreover, they have also taken a step in informing risk theory and the objectives that those managing the risks related to life insurance companies using genetic test results should aim for in future risk management efforts. Although this study has achieved its objectives, it may have been more productive if several limitations had been overcome. Section 7.4 illustrated some criticisms of the present study however, it is important to also note the limitations that were observed throughout the course of the research process. The following section shall explore these limitations.

8.3 Limitations

Perhaps one of the most prominent limitations in the present study relates to the question designed to measure the trust dimension – reliability. The question asked whether or not the respondent believed that the Government or the insurance industry changes policies regarding their management of insurers' use of genetic test results without good reasons. This question was piloted on three occasions prior to the survey being conducted and there did not appear to be any problems associated with the wording, which reflected that used by Poortinga and Pidgeon (2003). In addition, 180 participants responded to this question when surveyed and not one individual indicated any difficulties responding to the question. However, during the interview process, one individual commented that good reasons may be deemed differently depending on whether it is from the Government's point of view, given that "they could distort the facts and that necessarily wouldn't be for the public good" or from the public's point of view. The question does not state "whose good" is being questioned. This is a valid argument and suggests that there may be some inherent problems with the wording of this question. This finding poses a limitation for the present study, in that the wording was not corrected for the surveys and for the majority of interviewees, hence each individual may have understood the question differently. Therefore, the findings regarding the trust dimension 'reliability' may be somewhat tainted.

A further limitation identified throughout the course of the research process is that socio-demographic and cultural conditions were not accounted for when selecting cases for interview. Therefore, the sample chosen may not be representative of the total population available for interview (see Section 6.2). The reasons for this phenomenon were discussed in Section 6.2. Individuals were randomly selected for further contact from a list of the divided cases available for further interview. Some individuals contacted no longer wished to partake in further interview, a factor that limited the choice of interviewees somewhat. These factors and chance each contribute to the imbalanced case portfolio. It is advisable that future research should have more demographically representative cases. The increased representativeness of the cases would benefit investigations into the role that gender, age and economic status has upon risk perceptions and risk acceptance. Studies comparing how trust operates across various cultures is certainly deserving of increased scholarly work, given trans-national relations are now the norm across a range of endeavours, such as the management of insurers' use of genetic test results.

8.4 Suggested Further Research

Studies of risk acceptance and risk perception shall continue to be fruitful and informative. The factors and conditions that influence knowledge, social trust, risk perceptions, benefit perception and risk acceptance will vary according to the subject matter and the society in which future studies may take place. It is unlikely that the same findings from the present study shall apply in exactly the same manner in other studies examining insurers' use of genetic test results. However, the study does offer a stepping-stone from which future analyses can be compared. This research has only taken the first step in understanding and evaluating risk acceptance of the contentious issue of insurers using genetic test results. No research prior to this study has been completed in this area. This indicates that there is much room, and a need, for further research on this topic, given the powerful position life insurance occupies in society and the wider economy and continuous developments in genetic technology.

This research has taken a relatively small sample populations' viewpoint but considerations should be provided to completing a similar study with a larger population sample. Rather than carrying out further research on a local basis, it is advisable that future study on a national basis should be examined. When doing such research, it would be advisable to conduct exploratory analysis of knowledge, trust, risk perceptions, benefit perceptions and risk acceptance prior to quantitatively exploring the relationship between these factors (see Section 7.4 for reasoning).

Future analysis of trust should explore whether or not 'openness' can be considered a dimension of trust. This dimension was not included in the statistical analysis of general trust in the present study. Principal components analysis revealed how this item did not appear to measure the same condition as the other dimensions of trust. However, qualitative analysis later revealed that this factor was bound to a definition of trust and was linked to several other dimensions of trust. Therefore, openness should perhaps have been considered for inclusion in the study of general trust. The factors and conditions affecting the results of the present study are likely to be different to those in future studies, hence, a full analysis of each of the dimensions of trust should be carried out before any decisions regarding whether or not certain dimensions should be excluded from study.

Regarding the dimensions of trust, an exploration as to whether or not an organisation is considered as being part of a power elite in society should be conducted. Further analysis should be conducted regarding its affect upon trust attitudes. Future studies could explore whether or not this factor can be considered a dimension of trust and its role within the components of trust.

The question designed to measure the trust dimension 'reliability' of the Government and the insurance industry should be reworded so that it is understood that a changing of policies without good reasons applies to the reasoning of the individual and the entity under question. Future studies could take this issue into account in order to obtain a more accurate understanding of trust evaluations.

Further study should be carried out on the affects that the perceived lack of transparency of underwriting may have upon risk perceptions and benefit perceptions of insurers using genetic test results. The lack of transparency appears to generate a negative view of insurers and this could be analysed in more detail and the affects that this may have on life insurance purchasing.

It may be beneficial for the Irish Government to conduct regular research into the risks related to insurers using and not using genetic test results. The findings of such research could be made known publicly, along with the Government's position as the authority governing insurers' use of genetic test results, in an effort to provide peace of mind to members of society regarding this matter. It may also prove valuable if the methods used to manage the risks related to insurers using genetic test results were reviewed and the benefits from including the public in such decision-making discussed. If the public were to be involved, it would be advisable for the findings from any analysis relating to this matter to be revealed to society so that the public feel that they have the knowledge and the means to make decisions regarding this matter.

It may be of interest for future research to consider the effect that early claims have upon critical illness, income protection and long-term care insurance policies and not just the impact on policies specifically designed to cover premature death. In addition, an understanding of society's perception of these products and their 'need' for these policies could be explored, to further understand society's perceptions of the risks and benefits related to insurers using genetic test results to underwrite such policies. It would be interesting to note if the risks perceived from insurers using genetic test results differs in accordance with the type of policy under question.

8.5 Summary

This chapter presented several conclusions based upon the findings of the present study. The overall findings from the study were presented and discussed and it was demonstrated how the research objectives were achieved. The contribution that this study has provided to the field of risk perception research was briefly discussed, as were the implications for the insurance industry regarding their use of genetic test results in underwriting life insurance applicants. This discussion was followed by an outline of the limitations of the research that became apparent during the progress of the study. Finally suggestions were made regarding avenues for future research in this area.

This research has provided a platform on which to build further understanding of risk acceptance and the factors and conditions that influence risk acceptance evaluations and attitudes. Although, the study was mainly exploratory in nature, it has, nonetheless, provided a basis for future research into not only the trust and risk evaluations of individuals, but also the role of social reality in influencing the formation of society's beliefs and decision-making. In addition, the present study has made a contribution to risk theory by investigating the risks perceived as relating to insurers using genetic test results and the nature of the relationship between incomplete knowledge, social trust, risk perceptions, benefit perceptions and risk acceptance. The findings also demonstrate the advantages for risk managers in taking account of the knowledge, trust evaluations, risk perceptions, benefit perceptions and risk acceptance of individuals when making decisions that influence the welfare of society. Further understanding of these evaluations and decisions and the nature of their relationship with each other may improve the effectiveness and appropriateness of risk management decisions.

BIBLIOGRAPHY

Alhakami, A.S., & Slovic, P. (1994). A psychological study of the inverse relationship between perceived risk and perceived benefit. *Risk Analysis, 14*, 1085-1096

Alper J.S., Geller LN, Barash CI, Billings PR, Laden V and Natowicz MR. (1994). "Genetic discrimination and screening for hemochromatosis" Journal of Public Health Policy, 15, 345 In Lemmens, T. (2000) Selective justice, genetic discrimination and insurance: should we single out genes in our laws?, *McGill Law Journal,* 347-355

Alper, J.S. and Beckwith, J. (1998). Distinguishing Genetic From Non-genetic Medical Tests: Some Implications for Antidiscrimination Legislation, *Science Engineering Ethics, 4*, 141-145

Al Zeera, Z. (2001). Wholeness and holiness in education: An Islamic perspective. Herndon, VA: International Institute of Islamic Thought In Krauss, S.E. (2005), Research Paradigms and Meaning Making: A Primer, *The Qualitative Report 10* (4), 758-770

American Academy of Actuaries (1998) *Genetic Information and Voluntary Life Insurance*, http://www.actuary.org/pdf/life/genet.pdf (last assessed January, 2005)

American Research Group, Margin of Error Calculator, http://americanresearchgroup.com/moe.html (last assessed March 2004)

Archer, M. (1995). Realist social theory: the morphogenetic approach. Cambridge: Cambridge University Press, In Dobson, P.J. (2002). Critical realism and information systems research: why bother with philosophy?, *Information Research*, *7* (2)

Archer, M., Bhaskar, R., Collier, A., Lawson, T, and Norrie, A. (eds.) (2000). For structure: Its reality, properties and powers: A reply to Anthony King, *Sociological Review* 48: 464–72

Athaus, C.E. (2005). A disciplinary perspective on the epistemological status of risk, *Risk Analysis*, *25*, (3), 567-588

Ayer, A.J. (1936/1990). Language, Truth and Logic. Harmondsworth, Penguin

Bachrach, P. and Baratz, M.S. (1970) *Power and Poverty: Theory and Practice,* New York: Oxford University Press

Baird, B.N.R. (1986). Tolerance for environmental health risks: The influence of knowledge, benefits, voluntariness, and environmental attitudes. Risk Analysis, 6, 425-435 In Siegrist, M., Earle, T.C., Gutscher, H. (2003). Test of a trust and confidence model in the applied context of electromagnetic field (EMF) risks, *Risk Analysis*, *23* (4), 705-716

Barber, B. (1983). The Logic and Limits of Trust. New Brunswick, NJ: Rutgers University Press *In* Poortinga, W. and Pidgeon, N.F. Exploring the dimensionality of trust in risk regulation, *Risk Analysis*, *23* (5)

Barke, R.P., Jenkins-Smith, H. and Slovic, P. (1997). Risk perceptions of men and women scientists, *Social Science Quarterly*, 78, 167-176

Beck, U. (1992). *Risk Society: Towards a New Modernity,* London: Sage

—. (1993). *Risk: A Sociological Theory*, New York: Walter de Gruyter

—. (1994). The reinvention of politics: towards a theory of reflexive modernisation, in U. Beck, A. Giddens and S. Lash (eds) *Reflexive Modernisation*, 1-55. Stanford: Stanford University Press

—. (1995a). *Ecological Enlightenment: Essays on the Politics of the Risk Society,* New Jersey: Humanities Press

—. (1995b). Ecological Politics in the Age of Risk, Cambridge, Polity Press, Cambridge in G. Gaivoronskaia, and K.E. Solem (2004) Genetic testing: affected parties and decision-making, *Journal of Risk Research* **7** (5), 481-493

—. (1997). *The Reinvention of Politics*, Cambridge: Polity Press

—. (1999). *World Risk Society*, Cambridge: Polity Press

Beck, U., Giddens, A., Lash, S. (1994). *Reflexive Modernisation,* Cambridge: Polity Press

Becker, J., and Niehaves, B., (2006). An Epistemological Framework for Analyzing IS Research Approaches, *Information Systems Journal* (16) [to appear]

Beckwith, J. and Alper, J.S. (1998). Reconsidering Genetic Antidiscrimination Legislation, *Journal of Medicine and Ethics*, 26, 205-206

Benn, S. (2004). Managing toxic chemicals in Australia: a regional analysis of the risk society, *Journal of Risk Research,* **7** (4), 399-412

Berman, M. (1988). *All that is Solid Melts into Air: The Experience of Modernity*, New York: Penguin Books

Bernstein, P. (1996). *Against the Gods, the Remarkable Story of Risk*, New York: Chichester, Wiley

Bhaskar, R. (1979). *The Possibility of Naturalism*, East Sussex: Harvester-Wheatsheaf, Brighton

—. (1986). *Scientific Realism and Human Emancipation*, London: Verso

—. (1993). *Dialectic: The Pulse of Freedom,* London: Verso

Biel, A., and Dahlstrand, U. (1995). Risk perception and the location of a repository for spend nuclear fuel. Scandinavian Journal of Psychology, 36 In Viklund, M.J. (2003), Trust and risk perception in Western Europe: A cross-national study, *Risk Analysis, 23* (4)

Billings, P. , Kohn, M.A., de Cuevas, M., Beckwith, J., Alper, J.S. and Natowicz, M.R. (1992) Discrimination as a Consequence of Genetic Testing, American Journal of Human Genetics, 476 In Lemmens, T. (2000). Selective justice, genetic discrimination, and insurance: Should we single out genes in our laws?, *McGill Law Journal*, 347-355

Black, J. (1998). Regulation as facilitation: Negotiating the genetic revolution, *Law and Human Genetics: Regulating a Revolution*, Oxford: Hart Publishing

Black, K. and Skipper, H., (1994). *Life Insurance,* 12th Edition, U.S.: Prentice Hall

Bogdan, R.C. and Biklen, S.K., (1982). *Qualitative Research for Education: An Introduction to Theory and Methods*, Boston: Allyn and Bacon

Bond, S. (1993). Experimental research nursing: necessary but not sufficient. In: Kitson, A. (Ed) *Nursing, Art and Science*. London, Chapman and Hall

Bord, R., and Connor, R.E. (1997). The gender gap in environmental attitudes: The case of perceived vulnerability to risk, *Social Science Quarterly*, 78, 831–840.

Brackenridge, R.D.C. and Elder, W.J. (1998). *Medical Selection of Life Risks, 4th Edition,* Stockton Press

Bradbury, J.A., Branch, K.M. and Focht, W. (1999). Trust and Public Participation in Risk Policy Issues, in Cvetkovich, G., Lofstedt, R.E., *Social Trust and the Management of Risk*, London: Earthscan

Brenot, J., Bonnifous, S. and Marris, C. (1998), Testing the cultural theory of risk in France, *Risk Analysis, 18,* (6), 729-740

Brewer, J. and Hunter, A. (1989). *Multi-method Research; A Synthesis of Styles*, Newbury Park, California: Sage

Bridge, P.D. and Sawilowsky, S.S. (1999). Increasing physicians' awareness of the impact of statistics on research outcomes: comparative power of the t-test and and Wilcoxon Rank-Sum test in small samples applied research. *Journal of Clinical Epidemiology* 52, 229-235

Brownsword, R., Cornish, W.R. and Llewelyn, M. (1998). Human Genetics and the Law: Regulating a Revolution, In Brownsword, R., Cornish, W.R. and Llewelyn, M. (1998).*Law and Human Genetics: Regulating a Revolution*, Hart Publishing, Oxford

Bryman, A. (1989). *Quantity and Quality in Social Research,* New York, Routledge

—. (2001). *Social Research Methods*, Oxford, U.K.: Oxford University Press

Carter, B. and New, C. (2004). *Making Realism Work: Realist Social Theory and Empirical Research,* Oxford: Routledge

Central Statistics Office, Dublin *Various Publications*

Churchland, P. M. (1979). *Scientific Realism and the Plasticity of Mind.* Cambridge: Cambridge University Press

Cohen, B.L. (1983). Before its too late: A scientist's case for nuclear energy, New York: Plenum In Rosa, E.A. (1998). Metatheoretical foundations for post-normal risk, *Journal of Risk Research*, (1), 15–44

Coll, R. K. and Chapman, R. (2000). Choices of methodology for cooperative education researchers. Asia-Pacific Journal of Cooperative Education, *1*, 1-8. Retrieved April 23, 2003, from http://www.apjce.org/volume_1/volume_1_1_pp_1_8.pdf In Krauss, S.E. (2005), Research Paradigms and Meaning Making: A Primer, *The Qualitative Report 10* (4), 758-770

Comte, A. (1853/1971). The positive philosophy. In Thompson, K. and Tunstall, J. (Eds) *Sociological Perspectives*. Harmondworth, Penguin

Connelly, J. (2004). Realism in evidence based medicine: interpreting the randomised controlled trial, *Journal of Health Organization and Management, 18* (2), 70-81

Consumer Credit Act 1995, http://www.irishstatutebook.ie/ZZA24Y1995.html (last assessed March 2006)

Convention for the Protection of Human Rights and Dignity of the Human Being with regard to the Application of Biology and Medicine: Convention on Human Rights and Biomedicine, Oviedo4.IV.1997

Coote, A. (1998). Risk and Public Policy: Towards a High Trust Democracy, in Franklin, J. (1998). *The Politics of Risk Society*, Cambridge: Polity Press

Council of Europe Committee of Ministers, Recommendation No.R(92)3 of the Committee of Ministers to Member States on Genetic Testing and Genetic Screening for healthcare purposes, *(Adopted by the Committee of Ministers on 10 February 1992 at the 470th meeting of the Ministers' Deputies)*

Cousins, C. (2002). Getting to the "truth": Issues in contemporary qualitative research. Australian Journal of Adult Learning, 42, 192-204 In Krauss, S.E. (2005), Research Paradigms and Meaning Making: A Primer, *The Qualitative Report 10* (4), 758-770

Cumbo, J. (2005). Insurers to go easy on gay questioning, *Financial Times*, Sept. 30, http://news.ft.com/cms/s/ff6106e6-31bd-11da -9c7f -00000e2511c8.html

Cutter, S.L., Tiefenbacher, J., and Solecki, W.D. (1992) En-Gendered Fears: Femininity and Technological Risk Perception, *Industrial Crisis Quarterly, 6,* 5-22

Cvetkovich, G. and Lofstedt, R. (1999). *Social Trust and the Management of Risk*, London: Earthscan

Daar, A. and Mattei, J.F. (1999). Draft World Health Organization (WHO) Guidelines on Bioethics, online: *Nature: International Weekly Journal of Science* http://helix.nature.com/wcs/b23a.html (date assessed: 6 January, 2006)

Dake, K. (1991). Orienting dispositions in the perception of risk: An analysis of contemporary worldviews and cultural biases, *Journal of Cross-Cultural Psychology*, 22, 61-82

Dancey, C.P. and Reidy, J. (2002). *Statistics Without Maths for Psychology using SPSS for Windows*, London: Prentice Hall

Danermark, B., Ekstrom, M., Jakobsen, L. and Karlsson, J.C. (2002) *Explaining Reality; Critical Realism in the Social Sciences*, London: Routledge

Darke, P., Shanks, G., and Broadbent, M., (1998). Successfully completing case study research: combining rigour, relevance and pragmatism, Information Systems Journal (8:3), 273-289, In Niehaves, B. and Stahl, B.C., (2006). Criticality, epistemology and behaviout vs. design – information systems research across different sets of paradigms *Information Systems Journal* (to appear)

Data Protection Act, http://www.irishstatutebook.ie/ZZA25Y1988.html (last assessed February 2006)

Data Protection (Amendment) Act 2003, http://www.irlgov.ie/bills28/acts/2003/a603.pdf (last assessed February 2006)

Data Protection Commissioner, http://www.dataprivacy.ie/docs/Case_Study_2/99_Life_Insurance_ Company/137.htm (last assessed February 2006)

Davidson, D., and Freudenburg, W. (1996). Gender and environmental risk concerns: A review and analysis of available research. *Environment and Behaviour*, 28, 302–339

Daykin, C.D., Akers, D.A., Macdonald, A.S., McGleenan, T., Paul, D. and Turvey, P.J. (2003) Genetics and Insurance: Some Social Policy Issues, *Presented to the Institute of Actuaries*, 24 February, and to the *Faculty of Actuaries*, 17 March

Dean, H. (2000). Managing Risk by Controlling Behaviour: Social Security Administration and the Erosion of Welfare Citizenship, in Taylor-Gooby, P. (ed.), *Risk, Trust and Welfare*, London, Palgrave

Denzin, N. (1989). *The Research Act: A Theoretical Introduction to Sociological Methods*, New Jersey: Prentice Hall

Diamantopoulos, A. and Schlegelmilch, B.B. (2000), *Taking the Fear out of Data Analysis*, London: Thompson Learning

Dietz, T., Scott Frey, S. and Rosa, E.A. (1996). 'Risk, Technology, and Society, in R.E. Dunlap and W. Michelson (eds) *Handbook of Environmental Sociology*, Westport, CT: Greenwood Press.

Disability Bill, 2001, initiated as No.68 of 2001, Sponsored by; Minister for Justice Equality and Law Reform

—. 2004, initiated as No. 39 of 2001, Sponsored by; Minister for Justice Equality and Law Reform

Disability Act 2005, enacted as No. 14 of 2005

Dobson, P. J. (2002). Critical realism and information systems research: Why bother with philosophy? Information Research – An International Electronic Journal, 7(2). In Krauss, S.E. (2005), Research Paradigms and Meaning Making: A Primer, *The Qualitative Report* 10 (4), 758-770

Dosman, D.M., Adamowicz, W.L. and Hrudey, S.E. (2001). Socio-economic determinants of health and food safety related risk perceptions, *Risk Analysis: An International Journal, 21*(2), 307-17

Douglas, M. and Wildavsky, A. (1982). *Risk and Culture: The Selection of Technological and Environmental Dangers*, Berkeley, CA: University of California Press.

Douglas, M. (1978). Cultural Bias. Occasional Paper no. 35, Royal Anthropological Institute of Great Britain and Ireland

Drottz-Sjoberg, B.M., & Sjoberg, L. (1991). Adolescents' attitudes to nuclear power and radioactive wastes. *Journal of Applied Social Psychology,* 21, 2007-2036

Earle T.C. and Cvetkovich G.T., (1995). *Social Trust: Toward a Cosmopolitan Society*. Westport, CT:Praeger

Earle, T.C. and Cvetknvich, G. (1997), Culture, cosmopolitanism, and risk management, *Risk Analysis*, 17, 55-65

Earle, T.C. and Cvetkovich, G. (1998), Determining determinants of trust, *Risk Analysis,* 18, 231-232

Easterby-Smith, M., Thorpe, R. and Lowe, A. (2002). *Management Research: An Introduction*, Sage Publications

Eiser, J. R., Miles, S. and Frewer, L. (2002). Trust, perceived risk, and attitudes toward food technologies. *Journal of Applied Social Psychology, 32*, 1-12.

ELSI (2003). Genetics, Privacy and Legislation, http://www.ornl.gov/sci/techresources/Human_Genome/elsi/legislat.shtml (last assessed January 2005)

Enoch, M.A. and Goldman, D. (2001). Genetics of alcoholism and substance abuse, *Current Psychiatry Reports,* 3 (2) 144-151

Fenn, P. (2004). Genetic information and the sustainability of pooling insurance contracts, *Insurance Genetics and Fairness II Filling the Information Gap*, Conference, Imperial College, London, 13th July

Field, A. (2004). *Discovering Statistics using SPSS for Windows*, Sage Publications, London

Figueiredo, C.J. and Drotz-Sjoberg, B.M. (2000). Perceived control, voluntariness and emotional reactions in relocated areas of Russia, Ukraine and Belarus, Risk: Health, Safety and Environment , 11, 233-242 In Siegrist, M., Earle, T.C., Gutscher, H. (2003), Test of a trust and confidence model in the applied context of electromagnetic field (EMF) risks, *Risk Analysis, 23,* (4), 705-716

Finucane, M., Slovic, P., Mertz, C., Flynn, J. and Satterfield, T. (2000). Gender, race and perceived risk: the 'white male' effect, Health, *Risk and Society*, 2, 19-172

Fischer, E.P. and Berberich, K. (1999). *Impact of modern genetics on insurance.* Publications of the Cologne Re, 42

Fischoff, B., Slovic P. and Lichenstein S. (1982). Lay foibles and expert fables in judgements about risk, *American Statistician, 36,* 240 – 255

Flynn, J., Burns, W., Mertz, C. and Slovic, P. (1992). Trust as a determinant of opposition to a high-level radio active waste repository: Analysis of a Structural Model, *Risk Analysis,* 12, 417-429

Flynn, J., Slovic, P. and Mertz, C.K. (1994). Gender, race, and perception of environmental health risks, *Risk Analysis*, *14* (6), 1101-1108

Frankfort-Nachmias, C. and Nachmias, D. (1996). *Research Methods in the Social Sciences*, 5th Edition, London: Arnold

French, S., Maule, J., Mythen, G. and Wales, C. (2002). *Trust and Risk Communication (Technical Report)*. Leeds/Manchester: Manchester Business School/Leeds University Business School. 485.

French, J. R. P. and Raven, B. H. (1959). The bases of social power. In D. Cartwright (Ed.), *Studies in Social Power* (150-167). Ann Arbor: University of Michigan.

Freudenburg, W.R. (1989). Perceived risk, real risk: social science and the art of probabilistic risk assessment, *Science* 242, 44–49

—. (1993). Risk and Recreancy, *Journal of Social Issues*, 71 (4), 909-932

Frewer, L. (Sept. 1999). Risk perception, social trust, and public participation in strategic decision-making: implications for emerging technologies, *Ambio*, *28*, 6

Frewer, L.J., Howard, C., Hedderley, D. and Shepherd, R. (1996). What determines trust in information about food-related risks? Underlying psychological constructs. *Risk Analysis*, *16,* (4), 473-485

Frewer, L.J., Howard, C. and Shepherd, R. (1997). Public concerns in the United Kingdom about general and specific applications of genetic engineering: risk, benefit, and ethics. *Science, Technology and Human Values*, 22, 98-124.

Frewer, L.J., Howard, C. and Shepherd, R. (1998). Understanding public attitudes to technology. *Journal of Risk Research, 1*, 221-235

Frey, J.H. (1989). *Survey Research by Telephone*, 2nd Edition, London: Sage

Funtowicz, S. O. and Ravetz, J. R. (1992). Three types of risk assessment and the emergence of post-normal science. In S. Krimsky & D. Golding (Ed s.), (1992). *Social theories of risk* (251-274). Westport, CT: Praeger

Gaivoronskaia, G. and Solem, K.E. (2004). Genetic testing: affected parties and decision-making, *Journal of Risk Research, 7* (5), 481-195

Garcia, L. and Quek, F. (1997). Qualitative research in information systems: time to be subjective? In A.S. Lee, J. Liebenau and J.I. DeGross, eds. Information systems and qualitative research, 542-568. London: Chapman and Hall, In Dobson, P.J. (2002). Critical realism and information systems research: why bother with philosophy?, *Information Research*, *7* (2)

Gentner, D. and Collins, A. (1981). Studies of inference from lack of knowledge. *Memorandum Cognition,* 9, 434–443 In Rock, E.M., Ireland, M., Resnick, M.D. and McNeely, C.E. (2005), A Rose by Any Other Name? Objective Knowledge, Perceived Knowledge, and Adolescent Male Condom Use, *Pediatrics,* 115, 667-672

Geller, L.N., Alper, J.S., Billings, P.R., Barash, C.I., Beckwith, J. and Natowicz, M.R. (1996), Individual, Family, and Societal Dimensions of Genetic Discrimination: A Case Study Analysis, *Science and Engineering Ethics*, 2

Gergen, K. (1999). *An Invitation to Social Construction,* London: Sage

Gardner, C.B. (1990). Safe Conduct: Women, Crime, and Self in Public Places *Social Problems* 37, 311-328

Giddens, A. (1978). Positivism and Its Critics, in T. Bottomore and R. Nisbet (eds) *A History of Sociological Analysis,* 237–86, New York: Basic Books

—. (1984). *The Constitution of Society*, Cambridge: Polity

—. (1990). *The Consequences of Modernity*, Cambridge: Polity Press

—. (1991). *Modernity and Self-Identity. Self and Society in the Late Modern Age*, Cambridge, Polity Press

—. (1998). *The Third Way: The Renewal of Social Democracy*, Cambridge, Polity Press

—. (2000). *Runaway World.* 1st Edition, New York: Routeldge

Glass, K.C., Weijer, C., Cournoyer, D., Lemmens, T., Palmour, R.M. and Shapiro, S.H., (1996). Structuring the Review of Human Genetics Protocols: Gene Localisation and Identification Studies, A Review of Human Subjects Research In Lemmens, T. (2000) Selective Justice, Genetic Discrimination, and Insurance: Should We Single Out Genes in Our Laws?, *McGill Law Journal,* 347

Glenberg, A.M., Wilkinson, A.C. and Epstein. W. (1982). The illusion of knowing: failure in the self-assessment of comprehension. *Memorandum Cognition.* 10, 597–602 In Rock, E.M., Ireland, M., Resnick, M.D. and McNeely, C.E. (2005), A Rose by Any Other Name? Objective Knowledge, Perceived Knowledge, and Adolescent Male Condom Use, *Pediatrics,* 115, 667-672

Gostin, L. (1991). Genetic Discrimination: The Use of Genetically Based Diagnostic and Prognostic Tests by Employers and Insurers", 17 American Journal of Law and Medicine. 109 129 In Lemmens, T. (2000) Selective Justice, Genetic Discrimination, and Insurance: Should We Single Out Genes in Our Laws?, *McGill Law Journal*, 347, 355

Greenford, B.C. (2001). *The Management of Personal Injury Claims by Insurers in England and Ireland,* PhD available from University of Limerick

Greenford, B.C., Mullins, M., Garvey, J., Morris, L. and O'Meara, L. (2005). *International Insurance Markets. Between Global Dynamics and Local Contingencies*, Wharton University

Gregory, R. and Mendelsohn, R. (1993). Perceived risk, dread, and benefits. *Risk Analysis, 13,* 259-264

Grobe, D., Douthitt, R. and Zepeda, L. (1999). A model of consumers' risk perceptions toward recombinant bovine growth hormone (rbGH): The impact of risk characteristics. *Risk Analysis*, *19*, 661–673.

Groothius P.A. and Miller, G. (1997). The role of social distrust in risk-benefit analysis: a study of the siting of a hazardous waste disposal facility, *Journal of Risk and Uncertainty,* 15

Guba, E. and Lincoln, Y. (1994). Competing paradigms in qualitative research, In Denzin, N.K. and Lincoln, Y. (Eds), *Handbook of Qualitative Research*, Newbury Park, C.A.: Sage Publications, 105-117

Gustafson, P.E. (1998). Gender Differences in Risk Perception: Theoretical and Methodological Perspectives, *Risk Analysis, 18, (6), 805-811*

Gutteling, J. M. and Wiegman, O. (1993). Gender-specific reactions to environmental hazards in the Netherlands. *Sex Roles*, *28*, 433–447

Hall, Mark A., McEwen, J.E., Barton, J. C. , Walker, A.P., Howe, E.G., Reiss, J.A., Power, T.E., Ellis, S.D., Tucker, D.C., Harrison, B.W., McLaren, G.D., Ruggiero, A. and Thomson, E.J. (2005). Concerns in a primary care population about genetic discrimination by insurers. *Genetics in Medicine. 7*, (5), 311-316

Hansjurgens, B. (2000). Symbolische umweltpolitik – eine erklarung aus sicht der neuen politschen okonomie, in B. Hansjurgens and G Lubbe-wolff (eds) Symbolische Umweltpolitik, pp.144-82. Frankfurt/Main: Suhrkamp in D. Matten (2004) The impact of the risk society thesis on environmental politics and management in a globalising economy – principles, proficiency, perspectives, *Journal of Risk Research*, *7* (4), 377-398

Harper, P. and Clarke, A. (1990). Should we test children for 'adult' genetic diseases? Lancet 335, 1205-6 In Gaivoronskaia, G. and Solem, K.E. (2004). Genetic testing: affected parties and decision-making, *Journal of Risk Research, 7* (5), 481-493

Haughton, J. (2005). Growth in Output and Living Standards In O'Hagan, J. and Newman, C. (Eds.), *The Economy of Ireland: National & Sectoral Policy Issues*, 107-135. Dublin: Gill & Macmillan

Hawkins, D. (1997). Dangerous legacies; new gene tests provide fresh grounds for discrimination, Health Guide, http://ghs.gresham.k12.or.us/science/ps/sci/soph/gendiscrim/10priv.htm (last assessed January 2004)

Healy, M. and Perry, C. (2000). Comprehensive criteria to judge validity and reliability of qualitative research within the realism paradigm, *Qualitative Market Research: An International Journal*, 3, (3), 118-126

Healy, M. and Perry, C. (2000). Validity and Reliability of Qualitative Research within the Realism Paradigm. *Qualitative Research Journal: An International Journal.* 3 (3), 118-126.

Hebel, A. (2002), Parametric versus non-parametric – when to use them and which is more powerful, http://www.umes.edu/sciences, (last assessed January 2005)

Hilgartner, S. (1992). The Social Construction of Risk Objects: Or, How to Pry Open Networks of Risk. in J.F. Short, Jr. and L. Clarke (eds) Organizations, Uncertainties and Risk, 39–53, Boulder, CO: Westview Press. In Rosa, E.A. (1998), Metatheoretical foundations for post-normal risk, *Journal of Risk Research,* (1), 15–44

Hoban, J. (1997). Consumer acceptance of biotechnology: an international perspective, *Nature Biotechnology,* 15, 232-234

Hofstede, G. (1980). *Culture's consequences: International differences in work-related values.* Newbury Park, CA: Sage

Hogan, D. (2001). Cork reels under closure decision, *The Irish Times*, 16 June

Hofstede, G. (2005). Cultural Dimensions, http://www.geerthofstede. com/geert_hofstede_ contrarian _position.shtml (last assessed March 2006)

Holmwood, J. (2001). Gender and Critical Realism: A Critique of Sayer, *Sociology, 35* (4), 947–965

Holzheu, F. and Weidemann, P.M. (1993). Perspectives on Risk Perception, In *Risk is a Construct,* Munich: Bayerische Ruck

House of Lords. (2000). Science and Society, 3rd Report. London: HMSO. http:www.parliament.the-stationery-office.co.uk/pa/ld199900 /ldselect/ldsctech/38/3801.htm. (last assessed April 2004)

Hovland, C.I., Janis, I.L. and Kelley, H.H. (1953). *Communication and Persuasion. Psychological Issues of Opinion Change.* New Haven: Yale University Press

Howell, D.C., (1994). *Fundamental Statistics for the Behavioural Sciences*, Pacific Grove: Duxbury Press

Hughes, J. (1994). *The Philosophy of Social Research.* Essex, Longman

Human Genome Project Information, http://www.ornl.gov/hgmis/project/about.html

Hume, D. (1784/1984), *Enquiry Concerning Human Understanding.* New York, Bobbs-Merril Co.

Hunt, S., Frewer, L.J. and Shepherd, R., (1999), Public trust in sources of information about radiation risks in the UK, *Journal of Risk Research*, 2, 167-180

Hunt, S. (1991). Modern Marketing Theory, South-Western Publishing Co, Cicinnatti, In Healy and Perry (2000). Comprehensive criteria to judge validity and reliability of qualitative research within the realism paradigm, *Qualitative Market Research: An International Journal*, Vol. 3, No. 3, p118-126

Husted, J. (1999). Insurance, Genetics and Solidarity, In McGleenan, T., Wiesing, U. and Ewald, F. (1999). *Genetics and Insurance*, Oxford: BIOS Scientific Publishers Ltd

Inhaber, H. (1979). Risk with energy from conventional and nonconventional sources, Science 203, 718–23. In Rosa, E.A. (1998), Metatheoretical foundations for post-normal risk, *Journal of Risk Research*, (1), 15–44

Introna, L., (1997). Management, Information and Power: A narrative of the involved manager, MacMillan, London In Niehaves, B. and Stahl, B.C., (2006). Criticality, epistemology and behaviour vs. design – information systems research across different sets of paradigms *Information Systems Journal* (to appear)

Irish Financial Services Regulatory Authority (2004). *Independent Consumer Guide to Life Insurance*

—. (2005). *Independent Consumer Guide to Mortgages*

Irish Insurance Federation, (2001). *Factfile* 2001

—. (2001). *Code of Practice on Genetic Testing*

—. (2003). Research Highlights Public Attitudes to Insurance Fraud, Personal Injury Awards, Legal Costs and Premium Levels, http://www.iif.ie/media.htm (last assessed February 2006)

—. (2004). *IIF Annual Report*

—. (2004). *IIF Submission on the Disability Bill 2004* to the Joint Committee on Justice, Equality, Defence and Women's Rights

—. (30/8/2006). *IIF Welcomes Publication of 2005 Insurance Statistics*, Press Release, www.iif.ie/news-latest-article (last assessed February 2006)

Jacobsen P.B., Valdimarsdottier, H.B., Brown, K.L. and Offit, K. (1997). Decision-Making About Genetic Testing Among Women at Familial Risk for Breast Cancer 59 *Psychosometrika Medicine,* 459-460

Jallinoja, P. and Aro, A. R. (2000). Does knowledge make a difference? The association between knowledge about genes and attitude toward gene tests. *Journal of Health Communication,* 5, 29-39

Johansson, R. (2002). Theory of Science and Research Methodology, *Nordic Journal of Architectural Research*, 8, 33-49

Johnson, P. and Duberly, J. (2000) *Understanding management research: An introduction to epistemology*. London: Sage

Jungermann, H., Pfister, H.R., and Fischer, K. (1996). Risk, trust and democratic theory. Risk Analysis, 16(2), 251-261 In Poortinga, W. and Pidgeon, N.F. (1991), Exploring the dimensionality of trust in risk regulation, *Risk Analysis, 23* (5), 961-972

Jungermann, H. and Slovic, P. (1993), Characteristics of Individual Risk Perception. In *Risk is a Construct,* Munich: Bayersiche Ruck

Jungermann, H. Pfister, H.R. and Fischer, K. (1996). Credibility, Information Preferences, and Information Interests, *Risk Analysis, 16* (2), 251-261

Kaiser, H.F. (1974). An index of factorial simplicity, Psychometrika, 39, 31-36 In Field, A. (2004), *Discovering Statistics using SPSS for Windows*, Sage Publications, London

Kant, I. (1964 [1781]). *Critique of Pure Reason,* London: Macmillan

Kates, R.W. and Kasperson, J.X. (1983) Comparative risk analysis of technological hazards (a review), *Proceedings of the National Academy of Sciences* 50, 7027–38.

Kasperson, R.E. (1992). The Social Amplification of Risk: Progress in Developing an Integrative Framework, In Krimsky, S. and Golding, D. (1992). *Social Theories of Risk*, London: Praeger

Kasperson, R.E., Golding, D. and Tuler, S. (1992), Siting hazardous facilities and communicating risks under conditions of high social distrust, *Journal of Social Issues*, 48(4), 161-187

Kasperson, R.E., Golding, D. and Kasperson, J.X. (1999), Risk, Trust and Democratic Theory, In Cvetkovich, G., Lofstedt, R.E. (1999) *Social Trust and the Management of Risk*, London: Earthscan

Kolakowski, L. (1972). *Positivist science.* Harmondsworth, England: Penguin Books.

Koné' D. and Mullet' E. (1994), Societal Risk Perception and Media Coverage, *Risk Analysis,* 14, 114-119

Kosseim, P., Letendre, M. and Knoppers, B.M. (2004) *Protecting genetic information: a comparison of normative approaches*, HumGen, 2 (1), http://www.humgen.umontral.ca/en/GenEdit.cfm, (last assessed January 2006)

Kramer, R.M. (1999). Trust and Distrust in Organisations: Emerging Perspectives, Enduring Questions, *Annual Review of Psychology*, 50, 569-598

Kuhn, T.S., (1962). *The Structure of Scientific Revolutions*. Chicago: University of Chicago Press

Krimsky, S. & Golding, D. (1992) *Social Theories of Risk*, London: Praeger

Lang, J.T. and Hallman, W.K. (2005). Who Does the Public Trust? The Case of Genetically Modified Food in the United States, *Risk Analysis, 25* (5), 1241-1252

Langford, I., Georgiou, S., Bateman, I.J., Day, R.J. and Turner R.K., (2000). Public Perceptions of Health Risks from Polluted Coastal Bathing Waters: A Mixed Methodological Analysis Using Cultural Theory, *Risk Analysis*, 20, (5), 691-704

Lawson, T. (1997). *Economics and Reality,* London: Routledge

Layder, D. (1993). *New Strategies in Social Research*, Cambridge: Polity Press

Lemmens, T. (2000). Selective Justice, Genetic Discrimination, and Insurance: Should We Single Out Genes in Our Laws?, *McGill Law Journal*.347, 355

Lemmens, T., Joly, Y. and Knoppers, B.M. (2004). *Genetics and life insurance: a comparative analysis*, HumGen, Volume 2, No. 2, http://www.humgen.umontral.ca/en/GenEdit.cfm (last assessed November 2005)

Lewis, J.D. and Weigert, A. (1985). Trust as a Social Reality, Social Forces, *Social Forces,* 63, 967-85

Lipset, S.M. and Schneider, W. (1983). *The Confidence Gap: Business, Labor, and Government in the Public Mind*, Baltimore: John Hopkins University Press

Locke, J. (1894). An essay concerning human understanding. A.c. Fraser (ed.): Vol. 1, Clarendon, Oxford. In Dobson, P.J. (2002). Critical realism and information systems research: why bother with philosophy?, *Information Research*, 7 (2)

Lomax, G.P. (2000). From Breeder Reactors to Butterflies: Risk, Culture and Biotechnology, *Risk Analysis*, 20 (5)

Lowrance, W. (1976). Of Acceptable Risk: Science and the Determination of Safety, Los Altos, CA:William Kaufmann. In Rosa, E.A. (1998). Metatheoretical foundations for post-normal risk, *Journal of Risk Research,* (1), 15–44

Luhmann, N. (1979). *Trust and Powe*r. Chichester, UK: Wiley.

—. (1988). Familiarity, confidence, trust: Problems and alternatives. In D. Gambetta (Ed), Trust: Making and Breaking Cooperative Realtions (pp. 94-108). Oxford: Basil & Blackewell *from* Viklund, M.J. (2003), Trust and risk perception in Western Europe: A cross-national study, *Risk Analysis, 23* (4)

—. (1993). *Risk: A Sociological Theory*, Berlin: Walter de Gruyter

Lythcott, J. and Duschl, R. (1990). Qualitative research: From methods to conclusions. *Science Education, 74,* 449-460

Macdonald, A.S. (1997). How will improved forecasts of individual lifetimes affect underwriting? Philosophical transactions of the Royal Society B, 352, 1067-1075, and (with discussion) *British Actuarial Journal,* 3, 1009-1025 and 1044-1058

—. (1999). Modelling the impact of genetics on insurance, *North American Actuarial Journal,* 3 (1), 83-101

—. (2001). Moratoria on the use of genetic tests and family history for mortgage-related life insurance. Genetics and Insurance Research Centre, *Heriot-Watt University: Research Report* No. 01/3

Macer, D.R.J. (2000). Ethical challenges as we approach the end of the Human Genome Project, *Eubios Ethics Institute*, New Mexico, U.S.A.

Magee. B. (1985). Popper, London, Fontana In Healy and Perry, (2000). Comprehensive criteria to judge validity and reliability of qualitative research within the realism paradigm, *Qualitative Market Research: An International Journal,* 3 (3), 118-126

March J.G., Olsen J.P. (1994). *Democratic Governanc*e. New York: Free Press

Matten, D. (2004). The impact of the risk society thesis on environmental politics and management in a globalising economy – principles, proficiency, perspectives, *Journal of Risk Research,* **7** (4), 377-398

Matthews, M. (1992). Old wine in new bottles: A problem with Constructivist Epistemology, Philosophy of Education, http://www.ed.uiuc.edu/EPS/PES-Yearbook/92_docs/Matthews.HTM (last assessed October 2006)

Marris, C., Langford, I.H. and O'Riordan, T. (1998). A quantitative Test of the Cultural Theory of Risk Perception: Comparison with the Psychometric Paradigm, *Risk Analysis,* 18, 635-647

McGleenan, T. (2001). *Insurance and Genetic Information,* Association of British Insurers Conference, London, 8 November

McGleenan, T., Wiesing, U. and Ewald, F. (1999). *Genetics and Insurance*, Oxford: BIOS Scientific Publishers Ltd

McGovern, M. (2005). *Interview with representative of IFSRA*, 21/12/2005

Metlay, D. (1999). Institutional Trust and Confidence: A journey into a Conceptual Quagmire In Cvetkovich, G. and Lofstedt, R.E., *Social Trust and the Management of Risk*, London: Earthscan

Micceri, T. (1989). The unicorn, the normal curve, and other improbable creatures. *Psychological Bulletin*, 105:156-166.

Miles, M.B. and Huberman, A.M. (1994). *Qualitative Data Analysis*, (Second Edition) London: Sage Publications. Inc. Thousand Oaks

Miller, R.L. and Brewer, J.D. (2003). *The A-Z of Social Research*, Sage Publications, London

Miranda, S.M. and Saunders, C.S. (2003). The Social Construction of Meaning: An Alternative Perspective on Information Sharing, Information Systems Research (14:1), 87 – 106, In Niehaves, B. and Stahl, B.C., (2006). Criticality, epistemology and behaviour vs. design – information systems research across different sets of paradigms *Information Systems Journal* (to appear)

Misztal, B.A. (1996). *Trust in Modern Societies: The Search for the Bases of Social Order*, Cambridge: Polity Press

Moran, D. (2000). *Introduction to Phenomenology,* London, New York: Routledge

Morgan, M.G. (1995). Risk Analysis: A personal assessment of the state of the . eld, Society for Risk Analysis 1995 Distinguished Contribution Award's Keynote Address, annual meeting of the Society for Risk Analysis, Honolulu, Hawaii (3–6 December). In Rosa, E.A. (1998). Metatheoretical foundations for post-normal risk, *Journal of Risk Research,* (1), 15–44

Moultrie, T. A. and Thomas, R. G. (1997). The Right to Underwrite? An Actuarial Perspective with a Difference (with discussion), *Journal of Actuarial Practice* 5, (1), 125–47 In Macdonald, A.S. (1999). Modelling the impact of genetics on insurance, *North American Actuarial Journal,* 3:1, 83-101

Mouzelis, N. (2000). The Subjectivist-Objectivist Divide: Against Transcendence, *Sociology* 34:741–62.

Murray, T.H. (1997). Genetic Exceptionalism and 'Future Diaries': Is Genetic Information Different from Other Medical Information? In M.A. Rothstein, ed., *Genetic Secrets: Protecting Privacy and Confidentiality in the Genetic Era* New Haven: Yale University Press

Natowicz, M.R., Alper J.K. and Alper, J.S. (1992). Genetic Discrimination and the Law, American Journal of Human Genetics, 465 In J.S In Lemmens, T. (2000) Selective Justice, Genetic Discrimination, and Insurance: Should We Single Out Genes in Our Laws?, *McGill Law Journal*, 347, 355

Neimeyer, R. A. and Raskin, J. D. (Eds.). (2000). Constructions of disorder: Meaning-making frameworks for psychotherapy. Washington, DC: American Psychological Association

Nielsen, J.B., Gyrd-Hansen, D., Kristiansen, I.S. and Nexse, J. (2003). Impact of socio-demographic factors on willingness to pay for the reduction of a future health risk, *Journal of Environmental Planning and Management*, *48* (1), 39-48

Okrent, D. (1975). Risk-Benefit Methodology and Application, UCLA-ENG-7598, Los Angeles:

University of California. In Rosa, E.A. (1998). Metatheoretical foundations for post-normal risk, *Journal of Risk Research,* (1), 15–44

Oltedal, S., Moen, B.E., Klempe, H. and Rundmo, T. (2004). Explaining Risk Perception, an Evaluation of Cultural Theory, *Rotunde*, 85

O'Neill, O. (1997). Genetic Information and Insurance: Some Ethical Issues, Philosophical Transactions of the Royal Society B, 352, 1087–93 In Macdonald, A.S. (1999). Modelling the impact of genetics on insurance, *North American Actuarial Journal,* 3 (1), 83-101

O'Neill, Baroness (1999). *When and Why is Discrimination Wrong?, UK Forum for Genetics and Insurance*, Report of a one-day seminar, 26 July, Royal Society London

Otway, H. (1992). Public Wisdom, Expert Fallibility: Toward a Contextual Theory of Risk, In Krimsky, S. and Golding, D. (1992), *Social Theories of Risk*, London: Praeger

Outhwaite, W. (1983). Toward a realist perspective. In G. Morgan (Ed.), *Beyond method: Strategies for social research*, 331-330, Beverly Hills, CA: Sage

Palmer, C.G.S. (1996). Risk Perception: An Empirical Study of the Relationship Between Worldview and the Risk Construct. *Risk Analysis*, 16, 717-723

Pardo, R., Midden, C., & Miller, J. D. (2002). Attitudes toward biotechnology in the European Union. *Journal of Biotechnology, 98*, 9-24

Park CW, Gardner MP, Thukral VK. (1998). Self-perceived knowledge: Some effects on information processing for a choice task. *American Journal of Psychology,* 101:401–424 In Rock, E.M., Ireland, M., Resnick, M.D. and McNeely, C.E. (2005), A Rose by Any Other Name? Objective Knowledge, Perceived Knowledge, and Adolescent Male Condom Use, *Pediatrics,* 115, 667-672

Parker, M., Chartered Insurance Institute (1989). *Life Assurance Practice, Book Production Consultants*, 47 Norfolk Street, Cambridge

Parkhe, A. (1993). Messy research, methodological predispositions, and theory development in international joint ventures, *Academy of Management Review*, 18, 2, 227-268 In Perry, C., Riege, A. and Brown, L. (1999). Realism's role among scientific paradigms in marketing research, *Irish Marketing Review*, December

Patton, M.Q. (1990). *Qualitative Evaluation and Research Methods,* 2nd Edition, Sage Publications, London

Pauly, M.V. (1974). Overinsurance and Public Provision of Insurance: The Roles of Moral Hazard and Adverse Selection, *The Quarterly Journal of Economics*, MIT Press, 88(1), 44-62

Pawson, R. and Tilley, N. (1997). *Realistic Evaluation*, London: Sage Publications

Perry, C., Alizadeh, Y. and Riege, A. (1997). *Qualitative methods in entrepreneurship research*, Proceedings, Small Enterprise Association of Australia and New Zealand, Southern Cross University, Coffs Harbour

Perry, C., Riege, A. and Brown, L. (1999). Realism's role among scientific paradigms in marketing research, *Irish Marketing Review*, December

Peters, E. and Slovic, P. (1996). The role of affect and worldviews as orienting dispositions in the perceptions and acceptance of nuclear power, *Journal of Applied Social Psychology*, 26, 1427-1453

Petersen, A. (1963). The philosophy of Niels Bohr, *Bulletin of the Atomic Scientists*, 19, 6-14

Petts, J. (1998). Trust and waste management information expectation versus observation, *Journal of Risk Research, 1* (14)

Pidgeon, N., Hood, C., Jones, D ., Turner, B . and Gibson, R. (1992). Risk perception. In Royal Society Study Group (Ed.), *Risk: Analysis, perception and management,* 89-134, London: The Royal Society

Pijawka, K. D. and Mushkatel, A. H. (1991). Public opposition to the sitting of the high-level nuclear waste repository: The importance of trust. *Policy Studies Review*, *10*, 180–194

Pisano, C. and Woods, M.D. (2002). Communicating food safety across cultures, issues of trust and credibility within diverse populations, Paper presented at the *Agricultural Communicators in Education (ACE) Conference*, Savannah, Georgia, USA

Pokorski, R.K. (1995). "Genetic Information and Life Insurance" 376 Nature 13 In Lemmens, T. (2000) Selective justice, genetic discrimination, and insurance: Should we single out genes in our laws? *McGill Law Journal*, 347

Poortinga, W. and Pidgeon, N.F. (2003). Exploring the dimensionality of trust in risk regulation, *Risk Analysis,* 23 (5), 961-972

Poortinga, W., Bickerstaff, K., Langford, I. H., Niewohner, J. and Pidgeon, N. F. (2004). The British 2001 foot and mouth cri-sis: A comparative study of public risk perceptions, trust and beliefs about government policy in two communities. *Journal of Risk Research, 7* (1), 73-90

Poortinga, W. and Pidgeon, N.F. (2005). Trust in risk regulation: Cause or consequence of the acceptability of GM food? *Risk Analysis, 25* (1), 199-209

Popper, K. (1972). Objective Knowledge, Oxford, Clarendon, In Smith, M.J., (1998) *Social Research in Question*, London: Sage Publications

Poumadere, M. (1995). Enjeux de la communication publique des riskques pour la sante et l'environment, European Review of Applied Psychology, 45 In Viklund, M.J. (2003), Trust and risk perception in Western Europe: A cross-national study, *Risk Analysis, 23* (4)

Radecki, C.M. and Jaccard, J. (1995). Perceptions of knowledge, actual knowledge, and information search behavior. *Journal Exp Sociolgy Psychology,* 31, 107–138 In Rock, E.M., Ireland, M., Resnick, M.D. and McNeely, C.E. (2005), A Rose by Any Other Name? Objective Knowledge, Perceived Knowledge, and Adolescent Male Condom Use, *Pediatrics,* 115, 667-672

Rayner, S. (1992). Cultural Theory and Risk Analysis, In Krimsky, S. and Golding, D. (eds). (1992). *Social Theories of Risk* Westport, CT: Praeger

Renn, O. (1992). Concepts of Risk: A Classification In Krimsy, S. and Golding, D. (1992). *Social Theories of Risk*, London: Praeger

—. (1998). Three decades of risk research: accomplishments and new challenges, *Journal of Risk Research* 1, 49–71

Renn, O. and Rohrmann B. (2000). *Cross-Cultural Risk Perception: A Survey of Research Results,* Dordrecht and Boston: Kluwer

Renn, O. and Levine, D. (1991). Credibility and trust in risk communication. In R. E. Kasperson and P.J.M. Stallen (Eds.), Communicating Risks to the Public, The Hague: Kluwer, In Poortinga, W. and Pidgeon, N.F. (2003). Exploring the dimensionality of trust in risk regulation, *Risk Analysis,* 23 (5), 961-972

Remenyi, D., Williams, B., Money, A. and Swartz, E. (1998). *Doing Research in Business and Management: An Introduction to Process and Method,* London: Sage Publications

Ring, P.J. (2004). Providing security in pensions – a critical analysis of UK government policy, *26th UK Insurance Economists Conference*, March 30-31

Rippl, S. (2002). Cultural theory and risk perception: a proposal for a better measurement, *Journal of Risk Research*, 5, 147-165

Robertson, J.A., Brody, B., Buchanan, A., Kahn, J and McPherson, E. (2002). Pharmacogenetic challenges for the health care system, *Health Affairs,* 21 (4)

Robson, C. (2002). *Real World Research: A Resource for Social Scientists and Practitioner Researchers* (Second Edition), Oxford, U.K., Blackwell Publishing

Rock, E., Ireland, M., Resnick, M.D.and McNeely C.A. (2005). A Rose by Any Other Name? Objective Knowledge, Perceived Knowledge, and Adolescent Male Condom Use, *Pediatrics,* 115, 667-672

Rosa, E.A. (1998). Metatheoretical foundations for post-normal risk, *Journal of Risk Research,* (1), 15–44

Rose, N. (2000). Community, Citizenship and The Third Way, in D. Meredyth and J. Minson, eds., *Citizenship and Cultural Policy,* London: Sage

Rothstein, M.A. (1994). Genetic Discrimination in Employment: Ethics, Policy and Comparative Law In Swiss Institute of Comparative Law, ed., *Human Genetic Analysis and the Protection of Personality and Privacy* (Zürich: Schulthess Polygraphisher Verlag) 129 at 129 In Lemmens, T. (2000). Selective Justice, Genetic Discrimination, and Insurance: Should We Single Out Genes in Our Laws?, *McGill Law Journal*.347

Rowe, W.R. (1977). An Anatomy of Risk, New York: John Wiley & Sons In Rosa, E.A. (1998), Metatheoretical foundations for post-normal risk, *Journal of Risk Research,* (1), 15–44

Rundmo, T. (1995). Perceived risk, safety status and job stress among injured and non-injured employees on offshore petroleum installations. *Journal of Safety Research*, 26(2), 87-97

Saparito, P. and Chen, C. (2001). *Trust's Multidimensional Form and Role in Bank-Small Firm Relationships*, Academy of Management Annual Meeting

Sarle, W.S. (1995). *Disseminations of the International Statistical Applications Institute*, 1(4), Wichita: ACG Press, 61-66

Saunders, M., Lewis, P. and Thornhill, A. (2003). *Research Methods for Business Students*, 3rd Edition, Essex, England: Pearson Education Ltd

Satterfield, T.A., Mertz, C.K. and Slovic, P. (2004). Discrimination, Vulnerability, and Justice in the Face of Risk, *Risk Analysis,* 24 (1), 115-129

Sayer, A. (1992). *Method in Social Science: A Realist Approach*, London: Routledge

Schmied, J. (1993). Qualitative and quantitative research approaches to English relative constructions In Souter, C. and E. Atwell, eds. (1993). *Corpus Based Computational Linguistics*, Amsterdam: Rodopi, 85-96

Schwandt, T.A., (1994). Constructivist, Interpretivist Approaches to Human Inquiry, in Denzin, N.K., and Lincoln, Y.S. eds., (1994).

Handbook of Qualitative Research, Thousand Oaks, CA: Sage Publications, Inc., 118-137

Sellars, R.W. (1908). Critical Realism and the Time Problem: II, *The Journal of Philosophy, Psychology and Scientific Methods,* 597-602, http://library.advanced.org/18617/data/types/dna.html (last assessed November 2006)

Shapiro S. (1987). Policing trust. In *Private Policing*, ed. CD Shearing, PC Stenning. Thousand Oaks, CA: Sage

Sheppard, B.H. & Sherman, D.M. (1998). The grammars of trust: a model and general implications, *Academy of Management Review*, 23 (3)

Shih, F.J. (1998). Triangulation in nursing research: issues of conceptual clarity and purpose, Journal of Advanced Nursing, 28, 3, 631-641, In Crossan, F. (2001) Research philosophy: towards an understanding, *Nurse Researcher,* 11 (1)

Short, J.F., Jr. (1984). The social fabric at risk: toward a social transformation of risk analysis, *American Sociological Review* 49, 711–25

Shrader-Frechette, K. (1991). Risk and Rationality: Philosophical Foundations for Populist Reforms, Berkeley and Los Angeles: University of California Press. In Rosa, E.A. (1998), Metatheoretical foundations for post-normal risk, *Journal of Risk Research*, (1), 15–44

Siegrist, M. (1999). A causal model explaining the perception and acceptance of gene technology, *Journal of Applied Social Psychology*, 29, 2093-2106

—. (2000). The Influence of Trust and Perceptions of Risks and Benefits on the Acceptance of Gene Technology. *Risk Analysis, 20,* 195-203

Siegrist, M., Cvetkovich, G. (2000). Perception of Hazards: The Role of Social Trust and Knowledge, *Risk Analysis, 20*, 713-719

Siegrist, M., Cvetkovich, G., Roth, C. (2000). Salient Value Similarity, Social Trust, and Risk/Benefit Perception, *Risk Analysis, 20*, 353-362

Siegrist, M. and Cvetkovic, G.T. (2001). Better negative than positive? Evidence of a bias for negative information about possible health dangers. Risk Analysis, 21 (1), 199-206

Siegrist, M., Earle, T.C. and Gutscher, H. (2003), Test of a trust and confidence model in the applied context of electromagnetic field (EMF) risks, *Risk Analysis,* 23 (4), 705-716

Siegrist, M., Earle, T.C. and Gutscher, H. (2005). Perceptions of risk: the influence of general trust, and general confidence, *Journal of Risk Research,* 8, 145-156

Siegrist, M., Keller, C. and Kiers' H.A.L. (2005). A New Look at the Psychometric Paradigm of Perception of Hazards, *Risk Analysis,* 25 (1), 211

Simon, S. (2005). *Parametric versus Non-Parametric Tests,* http://www.childrens-mercy.org (last assessed October 2006)

Sjöberg, L. (1996). *Risk perceptions by politicians and the public* Rhizikon: Risk Research Reports 23. Stockholm: Centre for Risk Research.

—. (1996a). A discussion of the limitations of the psychometric and cultural theory approaches to risk perception, Radiation Protection Dosimetry, 68219-225, In Siegrist, M., Cvetkovich, G., (2000). Perception of Hazards: The Role of Social Trust and Knowledge, *Risk Analysis, 20*, 713-719

—. (1997). Explaining risk perception: an empirical evaluation of cultural theory, *Risk Decision and Policy*, 2(2), 113-130

—. (1998). Risk Perception: Experts and the Public, *European Psychologist, 3*, 1-12

—. (1999). Perceived competence and motivation in industry and Government as factors in risk perception, 89-99, In G. Cvetknovich & R.E. Lofstedt (Eds) (1999). *Social Trust and the Management of Risk*, London, Earthscan

—. (1999a). Consequences of Perceived Risk: Demand for Mitigation, *Journal of Risk Research* , 2 (2), 129-149

—. (2000). Factors in risk perception, *Risk Analysis*, 20, 1-11

—. (2001). Limits of Knowledge and the Limited Importance of Trust, *Risk Analysis*, 21 (1), 189-198

—. (2002). Attitudes to technology and risk: Going beyond what is immediately given. Policy Sciences, 35, 379-400 In Viklund, M.J. (2003). Trust and risk perception in Western Europe: A cross-national study, *Risk Analysis*, 23 (4)

—. (2005). Gene Technology in the Eyes of the Public and Experts, Moral Opinion, Attitudes and Risk Perception, *Working Paper Series in Business Administration*, 1-63

Sjoberg, L. and Drottz-Sjoberg, B. (2001). Fairness, risk and risk tolerance in the siting of a nuclear waste repository, *Journal of Risk Research,* 4, 75-101

Skipper, H.D. (1998). *International Risk and Insurance: An Environmental – Managerial Approach*, Boston: Irwin McGraw-Hill

Slovic, P. (1992). Perception of risk: Reflections on the psychometric paradigm. In S. Krimsky & D. Golding (Eds.) (1992). *Social theories of risk*, 117-152, New York: Praeger.

—. (1993). Perceived risk, trust and democracy, *Risk Analysis,* 13, 675-682

—. (2000). Trust, emotion, sex, politics, and science: surveying the risk-assessment battlefield, *Risk Analysis*, 19, 689

Slovic, P., Finucane, M.L., Peters, E. and MacGregor, D.G. (2002). Risk as analysis and risk as feelings: some thoughts about affect, reason, risk, and rationality, *Risk Analysis*, 24 (2)

Smith, A.D. (1995). *Nations and Nationalism in a Global Era*, Cambridge, Polity Press

Smith, M.J. (1998). *Social Science in Question*, London, Sage

Society of Actuaries in Ireland (2004). *Briefing Statement on Insurance provisions in the Disability Bill 2004*, 1-8

Stake, R.E. (1995). *The Art of Case Study Research,* Thousand Oaks: Sage Publications

—. (1998). Case Studies, In Denzin, N.K. and Lincoln, Y.S. *Strategies of Qualitative Inquiry,* London: Sage Publications

Starr, C. (1969). Social benefit versus technological risk: what is our society willing to pay for safety, *Science* 165, 1232–38

Statsoft (2003). Electronic Textbook, http://www.statsoft.com/textbook (last assessed October 2006)

Stevens, J. (1997). Applied Multivariate Statistics for the Social Science (3rd Edition), Lawrence Erlbaum Associate In Dancey, C.P. and Reidy, J. (2002). *Statistics Without Maths for Psychology using SPSS for Windows*, London: Prentice Hall

Sweeney, S. (2006). The Data Protection Agency is not aware of any other cases specifically involving unlawful access to medical records held by underwriters, *Interview response* (20/6/06)

Tabachnick, B.G. and Fiddell, L.S. (1997). Using Multivariate Statistics (3rd Edition), Lawrence Erlbaum Associates In Dancey, C.P. and Reidy, J. (2002). *Statistics Without Maths for Psychology using SPSS for Windows*, London: Prentice Hall

The United Kingdom Parliament, (25 October, 2001), Genetics and Insurance, Column 139WH, http://www.parliament.the-stationery-office.co.uk/pa/cm200102/cmhansrd/vo011025/halltext/11025h01.htm (last assessed January 2005)

Thompson, M., Ellis, R., and Wildavsky, A. (1990). *Cultural Theory*, Boulder, CO: Westview Press

Trauth, E., and Jessup, L.M. (2000). Understanding Computer-Mediated Discussions: Positivist and Interpretive Analyses of Group Support System Use, MIS Quarterly (24:1), 43-79, In Niehaves, B. and Stahl, B.C. (2006). Criticality, epistemology and behaviour vs. design –

information systems research across different sets of paradigms *Information Systems Journal* (to appear)

Trauth, E. (2001). Choosing Qualitative Methods in IS Research: Lessons Learned, Qualitative Research in IS: Issues and Trends, E. Trauth (ed.), Idea Group Publishing, Hershey, 271 – 287 In Niehaves, B. and Stahl, B.C., (2006). Criticality, epistemology and behaviour vs. design – information systems research across different sets of paradigms *Information Systems Journal* (to appear)

Trochim, W. M. (2000). *The research methods knowledge base*. Retrieved November 13, 2003, from http://www.socialresearchmethods.net/kb/ (last assessed November 2006)

Tsoukas, H. (1989). The validity of ideograph research explanations. Academy of Management Review, 14(4), 551-561. In Krauss, S.E. (2005), Research Paradigms and Meaning Making: A Primer, *The Qualitative Report, 10* (4), 758-770

Turney, J. (1998), *Frankenstein's Footsteps: Science, genetics and popular culture*, London: Yale University Press

Tyler, A. (2004). *Seeking a Solution I and II,* Conference on Insurance, Genetics and Fairness II, Genetics and Insurance Committee and Human Genetics Commission, Imperial College, London, 13 July

Universal Declaration on the Human Genome and Human Rights, UNESCO, 29th Session, 29

U.S. Presidential / Congressional Commission on Risk Assessment and Risk Management (Risk Commission) 1997 Framework for Environmental Health Risk Management (GPO Publication no. 055-000-00567-2), http://www.riskworld.com/nreports/1996/risk_rpt/RR6ME001.HTM (last assessed April 2004)

Varey, R.J., Wood-Harper, T., and Wood, B. (2002). A Theoretical Review of Management and Information Systems Using a Critical Communications Theory, Journal of Information Technology (17:1), 229 – 239 In Niehaves, B. and Stahl, B.C., (2006). Criticality, epistemology and behaviour vs. design – information systems research across different sets of paradigms *Information Systems Journal* (to appear)

Viaene, S. and Dedene, G. (2004). Insurance fraud: issues and allegations, *The Geneva Papers on Risk and Insurance,* 29 (2), 313-333

Vickers, A.J. (2005). Parametric versus non-parametric statistics in the analysis of randomized trials with non-normally distributed data, *BMC Medical Research Methodology*, (5)

Viggiani, F.A. (1999). 'Doing the right thing' – Organisational structure and process for democratic governance in the firm, *Industrial Relations Journal* 30, 229-42

Viklund, M. (2002). An expectancy-value approach to determinants of trust, *SSE/EFI Working Paper Series in Business Administration*, 2002, 13

Viklund, M.J. (2003). Trust and risk perception in Western Europe: A cross-national study, *Risk Analysis,* Vol. 23, No.4

Von Glasersfeld, E. (1989). Cognition, Construction of Knowledge, and Teaching, *Synthese*, 80 (1), 121-140

Voss, G. (2000). *Report to the Human Genetics Commission on Public Attitudes to the Uses of Human Genetic Information* London: Human Genetics Commission.

Wad, P. (2001). Critical realism and Comparative Sociology, 5th IACR Conference RUC 17-19 August

Walsham, G. (1995). interpretative case studies in IS research: nature and method, European Journal of Information Systems, 4, 74-81, In Dobson, P.J. (2002). Critical realism and information systems research: why bother with philosophy?, *Information Research*, 7 (2)

Weber, R. (2004). The Rhetoric of Positivism Versus Interpretivism, MIS Quarterly, 28 (1), iii-xii. In Niehaves, B. and Stahl, B.C. (2006). Criticality, epistemology and behaviour vs. design – information systems research across different sets of paradigms *Information Systems Journal* (to appear)

Webster, N. (1983). *Webster's new twentieth century dictionary* (2nd). New York: Simon and Schuster.

Wester-Herber, M. and Warg, L.E. (2002). Gender and regional differences in risk perception: results from implementing the Seveso II Directive in Sweden, *Journal of Risk Research, 5* (1), 69-81

White, M.P., Pahl, S., Buehner, M.J. and Haye, A. (2003). Trust in risky messages: The role of prior beliefs. *Risk Analysis, 23* (4), 717-726

Whipple, C. (1992). Inconsistent Values in Risk Management, in S. Krimsky and D. Golding (eds) *Social Theories of Risk*, 343–54, Westport, CT: Praeger

Wildavsky, A. and Dake, K. (1990). Theories of risk perception: Who fears what and why? *Daedalus,* 119, 41-60.

Wilson, R. and Crouch, E. (1982). *Risk/Benefit Analysis*, Cambridge, MA: Ballinger Publishing Co.

Winner, L. (1991). Upon Opening the Black Box and Finding it Empty: Social Constructivism and the Philosophy of Technology In Pitt, J., and E. Lugo, eds. (1991). *The Technology of Discovery and the*

Discovery of Technology. Blacksburg: Society for Philosophy and Technology

Wittgenstein, L. (1953). *Philosophical Investigations,* Oxford, Blackwell

Wollin, A.S. (1995). A hierarchy-based punctuated-equilibrium model of the process of emergence and change of new rural industries. Unpublished doctoral thesis, Griffith University, Faculty of Commerce and Administration, Brisbane, Australia. In Krauss, S.E. (2005). Research Paradigms and Meaning Making: A Primer, *The Qualitative Report* 10 (4), 758-770

Wright, G., Pearman, A. & Yardley, K. (2000). Risk perception in the U.K. oil and gas production industry: are expert loss-prevention managers' perceptions different from those of members of the public?, *Risk Analysis*, 20, 681-690

World Medical Association, 44th Assembly (1992). *World Medical Association Declaration on the Human Genome Projec*t, Doc. 17 S/1

Wynne, B. (1992). Risk and Social Learning: Reification to Engagement, in Krimsy, S. & Golding, D. (1992). *Social Theories of Risk*, London: Praeger

Yamagishi, T. (2000). Session on Trust at the 27th International Congress of Psychology, Stockholm, Sweden, July 23-28, 2000; In Viklund, M.J. (2003). Trust and risk perception in Western Europe: A cross-national study, *Risk Analysis,* 23 (4)

Yesley, M.S. (1999). Protecting Genetic Difference, Berkeley Technology Journal, 653-662 In Lemmens, T. (2000). Selective Justice, Genetic Discrimination, and Insurance: Should We Single Out Genes in Our Laws?, *McGill Law Journal,* 347, 355

Yin, R. K. (1994). *Case Study Research, Design and Methods,* 2nd ed., Newbury Park: Sage Publications

Zimmerman, D. W. (1994). A note on the influence of outliers on parametric and nonparametric tests. *Journal of General Psychology, 121* (4), 391-401

—. (1995). Increasing the power of nonparametric tests by detecting and downweighting outliers. *Journal of Experimental Education, 64* (1), 71-78

—. (1998). Invalidation of parametric and nonparametric statistical tests by concurrent violation of two assumptions. *Journal of Experimental Education, 67*(1), 55-68

APPENDIX A

A.1 Questionnaire

UNIVERSITY *of* LIMERICK
OLLSCOIL LUIMNIGH

Aim of Survey:
This survey has been developed to assess Irish residents views about the use of genetic test results by life insurance companies to decide if they will insure an applicant. The information you provide is strictly confidential. The data will be aggregated and the responses of individuals will not be identifiable, unless explicitly requested by the individual.

Note: Definition of Genetic Tests:
Genetic tests examine the genetic information contained inside a person's cells to determine if that person has or will develop a certain disease or could pass a disease to his or her offspring.

Start Here

Q.1 The following questions relate to your knowledge about the use of genetic test results by the life insurance industry. *Please write in the number from the scale below that best represents your knowledge.*

1	2	3	4	5
No Knowledge				**Perfect Knowledge**

- How much do you know about genetic testing? ☐
- How much do you know about the use of genetic test results by life insurance companies to assess if they will insure an applicant? ☐
- How much do you know about current policies guiding life insurers use of genetic test results to decide if they will insure an applicant? ☐

Q.2 Here are a number of statements that an individual may make about their perception of risks and benefits of life insurance companies using genetic test results to decide if they will insure an applicant. *Please write in the number from the scale below closest to your views.*

Strongly Disagree	Disagree	Agree	Strongly Agree	No opinion or Don't Know
1	**2**	**3**	**4**	**5**

- Life insurers by using genetic test results to decide if they will insure an applicant expose Irish society as a whole to many risks ☐
- Life insurers by using genetic test results to decide if they will insure an applicant expose my family and I to many risks ☐
- There are many benefits for Irish society as a whole if life insurers use genetic test results to decide if they will insure an applicant ☐
- There are many benefits for my family and I if life insurers use genetic test results to decide if they will insure an applicant ☐
- The life insurance industry should be allowed to use genetic test results to decide if they will insure an applicant ☐

Q.3 Here are a number of statements that an individual may make about their trust in the Government to manage life insurance companies use of genetic test results to decide if they will insure an applicant.

Strongly Disagree	Disagree	Agree	Strongly Agree	No opinion or Don't Know
1	**2**	**3**	**4**	**5**

In relation to life insurers use of genetic test results to decide if they will insure an individual, *please write in the number from the scale above closest to your views.*

- The Irish Government is doing a good job of managing the above ☐
- The Irish Government has the skills to manage the above ☐
- The Irish Government is competent enough to deal with the above ☐
- The Irish Government provides all relevant information about the above ☐
- The Irish Government changes policies without good reason regarding the above ☐
- The Irish Government is acting in the public interest when managing the above ☐
- The Irish Government listens to concerns raised by the public about the above ☐
- The Irish Government distorts information concerning the above ☐
- The way the Irish Government makes decisions about the above is fair ☐
- The Irish Government is too influenced by the insurance industry regarding the above ☐

Q.4 Here are a number of statements that an individual may make about their trust in the Life Insurance Industry to manage life insurance companies use of genetic test results to decide if they will insure an applicant.

Strongly Disagree	Disagree	Agree	Strongly Agree	No opinion or Don't Know
1	**2**	**3**	**4**	**5**

In relation to life insurers use of genetic test results to decide if they will insure an individual, *please write in the number from the scale above closest to your views.*

- The Insurance Industry is doing a good job of managing the above ☐
- The Insurance Industry has the skills to manage the above ☐
- The Insurance Industry is competent enough to deal with the above ☐
- The Insurance Industry provides all relevant information about the above ☐
- The Insurance Industry changes policies without good reason regarding the above ☐
- The Insurance Industry is acting in the public interest when managing the above ☐
- The Insurance Industry listens to concerns raised by the public about the above ☐
- The Insurance Industry distorts information concerning the above ☐
- The way the Insurance Industry makes decisions about the above is fair ☐
- The Insurance Industry is too influenced by the insurance industry regarding the above ☐

Q.5 Thinking now about your opinion of decision-making regarding life insurers use of genetic test results to decide if they will insure an applicant for insurance. *Please indicate how far you agree with each of the following statements using this scale:*

Strongly Disagree	Disagree	Agree	Strongly Agree	No opinion or Don't Know
1	**2**	**3**	**4**	**5**

- I have a say on the content of such decisions ☐
- I have a say in how decisions regarding life insurers use of genetic test results are made ☐
- I have the means to hold the Irish Government accountable for their decisions regarding the use of genetic test results by life insurers to decide if they will insure an applicant ☐
- I have the means to hold the insurance companies accountable for their decisions regarding their use of genetic test results to underwrite an applicant for insurance ☐
- The Irish public should participate more in any decisions made regarding life insurance companies access to genetic test results ☐
- The insurance industry is capable of managing its own use of genetic test results ☐
- It is necessary for the Irish Government to create legislation to prohibit insurance companies from using genetic test results to underwrite an applicant for life insurance ☐

Q.6 Now for some personal details. At this point, I would like to remind you that the information you provide is strictly confidential. The data will be aggregated and the responses of individuals will not be identifiable, unless explicitly requested by the individual.

PART A

- Are you male or female?

- What year were you born?

- What is your current occupational status?

- How would you describe your current occupation?

- Have you lived outside of Ireland for a continuous period of five years or more in the last ten years?

If 'yes': **please answer the following 2 questions *(i, ii)***

If 'no': **please skip to *Part B***

i) What year did you last take up residence in the Republic of Ireland?

ii) In which country was your last previous residence?

PART B

- Do you currently own any life insurance product?

If 'yes': **please skip to *Part C***
If 'no': *please answer the following question* (i, ii)

i) Have you ever owned a life insurance product?

ii) To the best of your knowledge, do you think that you will purchase a life insurance product in the future?

PART C

▪ Would you be interested in taking part in a further discussion of the use of genetic test results by the life insurance industry?

If 'yes': please provide your name and contact details below

Name: ____________________

Address: ____________________

____

Telephone Number: ____________________

E-mail address: ____________________

Thank you sincerely for taking the time to complete this questionnaire. Your assistance in providing the information is very much appreciated. If you would like to make additional comments, please do so in the space provided below.

THANK YOU FOR YOUR TIME AND HELP

Please address any comments or questions to
Louise Morris,
Graduate Centre for Business,
University of Limerick,
Limerick.

E-mail: Louise.Morris@ul.ie

APPENDIX B

B.1 Research Process 'Onion'

See over

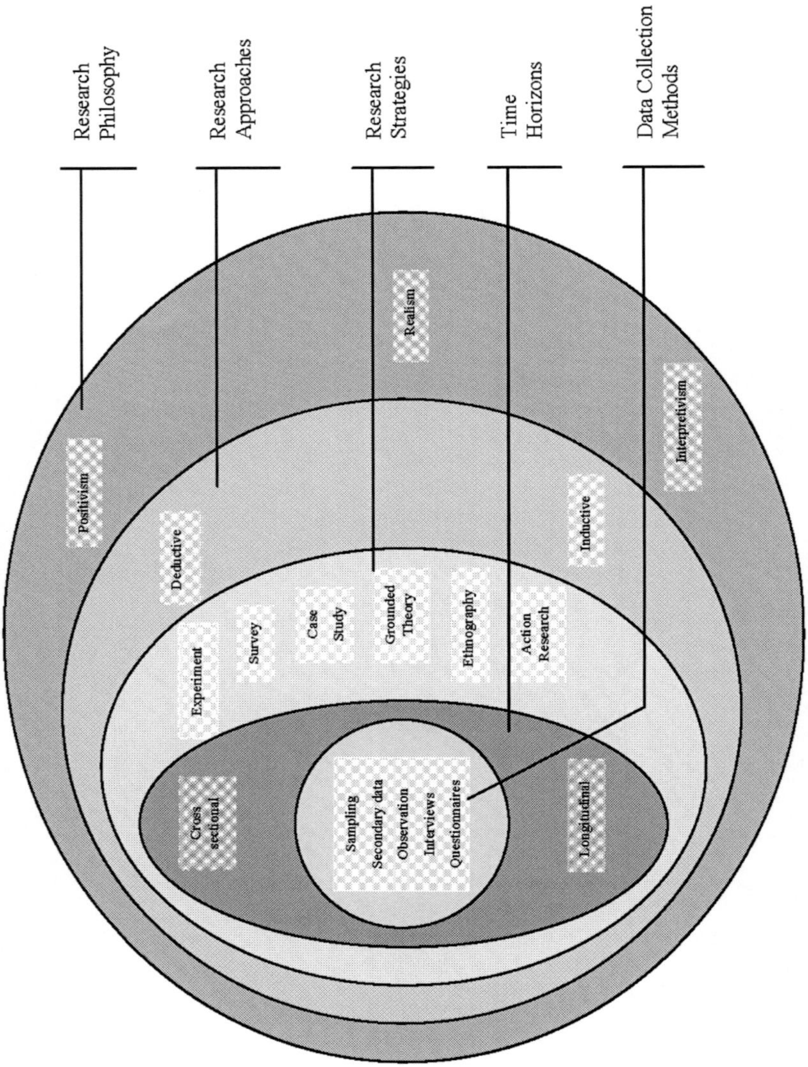

Adapted from Saunders, Lewis & Thornhill, 2003

B.2 Items Designed to Measure Knowledge, Social trust, Risk Perception, Benefit Perceptions and Risk Acceptance

Knowledge items:

1	2	3	4	5
No Knowledge				**Perfect Knowledge**

ITEM 1
How much do you know about genetic testing?

ITEM 2
How much do you know about the use of genetic test results by life insurance companies to assess if they will insure an applicant?

ITEM 3
How much do you know about current policies guiding life insurers use of genetic test results to decide if they will insure an applicant?

Trust in the Irish Government items:

Strongly Disagree	Disagree	Don't know	Agree	Strongly Agree
1	**2**	**3**	**4**	**5**

ITEM 1
The Irish Government is doing a good job with regard to managing the above

ITEM 2
The Irish Government is competent enough to deal with the above

ITEM 3
The Irish Government has the skills to manage the above

ITEM 4
The Irish Government distorts facts in its favour regarding the above

ITEM 5
The Irish Government changes policies regarding the above without good reasons

ITEM 6
The Irish Government is too influenced by life insurance companies regarding the above

ITEM 7
The Irish Government is acting in the public interest with regard to managing the above

ITEM 8
The Irish Government listens to concerns raised by the public about the above

ITEM 9
The Irish Government listens to what ordinary people think about the above

ITEM 10
I feel that the way the Irish Government makes decisions about the above is fair

ITEM 11
The Irish Government provides all relevant information about the above to the public

Trust in the Insurance Industry items:

Strongly Disagree	Disagree	Agree	Strongly Agree	No opinion or Don't Know
1	**2**	**3**	**4**	**5**

ITEM 1
The Insurance Industry is doing a good job with regard to managing the above

ITEM 2
The Insurance Industry is competent enough to deal with the above

ITEM 3
The Insurance Industry has the skills to manage the above

ITEM 4
The Insurance Industry distorts facts in its favour regarding the above

ITEM 5
The Insurance Industry changes policies regarding the above without good reasons

ITEM 6
The Insurance Industry is too influenced by life insurance companies regarding the above

ITEM 7
The Insurance Industry is acting in the public interest with regard to managing the above

ITEM 8
The Insurance Industry listens to concerns raised by the public about the above

ITEM 9
The Insurance Industry listens to what ordinary people think about the above

ITEM 10
I feel that the way the Insurance Industry makes decisions about the above is fair

ITEM 11
The Insurance Industry provides all relevant information about the above to the public

Perception of Risks and Benefits items:

Strongly Disagree	Disagree	Agree	Strongly Agree	No opinion or Don't Know
1	**2**	**3**	**4**	**5**

ITEM 1
Life insurers using genetic test results to decide if they will insure an applicant expose Irish society as a whole to many risks

ITEM 2
Life insurers using genetic test results to decide if they will insure an applicant expose my family and I to many risks

ITEM 3
There are many benefits for Irish society as a whole if life insurers use genetic test results to decide if they will insure an applicant

ITEM 4
There are many benefits for my family and I if life insurers use genetic test results to decide if they will insure an applicant

Risk Acceptance item:

Strongly Disagree	Disagree	Agree	Strongly Agree	No opinion or Don't Know
1	**2**	**3**	**4**	**5**

ITEM 1
It is necessary for the Irish Government to create legislation to prohibit insurance companies from using genetic test results to underwrite an applicant for life insurance

B.3 Information Sheet (1)

CONSENT FORM

UNIVERSITY *of* LIMERICK
OLLSCOIL LUIMNIGH

RISK RESEARCH GROUP

AIM OF SURVEY:
This survey has been developed to assess Irish residents views about the use of genetic test results by life insurance companies to decide if they will insure an applicant.

NOTE: DEFINITION OF GENETIC TESTS:
Genetic tests examine the genetic information contained inside a person's cells to determine if that person has or will develop a certain disease or could pass a disease to his or her offspring.

SURVEY CONTENT:
This survey is divided into 6 parts. The first five parts each present a number of statements about which you are asked to provide an opinion. At the top of each section there will be a number scale. Each number on the scale clearly represents a particular view. You will be asked to select the number from the scale closest to your views about each statement. After each statement there will be an empty box, inside which you can write the number from the scale closest to your opinion.

The sixth part of the questionnaire shall ask for some personal details, however, you do not have to provide your name and/or any contact details, unless you explicitly request to do so.

CONFIDENTIALITY:
The information you provide is strictly confidential.

The data will be aggregated and the responses of individuals will not be identifiable, unless explicitly requested by an individual participant.

CONSENT:
Would you be interested in taking part in this survey on the use of genetic test results by the life insurance industry?

Please note that you have the right to withdraw your consent at any stage during the survey.

Yes ☐ No ☐

Date:

B.4 Information Sheet (2)

UNIVERSITY *of* LIMERICK
OLLSCOIL LUIMNIGH

RISK RESEARCH
INFORMATION SHEET

Genetic Testing
Genetic tests examine the genetic information contained inside a person's cells to determine if that person has or will develop a certain disease or could pass a disease to his or her offspring.

Genetic testing involves looking for altered genes and for gene products that could signal an aberrant gene linked to a specific disease.

Not everyone with a gene linked to a disease will suffer the consequences or die from that genetic disorder. The presence of the mutation may merely only represent a slightly increased risk of developing a genetic disease. Genetic tests can determine if a person is a carrier of a particular gene, which then may lead to a determination of whether the person is predisposed to a particular disease or presymptomatic for a particular disease. However, a person may be neither predisposed nor presymptomatic to a genetic disease, in such a case, he or she is a carrier of a particular allele, which is simply a different form of the same gene.

Genes can mutate, therefore, a test taken many years ago may not accurately identify one's genetic make-up today.

There are various techniques for testing an individual's genetic characteristics, but normally a small blood sample is taken from an individual from which geneticists can retrieve an individual's genetic data.

The most common genetic tests used in the Republic of Ireland are for Hemochromatosis, cystic fibrosis and Down syndrome. However, genetic tests are available for a variety of other diseases, such as Huntington's disease, hereditary breast cancer, hereditary ovarian cancer, and familial Alzheimer's disease.

Life Insurance
Insurance is based on the principle of risk-defined equity. This principle describes how the premium paid by an insured individual is in line with their known level of risk. Therefore, the premium paid by each insured life is sufficient to cover the

risk which s/he brings to the life insurance fund. Risk-defined equity supports the insurers right to underwrite individuals' risks and to discriminate according to risks status that is the mortality or morbidity predicted. The result is that the severest risks are declined and an elevated premium is charged to those with higher than average risks commensurate with the extra risks they pose to the insurance fund. The results of genetic tests present the insurer with risk-status relevant information, leading to the insistence of insurers that genetic test results should be revealed to an insurance underwriter.

Genetic Testing & Life Insurance in the Republic of Ireland

The Council of Europe declared at a convention on Human Rights and Biomedicine that any form of discrimination against a person on grounds of his or her genetic heritage is to be prohibited, signalling that genetic testing can only be carried out for the purposes of health care or research linked to health purposes, and subject to appropriate genetic counselling. However, this convention has not yet been ratified by the Republic of Ireland, therefore, has no force until signed and implemented into national law.

In May 2001, the Irish Insurance Federation (IIF) issued a 'Code of Practice on Genetic Testing' that declared a moratorium on the use of genetic test results by all member companies. This exemption from the normal duty of full disclosure of material facts is intended to apply until 31st December 2005, allowing the insurance industry a 5 year period to evaluate the actuarial relevance of genetic tests. According to the Code of Practice, disclosure of the result of a genetic test is not required in new applications for life cover unless the sum insured on the new application exceeds €381,000 or the total of the sum insured on the new application and other policies, if any, taken out with any insurer between 1st April 2001 and 31st December 2005 exceeds €381,000. In the case of life cover exceeding this limit, insurers have been instructed to use only approved tests by the British Genetics and Insurance Council (GAIC) in the underwriting process. The GAIC have so far announced the reliability and relevance of only two genetic tests, both for Huntington's disease, as sufficient for insurance companies to use when assessing applications for life insurance. In the case of a proposal where genetic test results are not required to be disclosed, an applicant who has a negative result under an approved test can ask the insurer to take the result into account in the underwriting decision.

The Irish Government proposed a moratorium until 1st January 2010, on the commercial use of genetic information in the Irish Disability Bill 2001. The Bill outlined that a person shall not request, obtain, accept, or make use of the results of a genetic test of any applicant for insurance or any insured person. The Bill also outlined that information about the family history of an applicant for insurance may be sought and used by an insurer if the sum insured is above a sum to be determined from time to time by Minister for Enterprise, Trade and Employment in consultation with the Minister for Social, Community and Family Affairs or in the case of health related insurance, by the Minister for Health and Children. This

Bill has not this been enacted. It was withdrawn in February 2002 following negative reaction from disability groups, although it is likely to be reintroduced in the latter half of 2004.

Please address any comments or questions to:

Louise Morris
Graduate Centre for Business
University of Limerick
Limerick
Tel: (061) 202654
E-mail: Louise.Morris@ul.ie

If you have concerns about this study and wish to contact someone independent, you may contact:

The Chairman of the University of Limerick Research Ethics Committee c/o Vice President Academic and Registrar's Office
University of Limerick
Limerick
Tel: (061) 202022

APPENDIX C

DESCRIPTION OF THE DATA

C.1 Normal Curves

Mean Knowledge

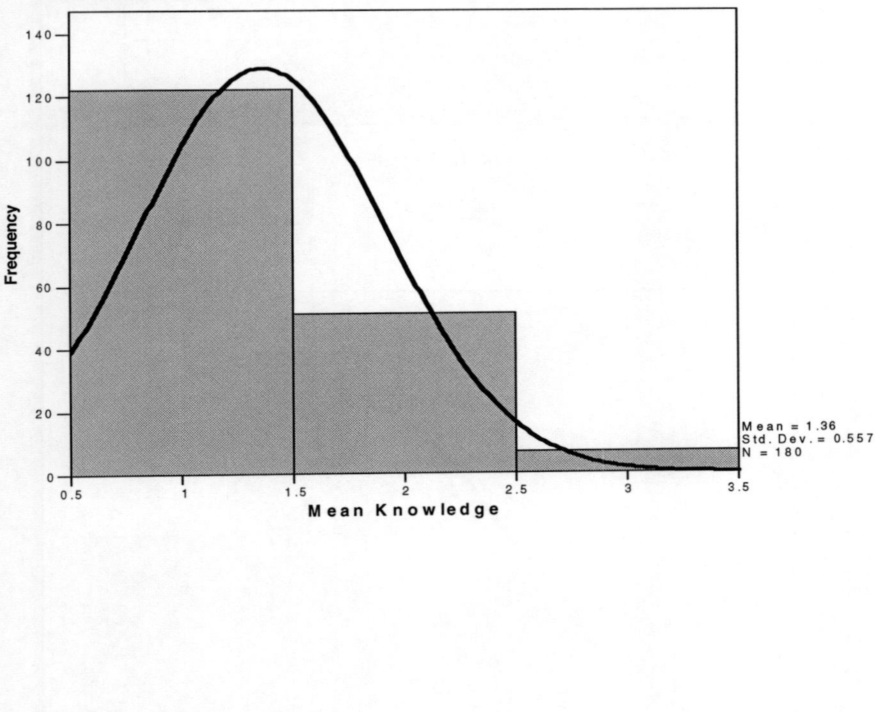

Overall Trust in Insurance Industry and Insurance Industry

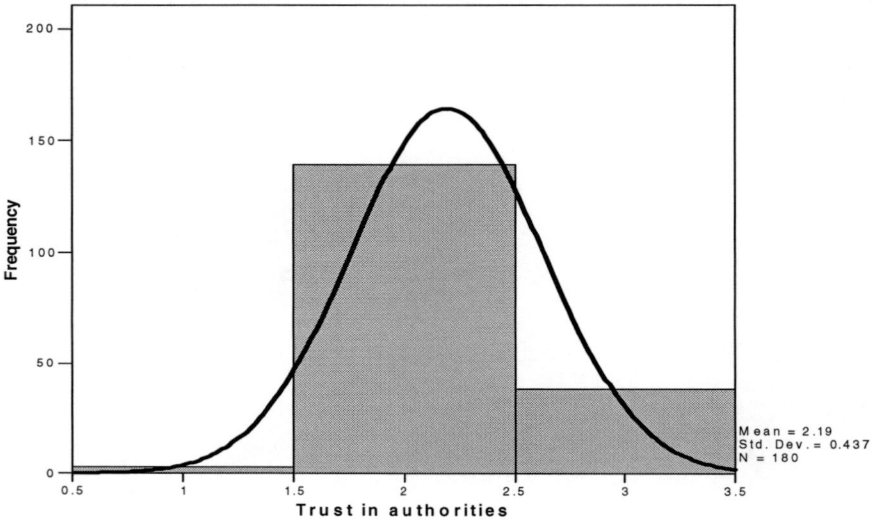

Risk Perceptions

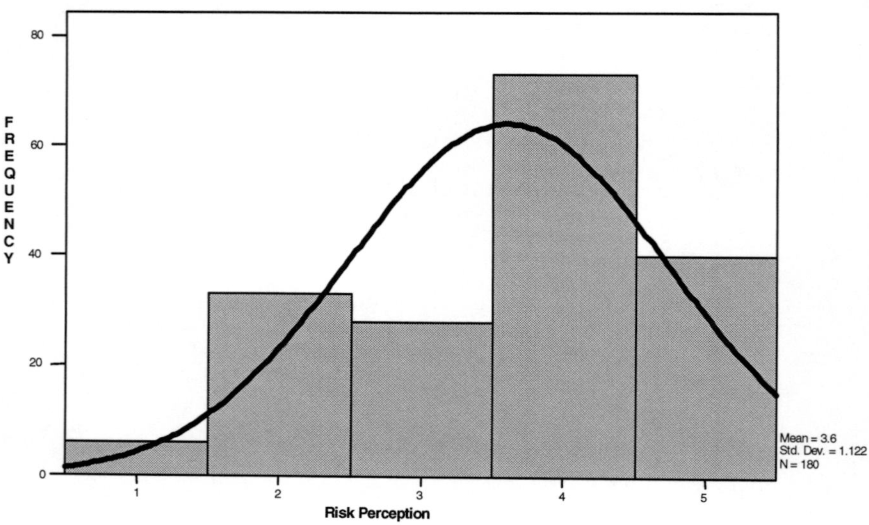

Benefit Perceptions

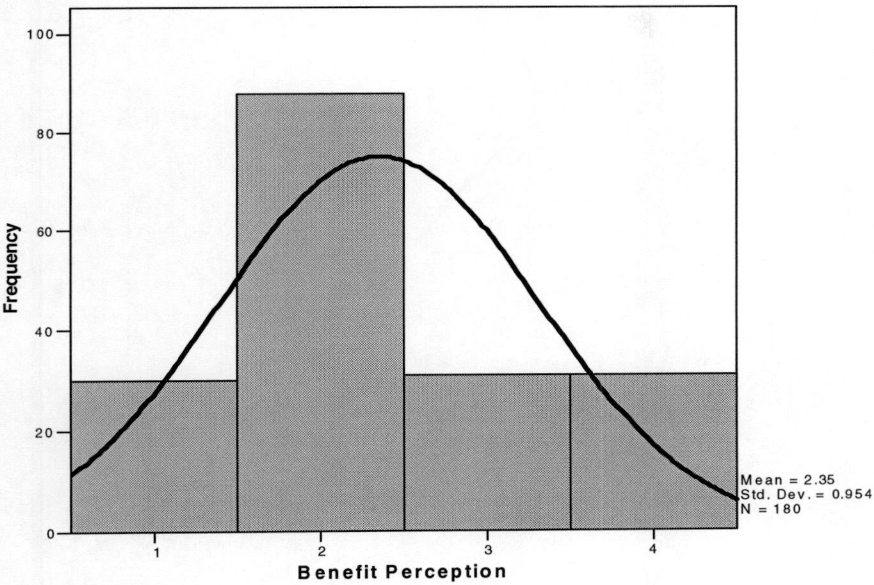

Risk Acceptance

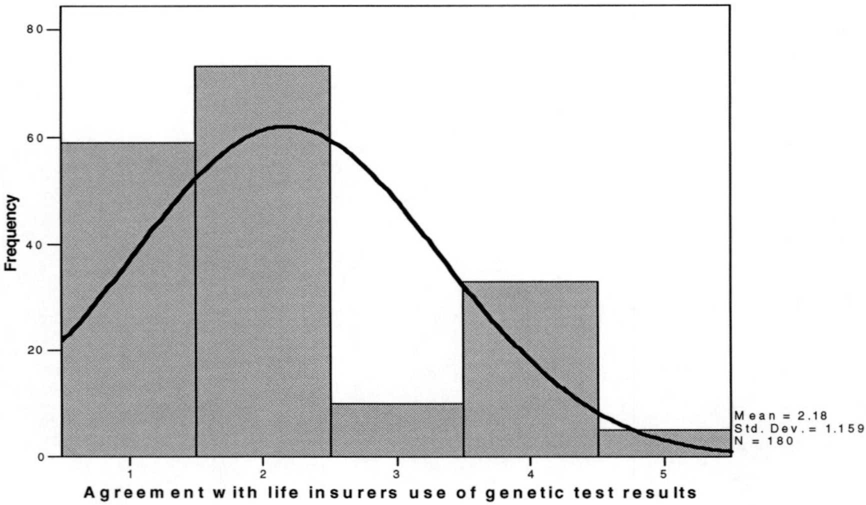

C.2 Box Plots

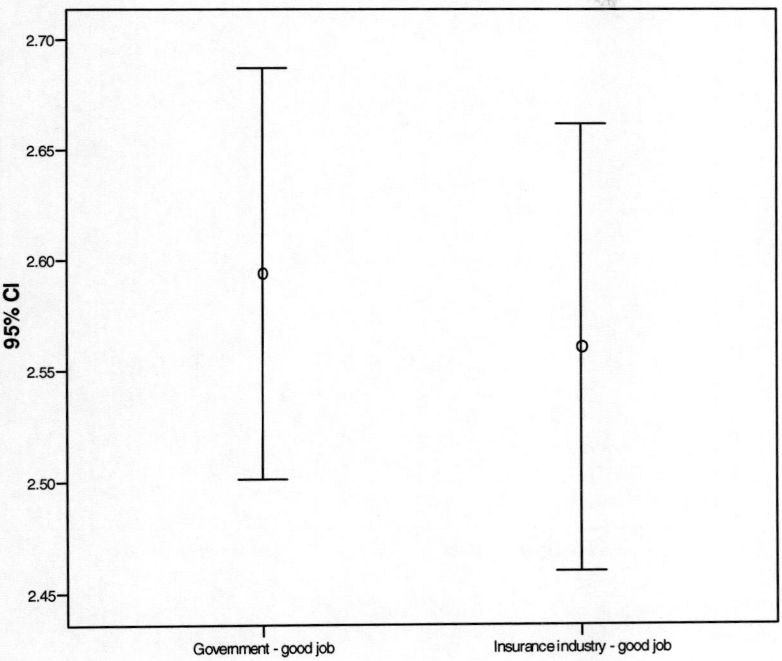
2.70
2.65
2.60
2.55
2.50
2.45
95% CI
Government - good job
Insurance industry - good job

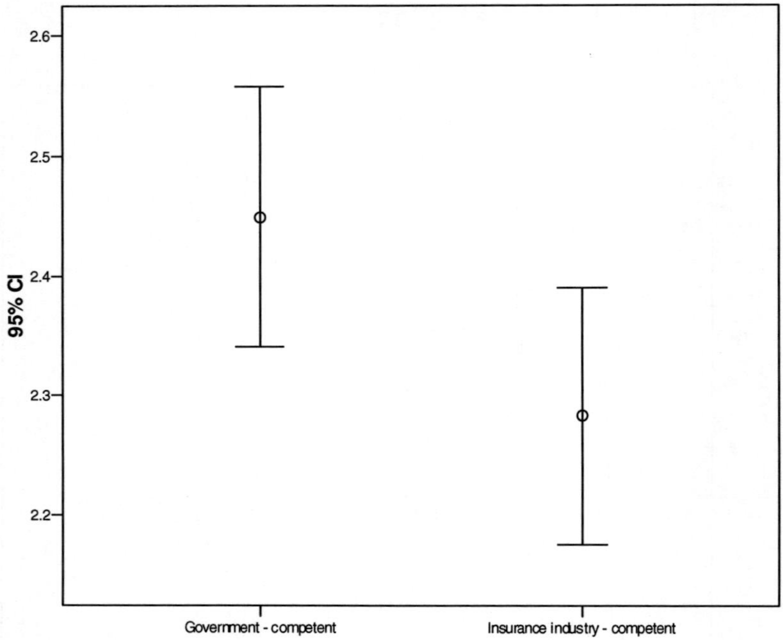
2.6
2.5
2.4
2.3
2.2
95% CI
Government - competent
Insurance industry - competent

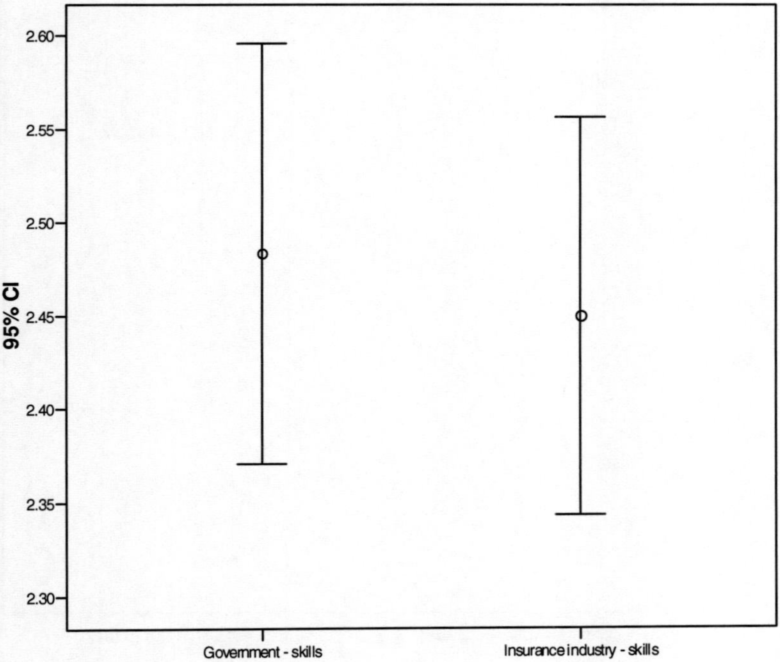
2.60
2.55
2.50
2.45
2.40
2.35
2.30
95% CI
Government - skills
Insurance industry - skills

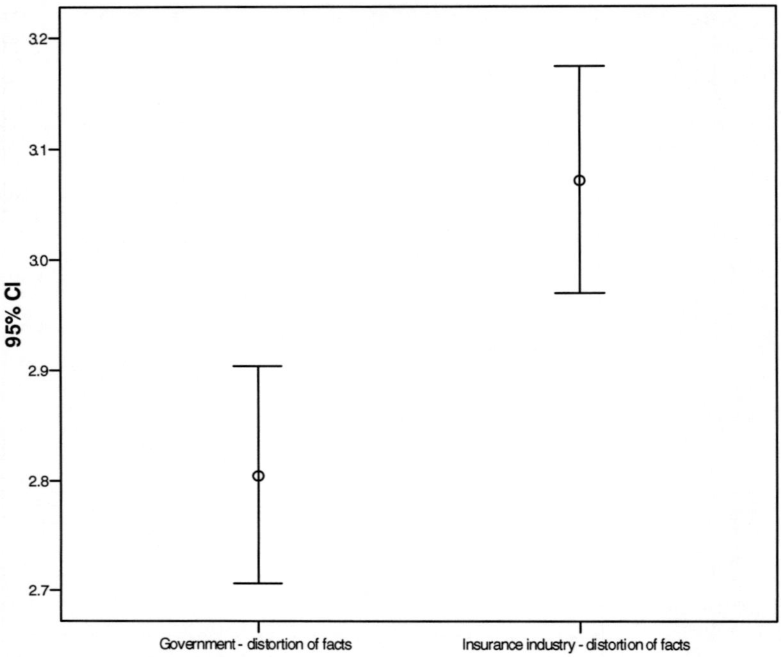
3.2
3.1
3.0
2.9
2.8
2.7
95% CI
Government - distortion of facts
Insurance industry - distortion of facts

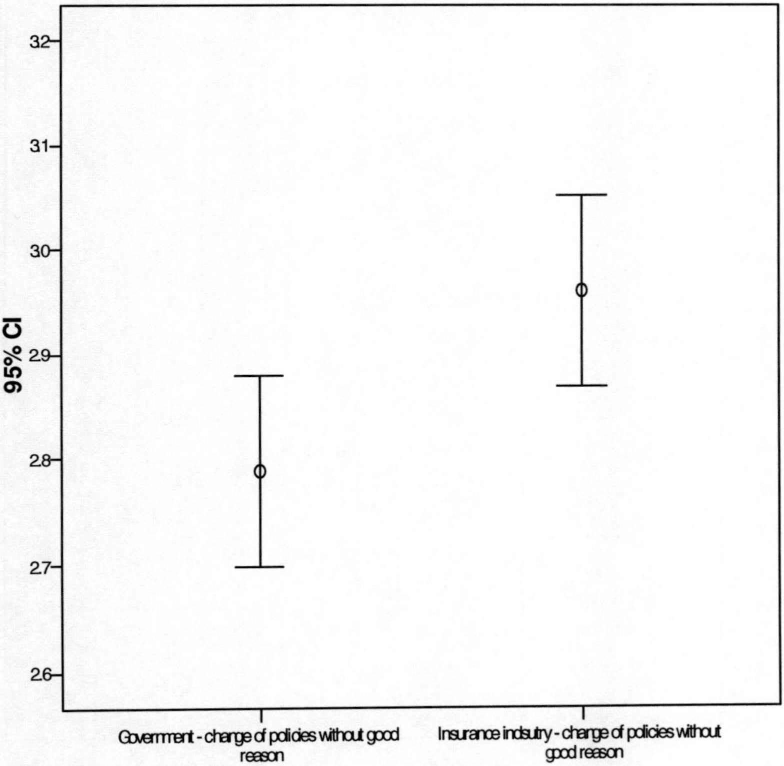
3.2
3.1
3.0
2.9
2.8
2.7
2.6
95% CI
Government - charge of policies without good reason
Insurance indsutry - charge of policies without good reason

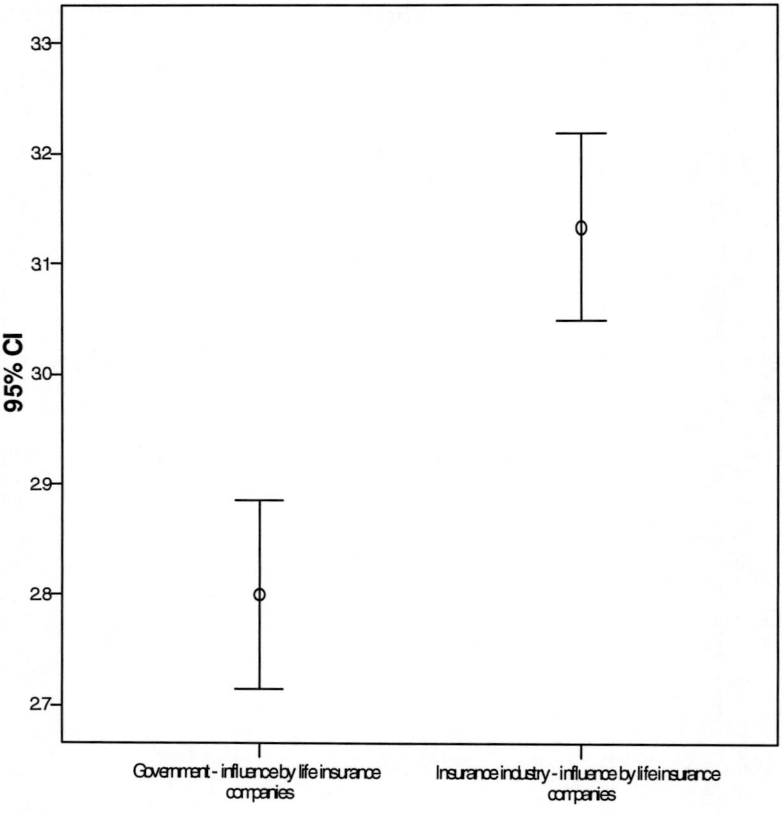
95% CI
33
32
31
30
2.9
2.8
2.7
Government - influence by life insurance companies
Insurance industry - influence by life insurance companies

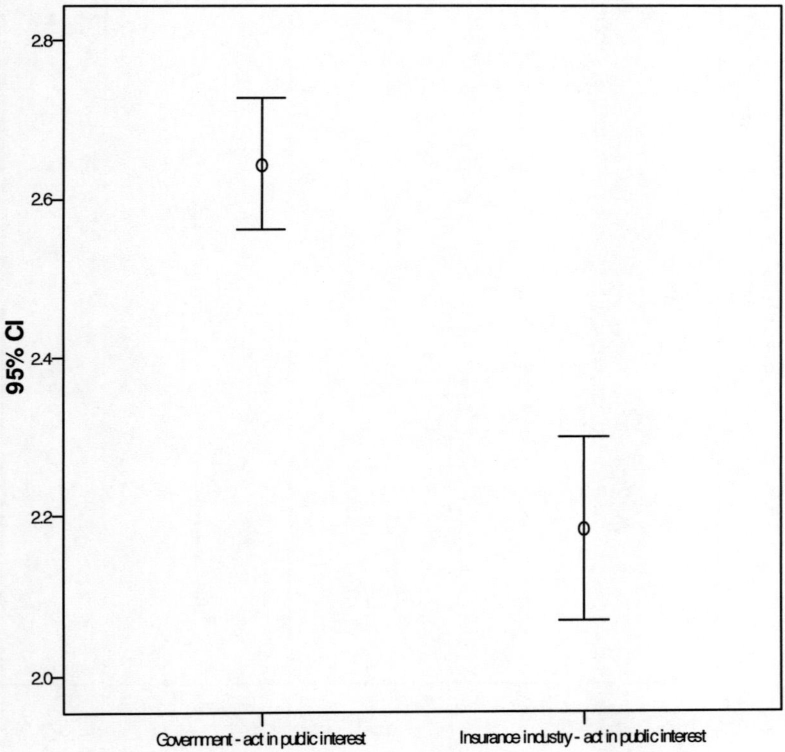
2.8
2.6
2.4
2.2
2.0
95% CI
Government - act in public interest
Insurance industry - act in public interest

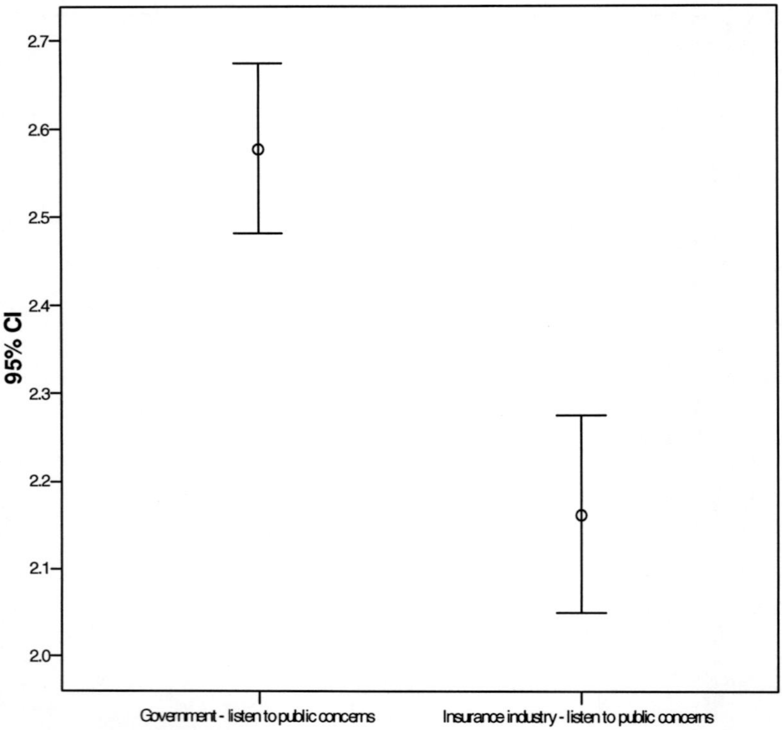
2.7
2.6
2.5
2.4
2.3
2.2
2.1
2.0
95% CI
Government - listen to public concerns
Insurance industry - listen to public concerns

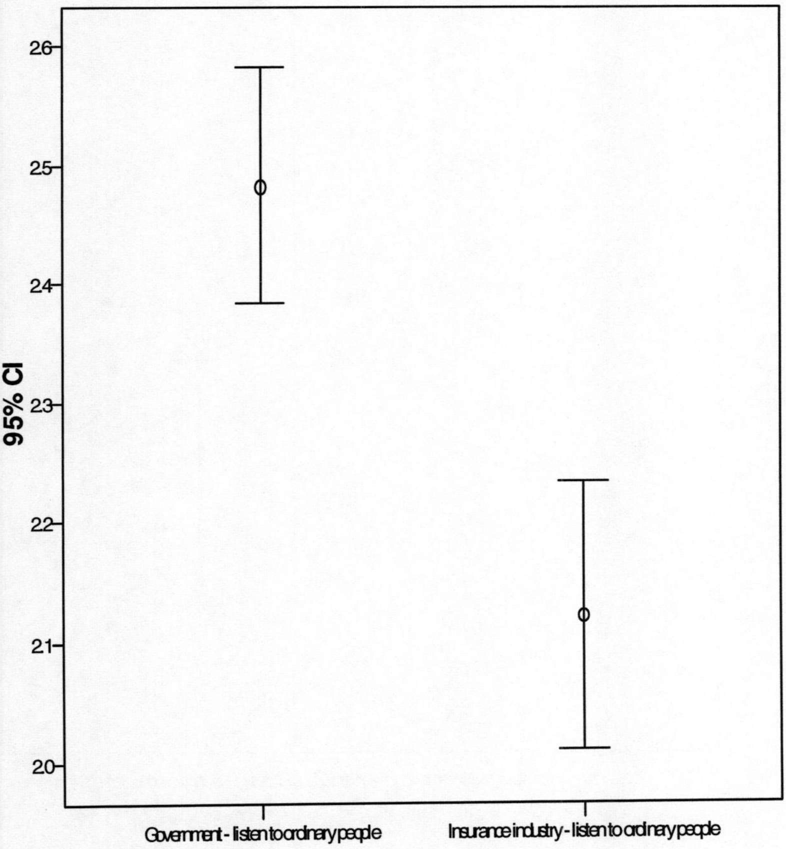

95% CI
26
25
24
23
22
21
20
Government - listen to ordinary people
Insurance industry - listen to ordinary people

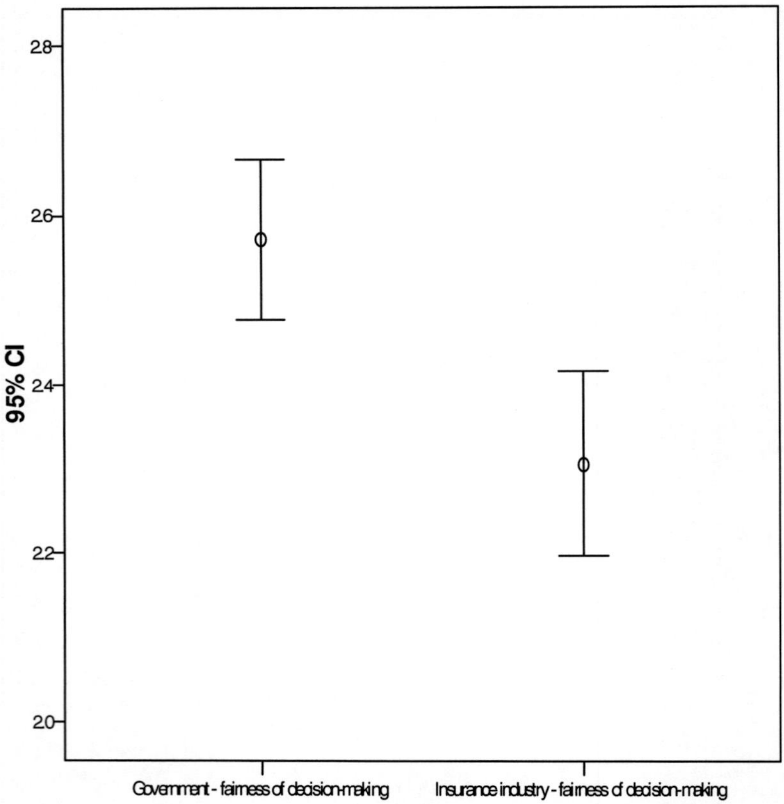

2.8
2.6
2.4
2.2
2.0
95% CI
Government - fairness of decision-making
Insurance industry - fairness of decision-making

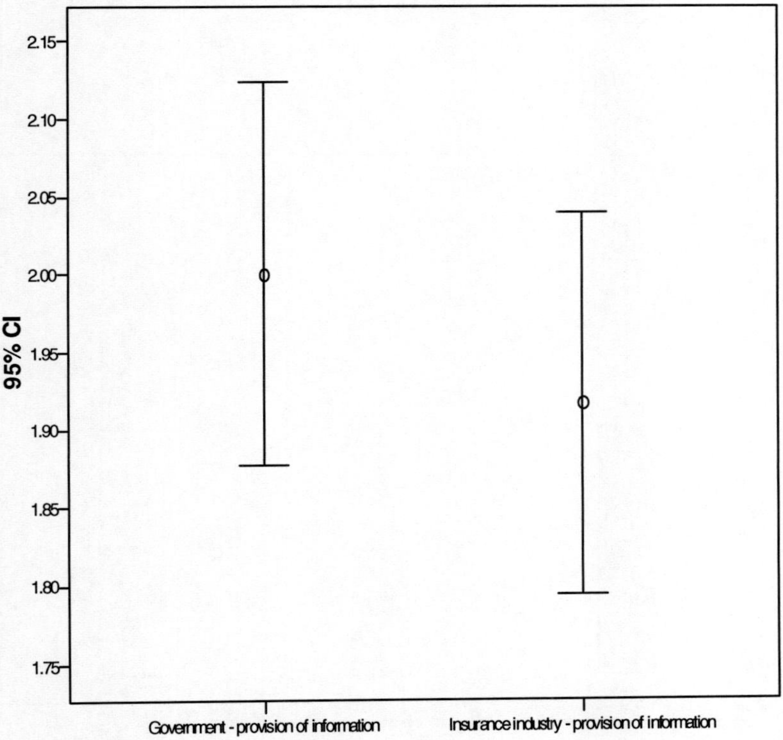
2.15
2.10
2.05
2.00
1.95
1.90
1.85
1.80
1.75
95% CI
Government - provision of information
Insurance industry - provision of information

C.3 Error Bars

Knowledge

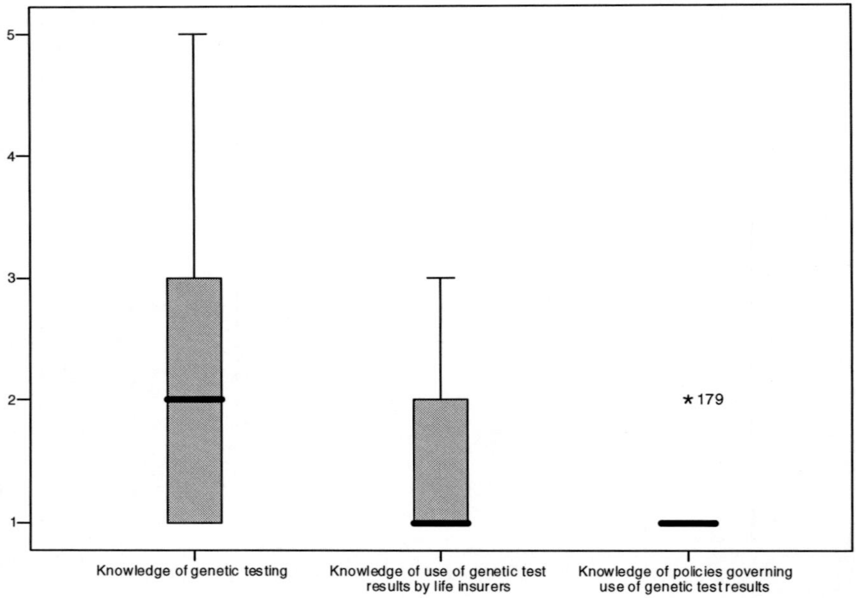

Trust in Insurance Industry

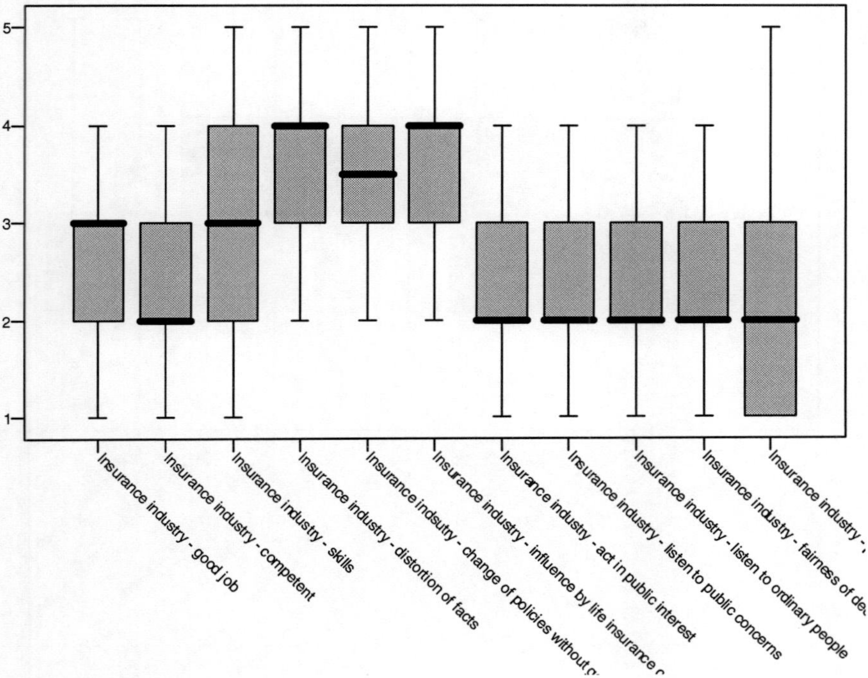

Trust in Insurance Industry

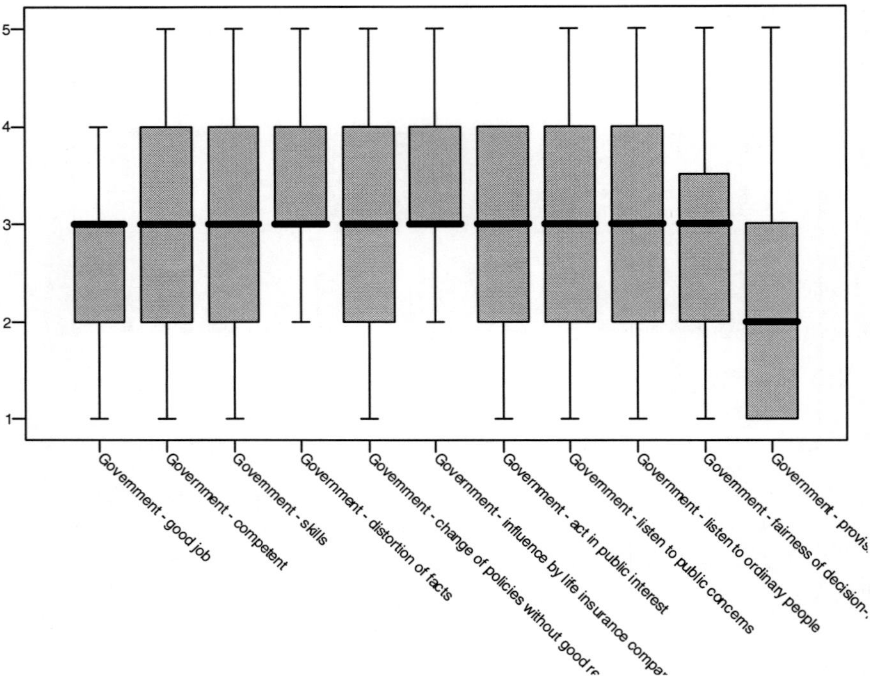

Risk and Benefit Perceptions

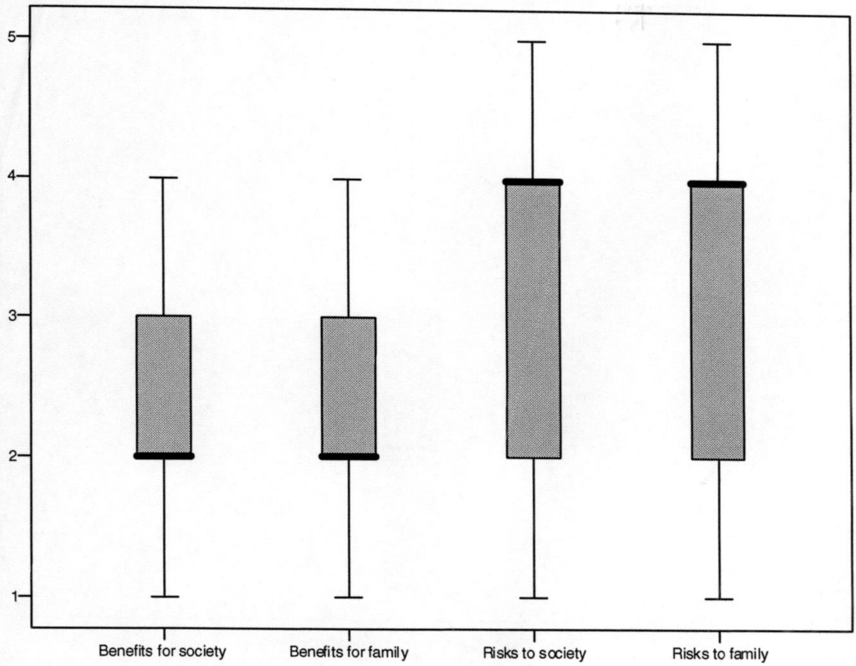

Risk Acceptance

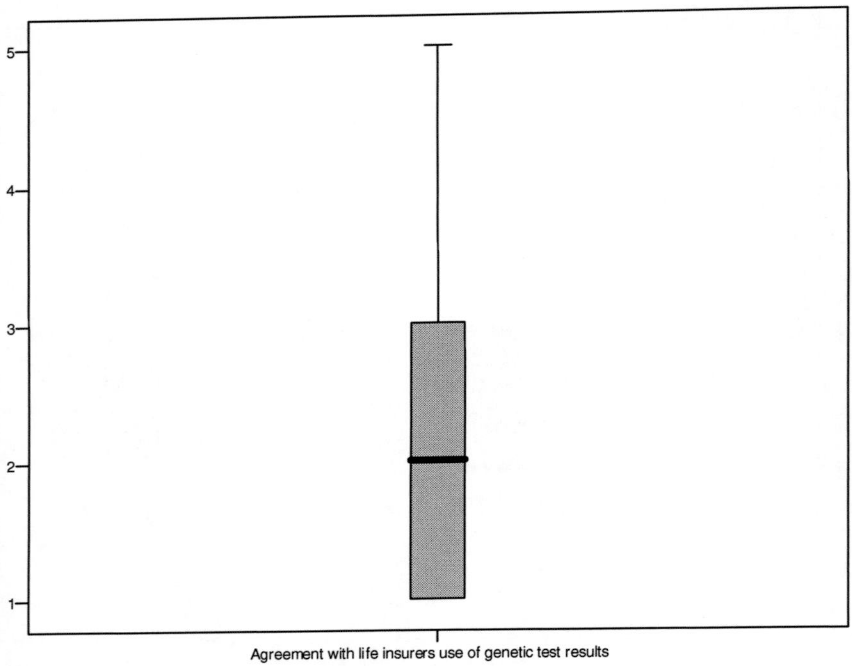

APPENDIX D

D.1 Interview Analysis

Table D-1 Interviewees knowledge of genetic testing based on questionnaire and interview findings

Subject	Questionnaire	Interview – How much do you know about genetic testing?	
		Initial Response	**Subsequent Response**
1	Some	"not a huge amount"	"genetic testing is used to find out what potential weaknesses people might have or… how likely someone might be to get a particular disease"
2	None	"They do blood tests, check if you've a history of… disease"	
3	Some	"I would only know a small amount"	"There are genetic tests now that can tell whether someone is vulnerable to a certain disease and whether it is family related"
4	Some	"they can test whether or not you may suffer from a disease, whether or not it's in your family, whether or not you might suffer from it in years to come"	
5	Some	"I really don't know much, no, not really"	"it's used for, to see for life insurance if someone's going to have a long life or a short life"

6	None	"Very limited knowledge"	"Genetic testing…is…taking a DNA sample from a person, analysing their genetic make-up"
7	Some	"Not much at all"	"it's testing for some genetic history, are you susceptible from different diseases"
8	None	"I don't know anything at all"	"testing for disease"
9	None	"Not a whole lot"	"for to say whether you have certain disease or not, like say Motor Neurons disease…you hear about Alzheimer's being carried in the genes…"
10	None	"Relatively little"	"you're trying to test for traces or genetic predispositions, are there diseases, are there illnesses"

Source: Analysis of Survey Data

Table D-2 Knowledge of genetic diseases and knowledge of insurers interest in gaining access to genetic test results prior to survey completion

Subject	Diseases Mentioned	Did you know about insurers using genetic test results prior to the survey?
1	Hemochromatosis	"No, I did not"
2	Huntington's disease	"No, no, not until I was told"
3	No mention of types of disease – not asked	"Not a huge amount"
4	Ovarian cancer	Not asked
5	No mention of types of disease – not asked	"Not much, actually, no"
6	Heart disease, thyroid problems	"No, not until you did the survey"
7	Alzheimer's, epilepsy	"During the survey was the first time I heard"
8	Cancer, heart problems	"No, no, I wasn't no"
9	Motor Neuron's Disease, Alzheimer's	"No", also agrees survey was first time to hear about insurers wanting to use genetic test results
10	Heart attack, hardening of arteries, diabetes, osteoporosis	"what I know is in relation to life assurance… I understand genetic testing may have an impact in relation to where and how and under what conditions life assurance may be issued in respect of clients"

Source: Analysis of Survey Data

Table D-3 Risk Perceptions

Subject	Survey	Risks Mentioned	
		Risks	**Discussion of Risks**
1	Some risk	• Selection of low risk individuals • Deny insurance • High premium	➢ "The risk that springs to mind is that they may use it to be selective about the people they insure and the kind of cover they provide for individuals, and may even use it to exclude certain individuals from cover altogether"
2	High risk	• Deny insurance • Privacy	➢ "I think excluding people, you know…they're going too deep into their…private lives" ➢ Genetic test results "are nobody else's business really" except for medics with "your health in mind" ➢ Insurers using "genetic testing is very much an invasion" (of privacy) ➢ "…just too invasive…"

3	Low risk	No risk • High premium	➢ "I think that is there a risk for society from genetic testing? I think no, it's an advantage. It's furthering science….it can enhance peoples lives" ➢ "I don't see the risks" ➢ "I don't see any risk for society as a whole" ➢ "From what I know, I didn't see any family risks ➢ Cannot see any moral dilemmas being introduced by insurers using genetic test results – "In that case, no" ➢ "I think…the information being known…by the insurance company would probably increase the payments, increase premiums and so on, and if they, if the individual already knows the result of the test, and the individual must make the information available to the insurer that probably pushes up the, the costs"

4	Some risk	• Discrimination • High premium • Deny insurance • Privacy	➢ "Well, firstly, the risk of discrimination against, families who may wish to take out life assurance but who probably have quite a few genetic disorders in their family, no, pretty much the risks of discrimination of families" ➢ "I myself could get breast cancer down the line or anything like that" Questions if could be "at risk to having been tested, or not being allowed to take out, or very, very high policy quote for taking out life assurance, and then I have a family along the line, they may be left without anything from me, to leave to them if I haven't been, if I have been told that I can't take out life assurance because of that" ➢ "In my mind, the only person who should hear about the results of those tests should be who I want to hear about it, other than doctors, I don't believe anyone should be notified of these tests that I've undergone at my own will, pretty much, I think it should only be people that I tell"

5	Some risk	Genetic testing is new, not possible to know negative effects • Privacy • Deny insurance	➢ "It's hard to know…I mean is a new, it's an extremely new thing, genetic, like variable, so like it's really hard to say whether it's going to be, like pose really negative effect. I'm not sure to be honest with you, I'm not sure" ➢ "…that question there like, whether I think it's going to have a bad effect on future generations or not, I can't really answer that like, we won't really know" ➢ "Just that your information about whether you're going to be healthy in a few years or not is, it's very kind of personal isn't it, information, and it shouldn't be given to an insurance company" ➢ "I think it's cheating… I mean, that's it for, for the company, if you, if they know, like that you're not going to make it, they're just going to say, okay not you… you're not going to make us money are you, they'll know that and that person won't be able to get life insurance and look after his kids, send them to college"

6	Low risk	No risks • Discrimination	➢ "I don't believe, well, I wouldn't believe that there'd be a, well let me think, en, no, I do not believe that there would be a major risk from it" ➢ "the risk would be that it would be used unethically and that people would be discriminated against because of their genetic make-up"
7	Some risk	• Privacy • High premium	➢ "it's not very fair on people, you know, it's very confidential information and they're going to have to pay a lot more for life insurance than everyone else"

8	Low risk	• Privacy • Discrimination	➢ "it's such personal information that you know I can, it's something that I suppose you don't want a company to, to know about you in detail, that's very private, or something you know that's life threatening, that you have to disclose this information to somebody else" ➢ "Outside of a family member or outside of…a medical profession…I think it delves too much into the person, I think it's too personal" ➢ "It's too personal, it's too private" ➢ "…you're discriminating against people for something that's not their fault. If you're born with a family, genetic history of disease, it's not your fault. ➢ "I think myself it's a very personal thing" ➢ "…genetic testing delves into too much, you know, something that might not happen……I consider too serious, too personal, too private"
9	Low risk	• High premium	➢ "they're obviously going to hype up the price if you have, you know say, anything different with your genes or even anything that is kind of, how would you say, anything that would be alarming, let's say"

10	Some risk	• High premium • Deny insurance	➢ "but now, having taken the test, we're not going to give you this, or …we're going to load you that" ➢ "information assessed from client by a company , the company love it, they have all the cards, they can say we will or we won't, no we can or we can't, or we're going to give it to you at such a price or indeed make you not want it, or it's too expensive…morally and ethically and every which way you're on a hiding to nothing"

Source: Abstracted from Analysis of Interview Transcripts

Table D-4 Other Risk Related Comments

Subject	Other Risk Related Comments
1	No further comment
2	***Punishment*** • "I think that it's bad enough for a person knowing…they would have, a genetic disease" • "I think that it's something you have to live with …and it shouldn't be used for…(insurance/commercial purposes) • "A certain thing which could be hereditary, it doesn't necessarily mean that you're going to go at the same time… I think it would be very wrong to punish….an individual who because of their family history says that…" ***Discrimination*** • "Everybody has a right…because none of us know but we could have something genetic, but it just hasn't come about yet" ***Need for insurance*** • Breadwinners take out life assurance "'cos the wife would be left with the children, the house, everything, so, you know, they felt there were the ones to have insurance" • "…everybody, you know, just wants to do their best for their family" • "I mean the reason you take out a policy in the first place is to look after, you know, who's left behind" ***New requirement*** • "I think that it would be very unfair to an individual who wanted to take out insurance , and the fact that, years ago there wouldn't have been any such thing as, as, things being hereditary, or genetic, or anything like that" ***Lack of privacy*** • "there's nothing stopping that information from being passed on elsewhere" (to other companies) "I think that would be totally unfair to the individual"
3	No further comment

4	***Need for insurance*** • "I have a family along the line, they may be left without anything from me, to leave to them if I haven't been, if I have been told that I can't take out life assurance because of that" ***Probability*** • "I would see it as unfair for them to use that, in anyway, because there's no saying that I'm definitely going to suffer, I could lead, I could decide from now on, I'm going to lead an extremely healthy life and put myself in the least possible risk to getting high cholesterol and high blood pressure, and therefore I could say that's extremely unfair, 'cos I've done everything in my power to stop myself from getting high blood pressure so…" • "…there's no definite answer that I would ever get ovarian cancer, I'm just at a more slight risk than another person, but then again another person, another woman could develop this in a year's time who took out an insurance policy and in a year's time she could actually get ovarian cancer, you know as in it could start a lot quicker, or anything like that. No, I don't think it would be fair at all" ***Affect future generations*** • "I would think the fact that it could affect in a few years time affect myself, affect my family, affect, if I have kids, it could affect them when they're a bit older"
5	***Psychological damage of tests*** • "…there probably could, yeah, probably against me, like if I go and do these tests and right now I feel , ah, I'm totally healthy, I'm happy out, and then I find out that oh well, actually that there's a good chance that when you hit 40 years of age, you know you're going to suffer from this and you're going to die like do you need to know that or would you live a happier life, just, just being lazy about the whole affair, you know like" ***Need for Insurance*** • "…they'll know that and that person won't be able to get life insurance and look after his kids, send them to college… And it's not fair, like, because if you're poor, that comes about, you'll have to save more money…" ***Effect on Future Generations*** • "…if (individuals who have revealed positive genetic test to insurer) their children are going for life insurance later on, well then they'll be blacklisted because they'll have found a trace of something"

	Physical Risks • "…maybe if people knew that" (insurers use genetic test results) they wouldn't go get tested" ***Other Risks*** • "it's a pretty advanced form, you know, it's pretty cool kind of thing, it's just as long as it's being used for good rather than being used for, for being used for, for bad things, that's probably the worry, you know that this kind of information can be extended to other aspects that might affect your life" • "it is dabbling in something that is, like, maybe we shouldn't"
6	***Faultless*** • "their genetic make-up…is…something that they cannot prevent and thus should not be a major issue in discerning insurance perhaps" • "A genetic disease is not something, as far as I can tell, is not something that a person brings on themselves, they inherit, therefore it isn't something they should have to pay extra premium because they cannot prevent having a genetic disease, therefore I do not believe that insurance companies should be allowed to use them to determine their premiums" • "The other person should be, should be, in general I believe that the other person should not have to pay more insurance premium for, for factors that is outside their control" • "I don't think a person should be discriminated against for something that they can't control"
7	***Faultless*** • "…you're discriminating against people for something that's not their fault. If you're born with a family, genetic history of disease, it's not your fault"
8	***Probability*** • "…going and saying are you going to be predisposed to such a thing that you may never get, but, or you could die of getting knocked down, you know you never know what's going to happen…" • "…genetic testing just delves too much into you know something that might not happen…"
9	No further comment

10	***Need for insurance*** • "Most people want insurance for themselves or their family" • "You're still looking for insurance…you're still wanting to protect your family" • "…employer wants them to get the thing…" (life assurance) • "an employer or family waiting out there saying, well, you must have this for the next job, or we really need this sort of way" ***Physical risks*** • "In terms of actual physical risks, they have to take a blood sample." This may cause physical harm "if it was done incorrectly" ***Persuasion*** • "we're not forcing you, but we're asking you if you want insurance" you should take a genetic test • "…the company says, hold up now, we can't without genetic tests, and the client goes out and gets a genetic test"

Source: Analysis of Interviews Transcripts

Table D-5 Benefits Mentioned

Subject	Survey	Benefits Mentioned	
		Benefits for Society	**Discussion of Benefits for Society**
1	Low benefit	Low benefit	➢ "no" awareness of any benefits related to insurers using genetic test results ➢ "there is a case for genetic testing for the benefit of those that don't actually carry genetic disorders"
2	Low benefit	No benefit	➢ "there can't be an advantage like…'cos they're using it for their own benefit like, why would they want the information in the first place? Only, you know, to stop them paying out so, such and such an amount… at a certain age. …the only reason they'd be doing it is to avoid the risks for them…" ➢ "they wouldn't be looking for your benefit in anyway, sort of like, they'd be lining their own pockets, wouldn't they really"
3	Some benefit	Fair premium	➢ "It could be a fairer assessment, at least" ➢ "Well, it's, it's like car insurance, if, if the costs, payments from insurance companies are generally reduced then the premiums can be reduced, and if you think of it like that, well maybe that's a benefit for society" ➢ "But only a financial benefit, whether there are moral benefits, I'm not too sure"

4	Some benefit	Fair premium	➢ "The most benefits would be that, you do know, you, people who would be maybe more likely to suffer cancer, or something like that, if they would see tell-tale signs of it, they might immediately take out a life assurance policy and therefore putting other peoples premiums up. That would be pretty much the benefit, just for the benefit for honest people who are coming along, taking out their own insurance policies, without having thought about it any particular way, that would be my… I'd imagine that would be the most benefit" ➢ "I don't know if my family, I don't know if it would be a negative fact that my family did have some genetic disorder that they didn't know about and then they did the genetic test but they were going to take out life insurance anyway and life insurance anyway, then yes, that would definitely be a bad thing, but otherwise, if my family are all going to stay healthy and they're not going to suffer from any, or if they don't have any genetic disorders, well then it wouldn't really have any effect in the future"

5	Low benefit	Low benefit Fair premium	➢ "Benefits for society, I'm not sure, actually what benefits there are, we're, we're kind of, well, like only for themselves" ➢ "only they can make more, kind of, money better" ➢ "if you're like absolutely 100% sure that you're going to be, you know, around for longer, I suppose, it could be to your advantage, you know, like you say some people when they're starting out, like, especially when they're tying to buy like a house, it wouldn't be fair, would it"? ➢ "…but maybe it works out real good, the advantages may be they'll find that this is an appropriate bracket for you and it's realistic, maybe I'll be optimistic about it in the future…"
6	Low benefit	No benefit	➢ "from an insurance point of view, I probably can't see any major benefits from it, no I don't think it would benefit"
7	Low benefit	No benefit	➢ "For society? No. It's only going to really help the insurance companies, cos they're going to know who to insure, or else, do you know, they're not going to really help society"

8	Low benefit	Fair premium	➢ "maybe the fact that your premium wouldn't be as high, you know you were fully healthy and you didn't have any hereditary diseases or predispositions to, to diseases, it probably would the fact that you're healthy and that you shouldn't be charged a premium in comparison to somebody else who has, a chance, a much higher chance of getting a disease that you don't, so I suppose in that way you benefit that from being assessed and, and in getting the right premium and not being over charged, so I suppose in that sense it's, it's a good thing for you, for people having to disclose that kind of information, other than that I'd say I don't know of any other benefits to society or, or to you personally, not really" ➢ "I can't see any major benefits to society or to people" ➢ "I mean like I don't see huge benefits for the person or society"
9	Low benefit	No benefit	➢ "well its only going to benefit the insurers like" ➢ "I wouldn't see any benefits" ➢ "It's only for the insurance company"

10	Low benefit	Fair premium	➢ "it's all very much on the insurance side" ➢ "I suppose for the individual it does out line where they're at in relation to their genetic history, where they're coming from, where they've come out of and potentially where they could go to now and for the future and if that gives them any safeguards going forward or any information, any help in insurance or no regards whether or not loading on"

Source: Abstracted from Analysis of Interview Transcripts

Table D-6 Other Benefit Related Comments

Subject	Other Benefit Related Comments
1	***Insurers Benefit*** • "I do see some arguments in favour of having them use genetic test results, I guess, but looking at it from the perspective of someone, as a client of an insurance company, I would prefer that they didn't. I feel that insurance companies would inevitably use it for their own financial gain first, and the welfare of the insured second, because I see them as a companies which are there to make a profit, to maximise their profits"
2	***Risk Averse Nature of Insurers*** • "the only reason they'd be doing it is to avoid the risks for them"
3	***Insurers Benefit*** • To "know the risks that would be involved for somebody in the future would be an advantage and it would be an advantage also from, from an insurance point of view" • "I can understand where there would be an advantage to, again, have that improved accuracy" ***Scientific Benefit*** • "it's an advantage, it's furthering science, it's, it can enhance peoples lives" • "I think it would be favourable. I would be in favour of the more information that you can gain from science the better" • "I think that it's a new science, I think that we should be interested definitely, we definitely should be pushing out the frontiers but I think they should be controlled" • "…it is a new science but at the same time, the DNA element of that science is well understood at this stage, it may only be an element of around five or ten years, but it's well understood at this stage. There's many, many, many benefits, but I think that if you look at the positive side of this situation, of genetic testing there is possible benefits as well" ***Risk Averse Nature of Insurers*** • "I think they will make decisions, looking at their future growth… And reduce their risk"
4	***Insurers Benefit*** • "in the other sense, well, you can kind of see where they're coming from"

5	***Other Benefits*** • "it's a pretty advanced form, you know, it's pretty cool kind of thing, it's just as long as it's being used for good rather than being used for, for being used for, for bad things, that's probably the worry, you know that this kind of information can be extended to other aspects that might affect your life" ***Insurers Benefit*** • "Very, very good for them to have it, to have that kind of information" • "Well, maybe they do need it, obviously, 'cos it's their business isn't it to know whether someone's going to have a long life or not, 'cos they'll lose out"
6	***Insurers Benefit*** • "I believe that insurance companies are collecting genetic information about people, and the only logical reason I can think if why they'd want to do that is to segregate their population and charge more to people who are prone to a genetic disease or who have a genetic disease"
7	***Insurers Benefit*** • "It's only going to really help the insurance companies, 'cos they're going to know who to insure"
8	***Insurers Benefit*** • It's "a business kind of thing for them" (insurers) "like the more information they get out of it the higher premiums they can charge" • "I think it's just a benefit to the insurance companies mainly than to anyone else if they have that kind of information"
9	***Insurers Benefit*** • "its only going to benefit the insurers" • "It's only for the insurance company, like and as I said any other company that would take on the results and would want to do something with them" ***Scientific Benefit*** • "I suppose there would be, you know a scientific advantage alright, you know you could see, say for example like the cancer genes or whatever, the ones that cause cancer, or possibly cause cancer, if other people had access to this information, it would show maybe areas where you know, it could possibly show environmental effect in different areas or that, so there would be advantages of I suppose genetic testing like that"

10	***Scientific Benefit*** • "I suppose genetic testing in a positive light would be to give an individual an indication of something in their genetic history that was you know a positive, a positive indicator such as that going forward if you're my age, you're 58 and getting old, you might do a genetic test to indicate if your parents had a predisposition towards early heart attack, that of whatever, hardening of the arteries, diabetes or anything like that, hopefully you would similarly, whether you would have, you know a positive or negative outlook in relation to those going forward, i.e., looking at your potential lifespan going forward you being at a certain age and you're trying to almost like pre-empt what might happen therefore you could put in train, now ultimately I suppose this is all negative 'cos it's related to illness, but you are trying to foresee the future and put in…progressive safeguards" • "you know at a certain age you try to progress or see forward the things, therefore you would put in breakers, you would put in good things that sort of you would put in positive medical interventions such that hopefully having found out that you wouldn't necessarily hopefully run into those things because you've put in…You know like osteoperosis, sort of like my mother, my grandmother, my great mother had severe osteoperosis by the time they were, now, can I at my age 20 or 30, can I drink more milk can I, that's very positive…" ***Insurers Benefit*** • "insurance companies have to guard themselves in relation to risk, and in relation to the premium like because the cost, they eventually will have some sort of payout at some point in time" • "…it's all very much on the insurance side…" • "insurance companies are in the business of risk, but risk aversion and risk aversion means looking at all the potential claims areas, angles, eventualities and everything in general whether they be car insurance house insurance, life assurance, you name it, risk insurance for aircraft, ocean liners, whatever, they look at all the risks, that means getting a test or an engineering assessment done or in our case looking at a genetic test, that gives the company all the cards again, they have all the information, they're at no loss at all, I know I understand that's the basis of insurance, risk balance, risk aversion, risk re-insurance, risk, you know, versus payment, so I understand, but they seem to have all

	the cards and more and more people are not necessarily happy at the power at the information and how that is used by the companies”

Source: Analysis of Interviews Transcripts

Table D-7 Definition of Trust

Subject	Definition of Trust
1	• “I would define it as… My level of confidence in another party, my trust in another party, my level of confidence in that party to make certain decisions or perform certain actions on my behalf and to do so with my best interests at heart”
3	• “I think the general population vote for the individual, the candidates. We trust the politicians, we vote for in a democracy, whether we vote for the individual winners or not, we’re a democracy, so we place our trust in our vote, in, in those individuals, and, they’ve got a job to work. And, in general they’re, they have rules and controls that confine them as well. And if we don’t like what they do, we kick them out the next time round.” • “Being able to place, being able to place either money, or, or a family member or, an idea with someone to look after it properly, to protect it to, whether it’s money, a family member or whether it’s, simply an idea you have, and to have confidence in them, and that they won’t, misuse the trust that you gave them”
4	• “Having faith in people” • “say you’re going to make a decision, say you actually trust someone, you’re going to like, allowing them to make a decision for you kind of, like do you know so, basically having faith in people”
6	• “a willingness to allow another person to, another person to act on your behalf or to make minor decisions for you so I suppose would be one way of describing trust” • “or placing value in another person’s opinion”
7	• “You’re confident that they’ll do the tight things and, you don’t need to worry about it”
8	• “I suppose trust is, when you trust somebody, you tell them something in confidence, and you hope that it’s not repeated you know to anybody else or it’s something that you can, you know you rely on somebody to carry out something that they said they would do and you hope that it’s done to the best that they can do

	it, so I suppose it'd be something that you kind of, you would give, I mean you would really trust, trust someone you would hope that they would fulfil something what they've said they've done, or if you tell them something you would hope that they wouldn't repeat it and you kind of rely on somebody al lot that you trust and you hope that they are true to their word"
9	• "well, if I was saying trust as regards a friend then I would define it as being able to confide in somebody, I suppose it can be taken in a good few meaning, it could be taken as to rely on somebody to so that you know I could rely on you to, to you know keep this say a certain amount of money for me you know, for a few days or whatever, you know, or a, I suppose on a deeper meaning it just means that you really can tells somebody anything and, and you know they won't you know tell anyone else or whatever, you know I don't know how that can be related to the Government or to insurance companies, but trust is I suppose you pay your premium at the start of the year, say for the insurance companies and I wouldn't have a whole pile of trust in them, you know I think they're always trying to do you out of, you know they're not going to let you make a profit out of them"
10	• "Trust is my understanding that the individual or institution or entity that I am dealing with will treat me fairly, honestly and will not penalise me financially, socially or any other way that might impact on me living within that society. In such a way that it is going to adversely affect what I have in my pocket or the way I have to act or interact within that society"

Source: Analysis of Interviews Transcripts

Table D-8 Trust in Government Risk Management of Life Assurance Companies' Use of Genetic Test Results

Subject	Survey		General Trust							
	General Trust	Sceptical Trust	Good Job	Competent	Skills	Public Interest	Listens Public	Listens Ordinary People	Fair decisions	Relevant Info.
1	Disagree	Unsure	Disagree	Disagree	Disagree	Disagree	Disagree	Disagree	Disagree	Strongly Disagree
2	Unsure	Unsure	Unsure	Agree	Disagree	Unsure	Disagree	Unsure	Unsure	Unsure
3	Agree	Disagree	Agree	Agree	Agree	Agree	Agree	Agree	Agree	Agree
4	Agree	Unsure	Unsure	Disagree	Agree	Unsure	Agree	Disagree	Unsure	Unsure
5	Unsure	Agree	Unsure	Unsure	Unsure	Unsure	Unsure	Unsure	Unsure	Disagree
6	Agree	Unsure	Unsure	Agree	Agree	Agree	Agree	Agree	Agree	Disagree
7	Agree	Unsure	Unsure	Agree	Agree	Agree	Disagree	Disagree	Unsure	Unsure
8	Unsure	Unsure	Unsure	Agree	Agree	Unsure	Unsure	Unsure	Unsure	Unsure
9	Agree	Disagree	Unsure	Agree	Agree	Agree	Agree	Agree	Unsure	Disagree
10	Disagree	Agree	Disagree	Disagree	Disagree	Disagree	Agree	Disagree	Disagree	Disagree

Subject	**Sceptical Trust**		
	Distorts Facts	**Changes Polices**	**Influenced by Insurers**
1	Unsure	Unsure	Unsure
2	Strongly Agree	Unsure	Unsure
3	Disagree	Disagree	Disagree
4	Agree	Unsure	Unsure
5	Agree	Strongly Agree	Unsure
6	Disagree	Unsure	Unsure
7	Unsure	Unsure	Unsure
8	Unsure	Unsure	Agree
9	Agree	Disagree	Disagree
10	Agree	Agree	Agree

Table D-9 Trust in Insurance Industry Risk Management of Life Assurance Companies' Use of Genetic Test Results

Subject	Survey		General Trust							
	General Trust	Sceptical Trust	Good Job	Competent	Skills	Public Interest	Listens Public	Listens Ordinary People	Fair decisions	Relevant Info.
1	Disagree	Unsure	Disagree	Disagree	Unsure	Disagree	Disagree	Disagree	Disagree	Disagree
2	Agree	Agree	Unsure	Unsure	Agree	Disagree	Agree	Agree	Unsure	Disagree
3	Disagree	Agree	Unsure	Disagree	Agree	Disagree	Agree	Agree	Disagree	Disagree
4	Agree	Agree	Unsure	Agree	Agree	Disagree	Agree	Disagree	Unsure	Unsure
5	Unsure	Unsure	Unsure	Unsure	Unsure	Disagree	Unsure	Unsure	Disagree	Unsure
6	Disagree	Unsure	Agree	Agree	Disagree	Strongly Disagree	Agree	Disagree	Disagree	Strongly Disagree
7	Disagree	Unsure	Unsure	Agree	Agree	Disagree	Disagree	Disagree	Unsure	Unsure
8	Unsure	Unsure	Unsure	Agree	Agree	Unsure	Unsure	Unsure	Unsure	Unsure
9	Disagree	Disagree	Unsure	Agree	Agree	Disagree	Disagree	Disagree	Unsure	Disagree
10	Disagree	Agree	Disagree	Disagree	Agree	Disagree	Disagree	Disagree	Disagree	Disagree

Subject	Sceptical Trust		
	Distorts Facts	**Changes Polices**	**Influenced by Insurers**
1	Unsure	Unsure	Unsure
2	Agree	Agree	Agree
3	Agree	Disagree	Agree
4	Agree	Unsure	Unsure
5	Unsure	Agree	Unsure
6	Strongly Agree	Disagree	Unsure
7	Unsure	Unsure	Unsure
8	Unsure	Unsure	Unsure
9	Disagree	Agree	Disagree
10	Agree	Agree	Agree

Table D-10 Trust in Government – Interview Results

Subject	Trust in Government	
	General Trust	**Sceptical Trust**
1	**Good Job, Competent** (The Government is not competent enough to deal with insurers using genetic test results) "perhaps not at the moment, I think that they're not doing a great job regulating certain parts of the insurance industry so, such as the car insurance sector, someone today quoted an example to me, a statistic to me, saying that there is 5 million people in Ireland, and there is 50 million perhaps or maybe ten times more in England, yet the profit made from the Irish people by the insurance companies is significantly more than made from the 50 million people in the UK, and I think that reflects badly on the Government's management of the insurance industry in Ireland and the way they set their charges, their insurance premiums. So, based on that, I would not have a high level of trust to manage the insurance companies and their use of genetic information" **Public Interest** "At the moment, I don't think that they are doing enough in the public interest" (because) "I just get	**Distorts Facts** "I think they have done so in the past, I think, however, a lot of other cases like that have been brought to light in a public forum, in recent years, they are less likely to do that going forward, so, in summary I think no" **Change Policies** "I think that good reasons would be required"

	that impression from the overall treatment of the health sector, that I just don't think they're putting enough focus on the health sector as a whole and I think that this is a part of that" **Fair Decisions** "I guess the way they make decisions on issues like this is a little unclear to me and that indicates to me that it is unfair, The fact that decisions are made and are not obvious to the public" **Relevant Information** "I'm not really aware at the moment the level of management they perform over this, I guess that's perhaps indicative to a lack of information to people like myself" (I don't believe the Government has provided enough information to the people). "That's what I think, yeah, However I have never sought that information either, so perhaps it is publicly available and but, I think it would be good to highlight it to people who are applying for life assurance"	

2	**Public Interest** “No” (I do not trust the Government) “because again, sort of like, you know, they work in their own interests” “No, I wouldn’t think so, I wouldn’t think” (that the Government represents the public interests) “Well, I think that they’d have to” (act in the public interest with regard to this matter), “you know, because, most people would like that information being, you know, if they were going for insurance and they had to give out information about genetics” **Listens to Public** “After a certain amount of they have to” (listen to concerns raised by the public) “because, do you know, I mean, you get the small fish somewhere trying to bring it up, but eventually then, sort of like, you know if it gets aired at all you’d people attention, well then I do think they have to sit up and listen then” **Relevant Information** “Oh, definitely, definitely” (think that the Government should provide more information to people about insurers using genetic test results)	**Distort Facts** “Well, they could” (distort facts), “but they’d be found out sooner or later, because it all comes out in the wash, especially after all the enquiries that there’s been lately, you know, I think they would be a bit scared of doing anything underhanded” **Influenced by Insurers** “They” (the Government) “could be” (swayed by insurance companies with regard to this matter). “Because if it mean a vote here or there, I’m sure that, you know because, again, everybody is out for their own benefit” “I don’t really know” (if the Government could be influenced by life assurance companies when making decisions on this matter). “Who can you trust these days sort of like, you know, a backhander here and a backhander there”

	(People would be better equipped to make their own decision if they had more information) "Absolutely, yeah, yeah"	
3	**Good Job** "My understanding from the briefing is that they're only beginning to look at it really, that, the insurance companies, at the moment are the ones that are governing practice, and that's conforming to practice in England rather than Ireland, so they don't seem to be doing it at the moment" **Competent** "I think that if they apply their, yeah, when they apply themselves to it, like that they, like, like everything else, they're suppose to be" **Skills** "Yes, I suppose they do, yes" (have the necessary skills to manage this) "Again, again just to clarify that" "they can either use their own independent, they can either use their own scientific laboratories, or they can use independent laboratories to get the information and the facts, but at the end of the day, they're there, they're there to write legislation, and again, and again I do believe that they have the capabilities to write the legislation"	**Distorts Facts** "I like to think that the Government has a duty to the people who elected it and they have to give proper information. The Government, information documents have to show both sides of the argument, so in preparation for the referendum and in preparation for any detail like that, the Government departments have to reveal all the information like that in a non-emotional, non..., informative fashion" "I think that when it comes to informing the public on legalities and laws, there are constraints put upon the Government to ensure that we get all of the information, the pros and cons of everything no matter what's under control" **Influenced by Insurers** "They're a lobby, they're a powerful lobby (pause) but I, I think that any government will step back and assess the situation with the general population in mind"

	Public Interest "I think we have to trust the Government. We don't really… You shouldn't think of them as who's in power at the moment or who's going…., or who was in power 20 years ago. The Government is who people choose, and the Government normally operates with its constituents in mind, and, and that's the way it should be" "I think that any government will step back and assess the situation with the general population in mind" **Listens to Concerns** "yes, yes, simply because we vote them in and we won't vote them in the next time if they don't listen" "Now, that doesn't just necessarily mean with respect to insurance, I mean it could be with respect to cost of living and everything, 'cos if we don't like what they do we won't vote" **Fair Decisions** "Slow, but fair"	

<table>
<tr><td></td><td>Relevant Information
“because I didn’t have the information, I didn’t know whether they provided it or not”

“It’s possible” (the fact that you haven’t received anything or heard anything about it, could be a reflection on the fact that they haven’t provided the information) “but, again, I, I don’t know, I haven’t gotten anything”

“I suppose you have to weigh out how much interest there is for one thing and if somebody is working in the field they’ll already know or go out of their way to find out about it, the general man on the street, or the person on the street will only, will, will probably get most of their info from news and from media”

“But it would be an advantage and of course the internet is there where I’m sure you could get information if you looked it up. But it is an advantage to know”

“I certainly think that the information should be out there and be available, yes”

“The information can come from either” (the Government or the insurance industry)</td><td></td></tr>
</table>

	"I like to think that the Government has a duty to the people who elected it and they have to give proper information. The Government, information documents have to show both sides of the argument, so in preparation for the referendum and in preparation for any detail like that, the Government departments have to reveal all the information like that in a non-emotional, non…, informative fashion"	
4	**General** "No" (I would not trust the current Government) "I wouldn't really, to be honest" **Good Job** "I don't know what the Government have to do with it, I'm not sure if, as, as I say I've never received any information about genetic testing before so, I'm not sure if they actually…" (are doing a good job) **Competent, Public Interest** "that's the same as the insurance company one, I'm not sure if, I'm sure they are competent enough, I'm sure they're all very intelligent and competent people to undergo, like, don't, I not be 100% sure that they would have our best interests at heart" **Skills** "Same thing, I do believe that they have the skills to manage it but…" (same as competence)	**Distort Facts** "like any, I've studied marketing myself, so I know the way you can twist words, you can twist phrases and everything to suit yourself so I do imagine that they're going to have somebody out there looking after, keeping their own name, do you know so, I don't know, I imagine that yes, they could actually distort facts to suit themselves and to make themselves sounds like, sound like they're going to look after everyone else and but, then again as I say, I believe that it's run like a business" "No, I actually couldn't" (remember any situation where this was the case, where they did distort information) "see now that's not backing myself up very well now is it? (laughs) But actually, I just know the way, like this morning, I was listening to the radio and I was listening for and against basically, cycling in a park, and the way the local councillor totally twisted what the councillor for it said, I, I can just see how that

	Public Interest "The same thing, if you ask me, the Government is pretty much a business, so they're going to be the exact same and they're always looking after their own rather than us" "Without good reasons, yes, definitely. Just as I say, I was listening to that radio show this morning and they completely disregarded a petition, they, the health of children, the heath of people, people cycling to, they say all the time about wanting people to be healthy and then there they are turning down people actually getting a good form of exercise in a public place. So, definitely and they took no regard for other peoples interests, so, yeah, definitely" "I wouldn't believe that, no" (that the that the Irish Government is acting in the public interest with regard to managing insurers use of genetic test results) "Well, just for the same reasons, I don't know whether I'd have full trust in the Irish Government so" **Listens to Public** (I believe that the Irish Government listens to concerns raised by the Irish public regarding this matter) "By the public, yes, because they're going, it's going to be heard about and there's going to be a	must be multiplied like on a huge scale to do with, to do with the bigger TD's and everything. That's just my belief. Like, like the stories you hear and they're all coming out years after the TD resigned whatever, and you hear all, we all know what's been going on in the, in the tribunals and everything like that so, I definitely believe that there's a lot more going on that we'll probably hear about in 20 years time, so I don't think it just ends" **Changes Policies** "I'm unsure what the Government actually do about it. I don't even know if the Government have any say in it, you know, I'm sure they do 'cos the Government has a say in all laws that are passed, but to what extent, I wouldn't be a 100%, and it's just, I've never heard it talked about, or I've never, I've never heard other people talking about what the Government would do towards genetic testing" "Without good reasons, yes, definitely. Just as I say, I was listening to that radio show this morning and they completely disregarded a petition, they, the health of children, the heath of people, people cycling to, they say all the time about wanting people to be healthy and then there they are turning down people actually getting a good form of exercise in a public place. So, definitely and they took no regard for other peoples

<table>
<tr>
<td></td>
<td>large majority of people saying this so if it's a group they're not going to be seen as if they just completely threw it out the window and no I don't want to do anything with this so…"

Listens to Ordinary People
"I took the ordinary people to mean a small group and the public to eb a large majority , kike I though of ordinary people as if it was just, just a group of people maybe saying this and that and I didn't think they would have much regard for it unless it was publicised. Unless, it was actually, put out in the newspapers or something where more people are going to hear about it and more people are going to be in disagreement with it, about it, but…"

"it depends how, if they were going to be fair in treating us to do with genetic testing? It depends big an issue it was, it depends how many people had raised awareness about it, and everything, it was kind of something that is known by just a few people, and there was a small group of people coming towards the Government to talk about it or whatever, well then no, I don't think they'd take much regard take into account, or take into account what that group of people thinks but I think that if there's a large body of people, then yes, I would think they'd take it into account"</td>
<td>interests, so, yeah, definitely"

Influenced by Insurers

"I don't know if they, if the insurance industry would funding the Government in any particular areas, they might, they might fund something in the health services, well then yes, if not, I'd say it'd be too much frowned upon by the public, and the Government would lose out on too many votes if they actually, went for, or took the side of the life insurance industry on the use of genetic testing. It depends also what the public vote, whether the public would think that genetic testing should or shouldn't be used"</td>
</tr>
</table>

	Fair Decisions, Public Interest "No" (I agree that the way the Government makes decisions is not always fair) "not always, I'm sure a lot of the time, yes, and certainly for well-known areas, definitely, because they don't want to be coming out as the under…., as be not thinking of their, like they're meant to be looking after us as number one so they don't want us to think that they're not. So, yes, if it's widely known issues, they do, but if it's going to be smaller issues, no, they don't. I'm sure they have a lot on their plates and I'm sure they don't always have time to about but that's what the local councillors and the local TD's are for, it's to look after the local people" **Relevant Information** "I have never received leaflets or pamphlets in the post or anything like that. I haven't even seen it in doctors, in surgeries or whatever, I have never really noticed it, or, but I certainly haven't been contacted through councillors, or local government or anything about it, or received leaflets in the post from it so" "I don't know if the Government have had information about it themselves to provide. If they haven't had information well then they obviously can't provide it to anyone, but if they have, maybe they want to decide what they're going to do first,	

	I'm not really sure why they wouldn't have provided information to the public yet" "if I knew more facts about genetic testing, I might be able to answer some of them more clearly, but to be honest I don't have enough grasp on the subject to be able to answer all of them with definite agree or disagree answers, so I wouldn't be a hundred per cent, so I, I just thought it'd be fairer to say that I don't know whether it should be allowed or not, so" "…they need to have the information before they can actually participate in decision-making and for the Gove… Whoever's kind of making the key decisions about this need to create awareness in the public and in the Irish people need to get known about, more information need to be spread about it basically, but it's definitely in the public interest to make decisions, the public should get information to make the decisions on this rather than just the Government"	
5	**General** (Perhaps you could say then that you might not trust the Government because of its previous history)? "Maybe, maybe, but I know that they haven't done anything with genetics, so and I'm still like, blind about a lot of things but, the initial feeling you get	**Distort Facts** "Yeah, they probably could, they probably could" "They could pay off certain members of the Government"

	from them, hmm, maybe suspicious, hmm, so maybe you've got about them being in control **Good Job** "I wouldn't know actually, like right now, I wouldn't know, I really don't know, but I hope that they are" **Competent** "I'd like to think so, yeah" (that the Government is competent enough to manage this situation) "I think that because it's such a new thing as well maybe it'd be great if the Government got experts in" **Skills** "I suppose they would" **Public Interest** "Yeah, I think they do, yeah" (act in public interest with regard to this matter) "'m not, I can't be sure but I'd like to have faith that they would" **Listen to Public** "Yeah, of course they would, yeah" **Fair Decisions, Listen to Public** "I do" (trust the Government to manage this situation) "yeah, I do at the moment because, like it seems that they really are trying to stay on top of everything,	**Influenced by Insurers** "…the Government should really be on top of the situation like that, 'cos it's really a private, is it, or is it the Government, or am I, is the Government really involved in this insurance as well, are they doing well from it"? (The Government could be swayed more so by the insurance industry than the public) "I'd say, it could be really paranoid of me, it's just yeah possibly, how are we to know, or say, how are, I mean things have come out before and they've just gone moving swiftly along, so maybe, if it was in their favour, they did see a way to move this to their advantage, pensions and other things, then they would" "They could pay off certain members of the Government"

	that's the way it seems at the moment, there's a lot of issues that have been kind of in their faces, and when it's brought to public attention, and once it's aired like that yeah, they deal with the situation, I think" **Relevant Information** "shouldn't the Government…be involved with the provision of information, to make sure that the information is being used appropriately I'd say, I don't know maybe the Government really more interested if they could use it as well against you like, I don't know" (If the Government knew about the situation and hadn't actually provided any information to the public, this would be bad, or a negative thing) "Yeah, definitely, they should tell the public really like, what they're moving on, 'cos the public put them where they are and they're supposed to do a job for the public" "I have no idea why" (the public have not heard about this) "it's a bit suspicious, maybe, maybe it's, maybe it's been like advertised, or you know, televised, or been talked about, the more you watch the more you hear. Maybe, it's been on the radio, but I haven't heard about it"	

	"there's a generation who really don't want to kind of think or do anything about it, 'cos they're kind of happy where they are, so like if something's landed on their lap, fair enough, and you even explain to them well look, there's this going on, but surely they'll kind of think oh that's not right or I agree with that. But look, there's no information on this thing"	
6	**General** "Yes, I believe, I would trust them to a large degree yeah" "I believe so at the moment, yes, I believe that the current Irish Government is fairly responsive to public opinion" **Good Job** "Well to be honest at the moment I know very little about it, I'm not sure as to whether or not they are managing it well, I'm not sure about that" **Competent** "I believe that if the situation arises, yeah, I believe they would be competent" **Listens to Public** "I believe so at the moment, yes, I believe that the current Irish Government is fairly responsive to	**Distorts Facts** "I believe that the Government is perfectly capable to doing that, yes" "I believe that they would to some degree" (do so with regards to this situation) "but I wouldn't believe that they would do it to a large degree. I wouldn't believe that they would try a complete cover-up" "No situation jumps into my mind at the moment, but no, nothing, no, I can't think of any situation right now" (when the Government distorted facts in its own favour) "no I don't thing they would distort facts to a high degree, they may more that distorting facts, both from the Government and the insurance point of view, I think they might more hold back on facts, rather than distort them, they may be more willing, I think at the moment in this issue, there is an awful lot of public

	public opinion" "I believe they do listen to them up to the point in that if they don't listen to them it would create political damage to the Government" **Public Interest** "not necessarily all the time no. I believe that the Government would more act in political interest than in public interest. I just hope that most of the time, they're one in the same thing" "again I believe that the Government would be highly influenced by insurance companies, perhaps a little bit more than by the public, I believe that there probably would be, that insurance companies would probably be a better interest group than the general public, but nevertheless I still believe that public opinion would be represented through equally strong interest groups on their side such as the media or through, health boards, so on so forth and through the Irish medical organisation, and other factors like this" **Relevant Information** "I believe that they probably haven't, because I've heard nothing about it anyway so no" "no I don't thing they would distort facts to a high	ignorance in this issue, and I think that they, the insurance industry and the Government are perfectly happy to allow the public to remain in that ignorant position as opposed to distorting facts they are just not providing the facts, I suppose" **Changes Policies** "Yes, I believe they could, but then it depends on what you mean by good reasons, from their own point of view, form a political point of view, I suppose they could distort the facts and that necessarily wouldn't be for the public good or for so on so forth, but I believe that it depends on what you mean by good. Whose good are you talking about?" **Influenced by Insurers** "I think that they would be influenced by insurance companies, yes" "again I believe that the Government would be highly influenced by insurance companies, perhaps a little bit more than by the public, I believe that there probably would be, that insurance companies would probably be a better interest group than the general public, but nevertheless I still believe that public opinion would be represented through equally strong interest groups on their side such as the media or through health boards, so on so forth and through the Irish medical

	degree, they may more that distorting facts, both from the Government and the insurance point of view, I think they might more hold back on facts, rather than distort them, they may be more willing, I think at the moment in this issue, there is an awful lot of public ignorance in this issue, and I think that they, the insurance industry and the Government are perfectly happy to allow the public to remain in that ignorant position as opposed to distorting facts they are just not providing the facts, I suppose" "the Irish public should be demanding to know at what level it currently stands, how much, what access do insurance companies at the moment have to this information and are they currently even using it. I'm unaware"	organisation, and other factors like this"
7	**General** "yes and no, I mean every Government's the same, I don't think any Government's better than another, they'll do some things right but they'll also pay for other things like. You trust them on average to do more good than bad" **Competent** "No, no" (the Government is competent enough to deal with insurance companies use of genetic test results) "You need experts in the field, politicians	**Distorts Information** "They don't withhold, they just highlight things that suit them, kind of, and don't tend to show, you know obviously if you dig deep enough in the freedom of information you are going to find things they don't want you to know about. But the things they actually make known" **Influenced by Insurers** "Probably not" (trust Government with regards to this situation) "because politicians, politicians can be

	there's only so much they know, they need to get the right people" **Listens to Public** "In general, they do, but again, if it's, they listen to concerns but if it doesn't really suit them they're not going to do much about it, they'll say they'll take it on board and do the usual, they'll look into it, and sometimes they do things about it but if it doesn't suit them then a lot of the time they don't" **Fair Decisions** "I can't think of a fairer system to be honest. You know it is fairly democratic, I know there's a lot of corruption, bribes and things like that, you can see that with all the tribunals but, you know there's no better way of doing it. So, I mean fair enough they will be, as I said, on average more fair than not" **Revealed Information** "I don't know anything about that now to be honest, but I just wouldn't know about that area, they could have but I don't know"	swayed usually, okay and if insurance companies wanted something, you know they'd just donate a lot of money, you know there are ways around to influence politicians" "They could be" (influenced quite heavily by life insurance companies) "The Government is always trying to push forward a positive thing, they want to stay in power, they want to be re-elected so they want to show positive things, or swing, just make things look more positive then they actually are, so they don't have a negative image" "If you just look at, I don't know, funding, for us, they don't talk about, if you're just looking at the roads problem or something, they talk about all the projects they have funded, but they don't talk about all the other problems, that are there at the time, they just keep changing the question back to we put this much money in" "I think that, that they could" (be too swayed by insurance companies) "I mean I'm not saying that the Government are all bad, but it just would really depend on who was chairing the committee overlooking this, how much power the insurance companies had"

8	**General** "In general, probably about 70%, I reckon, not 100% but I suppose I would trust them in general, I would but not fully, not 100%" "I don't (trust the Government, to manage insurers use of genetic test results, do you think that they can manage it successfully) "unless they have something like a specific task force, you know like people who are, experts themselves in that area, I don't know if they have, if they would employ a group of people, you know go to that kind of trouble and expanse of employing, the like of a committee or a task force to, to overview this kind of, the genetic testing. I, I suppose of it's, if it's, insurance companies, I mean I feel like insurance companies benefit more , they benefit more than society, or the person who's getting the insurance, so I think that, I kind of feel that the Government, I can't imagine that they would have a huge role in it or that they would really care whether or not, you know, that they, I kind of feel that they wouldn't put, be a hundred percent in putting effort into making sure that it's running properly and also that, I suppose they feel, you know, the whole expertise side of it as well, whether they would know enough about it or ploy people"	**Distort Facts** "I suppose they could yeah, easily, they do it all of the time really when you think about it, the tribunals and everything that's gone on, so I imagine if they wanted to I suppose they could, but I think these days everything, the truth always comes out, so maybe, maybe they wouldn't but I wouldn't be surprised if they did. You know I wouldn't be fully confident that they wouldn't distort facts" **Influenced by Insurers** "I think maybe they could be because, I mean, insurance is such a huge industry now with you know for all types of businesses, has, and I mean now, I mean it's huge, huge, you're talking about huge money, and huge job, I mean you're talking about a lot of jobs and companies that are set up here and that they kind of help the economy so much that I think maybe they would play ball with them as well, I think they might, kind of have to play ball with them as well"

	Competence, Skills "not, initially I wouldn't think that they would, but, then, then again would they bother to employ people who would, and pay those people to kind of monitor the situation , and you know kind of, report on it and things like that" "I'd imagine, I suppose they, the more it becomes a very common, topical thing, I, I think maybe they would have to, but I would say it'd be done at a push,, I don't think it'd be something that they would…." **Fair Decisions** "I don't really know anything about genetic testing, I haven't really heard anything about it lately or in general I don't know much about it" (In general) "I, mostly I do agree with them but in that sometimes I do kind of feel that that the things they make, the things that they do, they do are very incorrect and you know that, that they do have a lot to answer for but at the same time, I suppose they can only do as best as they can" **Listen to Public** "I think yeah, if that was the case then they'd have to, they'd have to react in some way, but like that it wouldn't be off their own bat, I think it'd be a matter of somebody pushing the matter on them and having	

	to maybe set up a committee or something" **Listen to Ordinary People** "I think, probably with this matter, I'm sure it would have to take somebody like a politician, a politician or somebody on a higher level to actually put the point forward and make them do, react, you know make them actually do something about it. But I think, you know, just the ordinary person, I don't think, see how the concern for the Government so I don't think they'd be too bothered unless they kind of had to react to it" **Relevant Information** "you see I don't know, I don't know, I have no life assurance or anything so I haven't really ever had to fill out forms or apply for it or look into it so" "suppose I kind of feel that if I was, was to go and apply for some, you'd imaging that most of the information, you know the information should be fully given by the insurance companies, if they're providing a service to a consumer well, then they should be, you know they should make them aware" "I suppose I would, you would hope that, like before, if you had a choice to vote that you would be informed exactly what you're voting for, or for or	

	against almost, or what it means, or what the consequences are, so I'd imagine, I think myself it's a very personal thing, if you think you want your results to be given to a company and then you based, you know everything is based on those, which, at the moment, I don't really agree with, I can't see any major benefits to society or to people, but, I suppose if I did have to make a choice I would hope that it would be informed and that you would be, you know, get your fact sheets get your for and against and why you would vote yes or no, so I'd hope that I should be informed before making a decision"	
9	**General** "Not a lot" of trust in Government (Why?) "about all the scandals and stuff unfolding in the last few years, you know Haughey and all these say nursing home things, and all those kind of things, you know, various scandals and stuff" "if there were strict enough guidelines, then I would yeah" (trust the Government to manage life assurance companies' use of genetic test results) **Competent** "I don't know it depends, you know, it depends, you know I don't know"	**Distorts Facts** "Yeah, I do yeah" (believe that the Government could distort facts in its own favour) "possibly, possibly, yeah" (with regard to this matter) **Influenced by Insurers** "Not so much so, you know I don't think that they would be you know"

	"I mean, you know I'm sure they could recruit, they can recruit anyone they want cos they have the money, so if they were to recruit a special team to do that, I'm sure they would be capable" **Listen to Public** "I think they would, if there's big enough public support then they would" **Fair Decisions** "well I think that you're never going to keep everybody happy, there'll be some decisions that I agree with, there'll be some that you know are for the good of society but you might not necessarily agree with them, but I think overall that they do make good decisions" **Relevant Information** "It's hard to say because I haven't seen anything"	

10	**General** "I would not necessarily, no, I would have to see one what they Government would be saying in relation to it, how they would manage it, and I would also have to look at what the insurance companies are going to say on how they would manage it in relation to this information, I would really have to see, willy-nilly I would not know, I would not necessarily accept the Government or the insurance companies. I would really want to see what they're outlining in relation to this before I would…" **Competent** "Again, unfortunately the answer would be no, in relation to a lot of stuff that we've seen in the last 10-15 years, and I think from the management of the blood bank, even this morning on the radio, we hear of the Government's management of properties and how, and also in the papers in the last number of days looking at, especially today, all the cobwebs that have happened, these are only the ones that get to the newspapers, how many more are out there, so no, I do not necessarily, I would have to see how they're going to do it, what sort of body they're going to put in relation to what sort of ombudspersons or specialist organisation, for example they're talking about the national treasury management for the money that we have for the pension or property, something that's	**Distort Facts** "Yes, I do" (believe that the Government could distort facts in its own favour with regard to this matter) "Well the blood transfusion and the Hepatitis C is a classic example. They buried they're head in the sand, but it was known in the blood bank that the blood was contaminated, and even when the information came out it still wasn't being watched, people still went on being affected" **Influenced by Insurers** "Yes, because they're a powerful lobby, they're a major commercial interest"

	upfront, efficient, well managed, well headed up by very competent trustworthy. I don't necessarily have trust for the Government or insurance companies" **Public Interest** "access to this information will only ever be used, on the basis of what's good for the company and not necessarily what's good for the client" "Only where the public keep a very wary and watchful eye over the Government and bring them to book where it hurts in relation to keeping them up, on their p's and q's and like everything from the media and concerned individuals, getting it out and telling it like it is, this is what's needed. We're watching you, don't think that we're not watching you"	

Table D-11 Trust in Insurance Industry – Interview Results

Subject	Trust in Insurance Industry	
	General Trust	**Sceptical Trust**
1	**General** "No" I would not place a lot of trust in the insurance industry" **Good Job** "I'm not sure, it's not fully clear to me at them moment, what, how much genetic testing they have available to them and how they use it. So, I couldn't comment on that" **Competent** "I don't believe so, I believe that, an independent third party should be involved to regulate their use" **Public Interest** "I feel that insurers would inevitably use it for their own financial gain first, and the welfare of the insured second, because I see them as companies which are there to make a profit, to maximise their profits"	**Distort Facts** "Yes" (I believe that the insurance industry could ever distort facts in its own favour regarding this situation, because) "again it comes back to my overall impression of the insurance industry, as an industry which is very much self centred, almost verging on greedy and I think that an attitude like that in any industry would certainly lead them to that type of behaviour" "I don't think they could" (distort facts in their own favour) "With all the laws and regulations, rules and regulations that are out there at the moment. Of course, with all things, you've got to read the small writing , so and I mean, that's very important, you know, because, I'm sure they have themselves covered left, right and centre. It's up to you if you're taking out a policy or whatever, to read the small writing in between"

	"I see them very much as a profit making industry and I basically believe that their first priority is their own pockets, that's why I don't trust them to take my considerations, or consideration of the Irish people in general…." "no, I don't think it's in the best interest in the public" **Listens Public Concerns** "Only if it doesn't take too much from their profits" "I haven't heard an awful lot about it" (the insurance industry listening to concerns raised by the public about their use of genetic test results) "so I can't that, I'd have to know more" (Insurers would not ignore public concerns) "No, I think that if, if enough people knew about it because, that's the first I heard about it now, this evening , so I'm sure if a lot of the public were more aware about it, like they would be up in arms" "if it was aired, you know, publicly, I'm sure they'd have to listen because, I'm sure, somebody or other in the Government, like, would…" "You know it'd be brought before the Dail"	

	Fair Decisions "No" (I don't think it's very fair the way they make these decisions) 'cos I don't think they should be doing it in the first place **Relevant Information** "No" (I do not believe that the insurance industry has provided enough information to the people regarding its use of genetic test results". "Yes, I would" (like to see more information)	
2	**General** (I would not trust the insurance industry) "Not really, not really" "Because they're there to make a profit" "They're certainly not your friend" **Public Interest** "they wouldn't be looking" (for genetic tests results) "for your benefit in anyway, sort of like, they'd be lining their own pockets" "They're doing a job like and I mean people who want insurance have to go somewhere for it, so I suppose they are working in the public interest" "Of course" (I think that they would look after their own interests before the public)	

3	**General** "It's not that I would mistrust them, it's just that I think that there should be a controlling body, independent of it" **Good Job** "I think that if they're following the constraints laid down by the UK, then they're doing their best. I think they're not abusing they system" **Public Interest** "I think it's a business and they can put their own slant on it, I think therefore there would be opportunities to abuse the situation. Maybe not even intentionally, but over a period of time there could be misuse, and it's better if consumers have an independent body" **Listens to Concerns** "I think they listen. I think they interpret and, like advertising, you can interpret what you want from it. And they take what they want from it" "I think that they will act on the information they gain, but whether it's, if, if, if, individual voices are concerned, it's an individual, and it's like the voice in the wilderness. If, it's, a thousand people, people	**Distort Information** "I think they can always accentuate the positive, and somehow forget some of the negative, you, you can put your slant on it" **Influenced by Insurers** "I think that just looking at it from their own slant, and therefore it could be possible in several years, oversights maybe in certain areas"

	pay more attention, if it's a hundred thousand people, then, I think they will pay more attention again. It just depends how strong the voice is and how loud the voice is, I think they listen. And again, determine whether it's necessary to act in something that, for instance, if they were going to go a certain direction, and there was an objection, a major outcry, I thin they would listen, and then they would assess how strong that outcry is and determine which way to go. Would they go in the original direction or would they change their, would they change their viewpoint. It depends on the level of expert, the level of complaints, or the level of agreement, and they would assess it on that basis" **Fair Decisions** "I've no reason not to believe it, but I think they will make decisions looking at their future growth" "And reduce their risk" **Relevant Information** "The information can come from either" (the Government or the insurance industry) "think that, companies put their slant on it"	

4	**General** "Oh yeah, definitely" (I have a lack of trust in the insurance industry) "Not just in life insurance, in any insurance industry, not just life assurance" **Good Job** (Unsure whether or not the insurance industry is doing a good job with regards to managing it and once again that's because you didn't know the situation?) "Because I didn't have enough knowledge" **Skills** "Yeah, I'm sure they do" **Competence, Skills** "I didn't know whether I'd have full trust in the Insurance Federation with actually managing it but, I'm sure they would be very capable of doing it but, the same as other question, I'm sure that they'd have every means possible to do and the key people who would be very, very capable and very skilled at doing their work" **Competence, Public interest** "I'm sure they would be competent enough, I'm sure they would actually be able to do it if they were	**Distorts Facts** "I'm sure they have the workers there to do it but I just wouldn't know if they'd do it in our interest" "I'm sure they have their statistics which they give out, but then again, as I say, marketing is all about making yourself, selling yourself, selling your own company so, yes, facts are twisted, you're almost taught how to do it, so, definitely, I'm sure they distort certain information to bring themselves out trumps" **Changes Policies** "I'm sure there's legislation, well I'm not sure if there is but I imagine there's legislation about whether or not they can change policies, so…" "I wouldn't know" (they've ever changed policies without good reasons) "Well, to do with driving I suppose, young, young drivers, they say that there's statistics out there, and there are quite high statistics of young drivers in accidents, but, there's also quite high statistics for other drivers, do you know, who would cause accidents so, yeah they might change policies"

	going to but I wouldn't be sure of how much I'd trust them to use it to the publics best interest. So that was pretty much.. I didn't mean that they are not going to be able to do this, I meant more so that I wouldn't full trust them as they're a business and they're going to be looking after their own before they look after any of us" "I'm sure they have the workers there to do it but I just wouldn't know if they'd do it in our interest" **Public Interest** "I wouldn't trust them in the sense that I wouldn't believe that they're going to look after us before they look after themselves, basically the same thing, I wouldn't fully think that they're going to say okay, look well we've got the public interest at heart, they may say that in their ad campaigns or slogans but they're not going to let themselves go bankrupt or anything just to look after us. That's why I wouldn't trust them" "I'm sure they have the workers there to do it but I just wouldn't know if they'd do it in our interest" "I wouldn't fully agree with that" (that the insurance industry is acting in the public interest with regard to managing insurers use of genetic test results) "no,	**Influenced by Insurers** "I'm sure, of course, that they're not going to do anything major to, it, so that the public are going to say well, you know, that they're going to lose out on clients and customers but you always hear of stories that are covered up and different areas that have been kind of pushed or brushed under a carpet so, I'm sure there will always be places like that. But then again, there's always going to be cases like that in everything, so it just, it depends I suppose"

	because I do believe that the insurance companies act with their own interest at, in their minds or whatever” (Those interests are) “Make money, basically make money, get more clients, and make more, make more people aware about their company, or those kind of areas you know, whatever, but, I do believe that yes to a certain extent they definitely will act in the public interest because it’s going to be very obvious if they don’t, but with, in other areas where they might get away with it, no, they definitely won’t. Areas where they are going to make more money, perhaps, without people re, realising that they’re doing things that they shouldn’t be, or whatever, you know” **Listens to Public** “I would believe that they would, when it’s raised by the public because they’re not going to want, um, large amounts of people, I, I took public again as meaning a bigger body of people than just ordinary people, so, I, I don believe they would when it’s raised by the public, ‘cos if they don’t it’s going to be very well publicised, it’s going to be in the papers, it’s going to be in the media, mass media, so, I don’t think they’d want their names brought through with negative information brought about them or whatever”	

	Fair Decisions, Relevant Information "Well, with regards genetic testing, I'm not too sure, because I don't know what way they would make decisions regarding genetic testing, in other areas, no. Probably not. To do with… You see, my only dealing ever to do with insurance companies was when I would have started driving, so, and we all know what it's like for young drivers trying to get insurance, it's very, very difficult and the premiums are extremely high, so, no with that regard, I don't believe it's fair. But, I wouldn't be able to answer without being informed, a proper, knowledgeable answer" **Relevant Information** "Well, I haven't received information from insurance companies or anything about it, but I'm sure that they have it in some sort of book somewhere, but they might not just have passed it out very adequately to the public, so I wouldn't" "I wouldn't be a hundred per cent sure basically" "…if it's a big query and the life insurance companies and the government are thinking about it, well then I wouldn't consider it very fair at all to not bring it to the pub.. the peoples attention"	

5	**General** "Well, it's a bit, you know, sneaky for them, isn't it? Very, very good for them to have it to have that kind of information" (genetic test results) (Never heard of anybody who's been treated unfairly by the insurance industry or would you have anything to base a lack of trust in the insurance industry on) "No, not really, just, I imagine that the size of the, that, kind of something like anything multi, huge sized (laughs) companies or organisations have one kind of purpose and it's really kind of the take up, you know to make money, I think that in anyway, but that's just my own opinion" **Good Job** "I'd like to think so" (that they do a good job with regard to managing insurers use of genetic test results) **Competent** "I don't know, but I'd like to, yeah, think so"	**Distorts Facts** (if the insurance industry were to provide this information that they could be biased) "Yeah, of course, it would be in their favour, surely like they'd have to make it sound really nice without giving you, like the downside to it, like… And, there's a lot of people of a certain age are kind of so easily lead into thinking that these big companies are really trustworthy" (The insurance industry could distort facts in its own favour regarding this matter) "Yeah, yeah, they could, I guess, I can't say, I don't know, and how would you know? …they could say they're giving you the facts, but you can't go to someone else about this, to check if it's true or not" **Change Policies** "If they're writing them, then they can change them"

	Skills "Yeah, they must do" (have the skills to manage this) "if they're, well, I'd like to presume so, if that's kind of the area that they're going to" **Public Interest** "But like that's their interest isn't it, as a big business like, they need the money" "I just think that it's like, it's a business, their, it's really in their interest isn't it to be good for themselves, and to say to people that we're actually, we're doing this for you, or, you know, I don't have too much experience with, like, insurance companies, anyway" "Well, they act in their interests" "So, like if you're really providing them with it so its in their interests" **Listens to Public** "Yeah, I think they would, because they're, they've got to listen, they are really relying on people at the end of the day to kind of keep then where they are, so, like if they're anyway a good company, a good organisation, of course, they're going to listen if their members are interested or worried about something and see how they can sort it out maybe"	

	Relevant Information “if the insurance industry were to provide this information that they could be biased” “No, I don’t think so, I haven’t seen any or heard anything about it. This is the first I’ve heard anyone talking about it” (the insurance industry has provided all relevant information about their use of genetic test results to the public)	
6	**General** “as much as I can trust them, yes I would trust them, well to be honest I suppose I trust them from the point of view that I can see no other system that I would trust more than the system that is already there” “yes, I suppose, I would not trust, if I found out that insurance companies were, using genetic testing to determine peoples premiums, I wouldn’t trust that” “The insurance industry, well it has in the past I suppose, with many different forms of insurance , that there has been a lot of excess profits made by Irish insurance companies in the past and that they have shown that they’re out there to make money , to make profits, if allowed they will charge excess on premiums”	**Distort Facts** “I believe it could, it could distort facts in its own favour” “no I don’t thing they would distort facts to a high degree, they may more that distorting facts, both from the Government and the insurance point of view, I think they might more hold back on facts, rather than distort them, they may be more willing, I think at the moment in this issue, there is an awful lot of public ignorance in this issue, and I think that they, the insurance industry and the Government are perfectly happy to allow the public to remain in that ignorant position as opposed to distorting facts they are just not providing the facts, I suppose”

	Competent "I believe that it is competent in running it, again, I believe that insurance industry are under pressure, especially with medical at the moment, medical issues at the moment are being scrutinised an awful lot by Irish public opinion at the moment and if the insurance industry did try and manipulate the situation they'd know that there'd be a very high chance of creating public scandal and that that pressure will create a certain level of competency or a certain level of caution" **Listens to Public** "I believe that in the longterm they would, yes, I believe that definitely at the moment that public opinion would be reflected through Irish media, yes that at the moment they would be listening to public opinion closely" **Fair Decisions** "actually at the moment, I'm unaware of how much access insurance companies have to genetic information, so I don't know whether it's appropriate or not"	**Influenced by Insurers** "I suppose, yeah they could be, they would be by the fact that they would have access to genetic information, yes"

7	**Competent** "No, you have to regulate it really, they're gonna, they want to, all they're worried about is their profits, so they're going to use, they're going to do everything they can to get those figures, so they can not insure those people or charge them a lot of money" **Public Interest** "I don't really know, I mean, they're companies, they're out to make profit, they don't, I've never really trusted companies that are out to make a profit, they're always going to do whatever suits, their, their margins, so I wouldn't trust them that way" "To an extent, but again, I just never, I just would never trust a company that's only worried about profit, they're always going to be" **Listen to Public** "They'd listen, but it mightn't affect them too much unless there's a huge problem and then they'll have to sort it out, but if it's a small one then they probably won't" **Fairness of Decisions** "No, because you're discriminating against people	**Distort Facts** "Again they're companies and companies often do" "Not really" (think of a time when the insurance industry has distorted facts in its own favour) "car insurance I have, when they talk about figures, they're figures, it's very high in Ireland and they're actually comparing their figures to the European and they're not really good really" "And they're talking about coming down in percentage terms when the percentage really means nothing, when you actually look at the abstract figures compared to the other so that kind of makes me think that, but I don't know about health insurance, that kind of think, life insurance" **Change Policies** • • "I hope not anyway, I don't know. It depends on who, I don't even know who sets the regulations"

	for something that's not their fault. If you're born with a family, genetic history of disease, it's not your fault. It's discriminating basically so…"	
8	**General** "I probably wouldn't a hundred percent, the fact that they're not really being regulated by the Government and especially with regard to genetic testing, something that I suppose, something would be relatively new to me, an idea, so I suppose then it's becoming more common as well, but I think, I wouldn't trust them a hundred percent if, yeah, if, I mean I'd probably go and maybe research it myself, or if there was terms in the contract or terms in the agreement that I didn't understand, you know that you are signing up to something that is extremely important in your life, you know, so I probably wouldn't trust them a hundred percent , especially if they're not being regulated or don't have to answer to anybody" **Competence, Skills** "I'd imagine that the would be, I imagine that they would be, you have an insurance company you hope that everyone, that people working there are skilled and that, they are competent enough to run a business and.. So I suppose I would trust them, or not trust them but I would that they would, the would be	**Distort Information** "I suppose if they don't have any Government really you know overlooking what they do and they don't have to answer to anyone, I suppose they could distort the information, like, like, I wouldn't be, it wouldn't surprise me if they did"

	competent to do their job, so I would think so, yea, yeah” **Listens to Public** “I suppose they, if they don’t want to listen and they really don’t have to, you know, that’s the thing, if they aren’t being regulated if they don’t have to answer to anyone they can do their own thing, but then I’m sure there’s, they don’t have to respond to the public if they don’t want to or it could be costly, they might have to reduce premiums, so I, I’d say then, no, I’d say they wouldn’t have to if they don’t want to, that the way I’d imagine” **Relevant Information** “I’m sure, I don’t know myself but, I, I;m sure they have to , if they’re selling something they have to sell a good or sell a service they have to, you know you’re signing a form, you’re signing up to a certain insurance, and insurance that’s going to cover x, y and z, I think that, I presume that they would have to provide information”	

9	**General** "I suppose you pay your premium at the start of the year, say for the insurance companies and I wouldn't have a whole pile of trust in them, you know I think they're always trying to do you out of, you know they're not going to let you make a profit out of them" "Not a lot" (of trust in insurance industry) **Skills** "well, I'm sure they do, you know say like they have all the, the skills to adjust for people who have had claims, on their let's say previous policies and that so, they could, it's only going to mean a price increase anyway for those people who, whose genetic testing shows up something bad, so, it's going to adjust upwards anyway so I'm sure they will take, they're not going to, I suppose, screw people either" **Public Interest** "I'd say not in the public interest, no, I'd say it's in their own interest" **Listen to Public** "No" (insurance industry would not listen to public concerns)	**Distort Facts** "Possibly distort but I wouldn't say gross distortion" "Well, I think they'd have to be (very clear about the relationship to the price increase and explain how much they're increasing it for), whether they would or not is another question, but, you know I suppose, yeah, you know that they would because on your renewal form every year for car insurance it says how much no claims discount and that so I'm guessing that they would have to give a break down to say this much extra is going on your policy because of…" **Change Policies** "Yeah" (the insurance industry could change its policies without good reasons) "change its policies, well, say for you know late 1990's, early 00, the early years or whatever, they just jacked up prices every year and you know still making, you know it wasn't as if they were making losses, they were still making profits, and the profits kept growing, growing and growing, so, and it was mostly say male driver or whatever that was suffering"

	Relevant Information "I can't say, I've never seen anything"	
10	**General** "Having worked in it and on the other side of it, no, the individual has, caveat emptor, let the buyer beware, you really have to be aware what you want when you want it, how you want it and then see what a particular insurance company or insurance companies would offer you in relation to any of the insurance types, you know you have to be very careful, they are big conglomerates dealing in a massive market, global risks, and in Ireland we're down to a handful, a half, dozen of really big insurance companies, you really have to be very careful, no I don't trust them" **Competent, Public Interest** "They probably would be competent enough to deal with the information they'd get alright, but I don't trust them because the information they get, they will use to their own best interests, so rather than competence, it's how they would deal with this information"	**Distort Facts** "Well, yes, I suppose is the answer, but presume on the basis that you're going to be honest enough to put it into a proposal form for insurance that they're equally going to put up there, and be honest and straight. No, from past experience, from dealing with them, insurance companies will do anything to reduce their risk, or get out of paying large settlements, they can get a large amount of legal help and settle on the steps of the court. If it's in court the only people benefiting are the legal side, as the costs are put down to the consumer."

	Public Interest "No, it wouldn't necessarily, they act in their own interest" "I wouldn't necessarily trust them or I wouldn't consider that they would, because they are vested interest, bottom-line profit, bottom-line shareholder, they are going to deal with the information in a way that I suppose secures their best interest and not necessarily the clients" **Listen to Public** "They would, when there's enough people out there getting onto Government and getting onto the opposite side of the lobby group to the industry, the Government would say hold on, because people are on the ball, they're interested, and watchful, they're listening and they're pulling the Government to book, or indeed enough of them are getting out there and saying we're watching you or we're not happy with this or no we're not going to pay anymore. But you tend to be caught over a barrier because the insurance company, you come to them looking for cover, whether it be mandatory or otherwise, but if you're watchful, and thoughtful and ask plenty of questions and keep the Government up to their job, and watching the life insurance companies"	

Table D-12 Relationship Between Risk Perception, Benefit Perception, General Trust and Sceptical Trust in the Government Based on Questionnaire Analysis

Subject	Risk	Benefit	Trust – General	Trust - Sceptical
1	Some risk	Low benefit	Disagree	Unsure
2	High risk	Low benefit	Unsure	Unsure
3	Low risk	Some benefit	Agree	Disagree
4	Some risk	Some benefit	Agree	Unsure
5	Some risk	Low benefit	Unsure	Agree
6	Low risk	Low benefit	Agree	Unsure
7	Some risk	Low benefit	Agree	Unsure
8	Low risk	Low benefit	Unsure	Unsure
9	Low risk	Low benefit	Agree	Disagree
10	Some risk	Low benefit	Disagree	Agree

Source: Analysis of Interviews Transcripts

Table D-13 Relationship Between Risk Perception, Benefit Perception, General Trust and Sceptical Trust in the Government Based on Interview Analysis

Subject	Risk	Benefit	Trust – General	Trust - Sceptical
1	High risk	Low benefit	Disagree	Unsure
2	High risk	No benefit	Unsure	Unsure
3	No risk	Some benefit	Agree	Disagree
4	High risk	Some benefit	Unsure	Unsure
5	Some risk	Low benefit	Unsure	Agree
6	No risk	No benefit	Agree	Unsure
7	Some risk	No benefit	Unsure	Unsure
8	High risk	Some benefit	Unsure	Unsure
9	Some risk	No benefit	Agree	Disagree
10	High risk	Some benefit	Disagree	Agree

Source: Analysis of Interviews Transcripts

Table D-14 Relationship Between Risk Perception, Benefit Perception, General Trust and Sceptical Trust in the Insurance Industry Based on Questionnaire Analysis

Subject	Risk	Benefit	Trust – General	Trust - Sceptical
1	Some risk	Low benefit	Disagree	Unsure
2	High risk	Low benefit	Agree	Agree
3	Low risk	Some benefit	Disagree	Agree
4	Some risk	Some benefit	Agree	Agree
5	Some risk	Low benefit	Unsure	Unsure
6	Low risk	Low benefit	Disagree	Unsure
7	Some risk	Low benefit	Disagree	Unsure
8	Low risk	Low benefit	Unsure	Unsure
9	Low risk	Low benefit	Disagree	Disagree
10	Some risk	Low benefit	Disagree	Agree

Source: Analysis of Interviews Transcripts

Table D-15 Relationship Between Risk Perception, Benefit Perception, General Trust and Sceptical Trust in the Insurance Industry Based on Interview Analysis

Subject	Risk	Benefit	Trust – General	Trust - Sceptical
1	High risk	Low benefit	Disagree	Unsure
2	High risk	No benefit	Unsure	Agree
3	No risk	Some benefit	Disagree	Agree
4	High risk	Some benefit	Unsure	Unsure
5	Some risk	Low benefit	Unsure	Unsure
6	No risk	No benefit	Disagree	Unsure
7	Some risk	No benefit	Unsure	Unsure
8	High risk	Some benefit	Unsure	Unsure
9	Some risk	No benefit	Disagree	Disagree
10	High risk	Some benefit	Disagree	Agree

Source: Analysis of Interviews Transcripts

Table D-16 Analysis of Risk Acceptance

Subject	Questionnaire	Interview
1	Insurers should not be allowed	• "I do see some arguments in favour of having them use genetic test results, I guess, but looking at it from the perspective of someone, as a client of an insurance company, I would prefer that they didn't" • (do you think that insurance companies should be allowed to use family history information and not genetic test information or do you think that they should be allowed to use the two or neither?) "I would prefer neither"
2	Insurers should definitely be allowed	• "I think that it would very unfair to an individual who wanted to take out insurance, and the fact that years ago there wouldn't have been any such thing as things as, as, thing being hereditary, or genetics, or anything like that" • (Genetic test results should be accessed "only, only by medical people, but only with your health in mind, like you know, and, and trying to help you, sort of like, but, I don't think it, it would be anybody else's business really" • "I don't think they should be doing it in the first place" (insurers using genetic test results) • "if they're going to put through this Disability Bill, well then, they must have got some information, and they're going to try and stop it which is only right"
3	Insurers should be allowed	• "I do think that there should be more of an over-riding authority on the insurers" • "I think you have to look at it from both sides, the, the individual knows the result, while the insurance company doesn't and is being asked to insure the person and, he has a right to know" • "it should be a fair assessment of the risk" • (do you believe that the insurance industry should be allowed to use genetic test results when underwriting individuals?) "I think so" "Yes, I think it would be favourable" • "I think that it's a new science, I think that we should be interested definitely, we definitely

		should be pushing out the frontiers but I think they should be controlled"
4	Don't know	• "I would see it as unfair for them to use that, in anyway, because there's no saying that I'm definitely going to suffer, I could lead, I could decide from now on, I'm going to lead an extremely healthy life and put myself in the least possible risk to getting high cholesterol and high blood pressure, and therefore I could say that's extremely unfair, 'cos I've done everything in my power to stop myself from getting high blood pressure so…" • "It's a tricky one because in, in once sense, yes, it's definitely unfair, because it means that I'm not going to be allowed to take out life assurance, for that reason, or else, they're going to out the quote so much over my limit that I'm not going to be able to take it out. But in the other sense, well, you can kind of see where they're coming from. If they took it in, but, if there was a public vote over it taking genetic testing, allowing life insurance companies to use genetic testing or not, I would probably vote not, that they would not, that they should not be allowed to use it" • "if it went to a public vote now, and based on my knowledge of it, I would say no, that they shouldn't be allowed to use genetic test results"
5	Insurers should not be allowed	• "Very, very good for them to have it (coughs) to have that kind of information, so yeah, no maybe I should agree with it" • (Do you think that the insurance industry should be allowed to use genetic test results to decide whether or not to insure an individual?) "No, I don't think so, I think it's cheating" "the company, if you, if they know, like that you're not going to make it, they're just going to say, okay not you" • (If insurance companies are denied the use of genetic test results, do you believe that this could have an undesired impact on future generations?) "No, no, I think it's be pretty good"

6	Insurers should not be allowed	• "all in all, I suppose, in determining premiums, no they shouldn't be. They shouldn't be" • "I believe that insurance should be, the basis of insurance is to share out risks amongst a large number of people and I believe that by following that insurance should be able to minimise the risk to them from genetic diseases which people pick up. A genetic disease is not something, as far as I can tell, is not something that a person brings on themselves, they inherit, therefore it isn't something they should have to pay extra premium because they cannot prevent having a genetic disease, therefore I do not believe that insurance companies should be allowed to use them to determine their premiums. I believe that insurance companies are collecting genetic information about people, and the only logical reason I can think if why they'd want to do that is to segregate their population and charge more to people who are prone to a genetic disease or who have a genetic disease" • "The other person should be, should be, in general I believe that the other person should not have to pay more insurance premium for, for factors that is outside their control" • "In the short run, yes, I believe it should not be use, yes, I believe the Government should intervene"; "In the long run, it would be optional but again I would not see a reason for it to be used, again for the overall reason that I don't think a person should be discriminated against for something that they can't control"
7	Insurers should not be allowed	• "I think they shouldn't, yeah, I think they shouldn't be allowed"
8	Insurers should not be allowed Insurers should not be allowed	• "I suppose it's, maybe they're right in, in the business that they're in, they should have some information about that person" • "I wouldn't be for it" • "I think that the risks are too high, I mean like I don't see huge benefits for the person or society, I see it more as a business entity, or a business kind of thing for them, like the more information they get out of it the higher premiums they can

		charge so I kind of feel like the benefits don't outweigh the risks of it, I think…"
9	Insurers should not be allowed	• (Do you think that the insurance industry should be allowed to sue genetic test results?) "If the person isn't forced to take them, then I see no, you know, there's no reason not to" • (Insurance companies should be allowed to access that information) "Yeah, if, it's up to the person, yeah, if the person isn't forced to take the test" • "No" (I don't think that the Government should prohibit insurance companies from using genetic test results) • "I think that if you're taking out a life assurance policy or a huge amount then possibly you should be charged more, like if it is an outrageous amount, whereas if it's to cover al mortgage pr something, well then like…" "A mortgage plus 50 or a hundred thousand" "Well then again you know, a mortgage in Dublin is different to a mortgage in, but like you know a mortgage anyway and then something else because you know that has to be paid off anyway" • "if I knew in the morning that there was something wrong with me, and I had to take out life assurance then, fair enough if you have to pay a little extra but if you're being really, really penalised for that then it's nothing that you can do about it, everyone has to get a mortgage, everyone has to, you know, to live as long as they can, you know so I don't believe anyone should be penalised, there's enough people in the world scheme of things to absorb any…"

10	Insurers should not be allowed	• "it's a catch 22 situation" • "I would say no, no don't, they should not be allowed" • "because companies are by their nature, because they're commercial organisations they'll use it to their own best advantage, they'll not necessarily use it benignly, they'll not necessarily use it ethically, morally or anything else because insurance companies are financial entities, they're based on shareholders, and in most cases even the most avid of mutuals, such as Standard Life, is going to demutualise shortly so there's going to be no mutual companies which would return the benefits to their clients, kind of venture holders, mutual company members, they're all plc's, bottom line share price, profit, that sort of way. An awful lot of ethics, morals, and everything else gets lost in that. And therefore, access to this information will only ever be used, on the basis of what's good for the company and not necessarily what's good for the client."

Source: Analysis of Interviews Transcripts

Table D-17 Overall Relationship Between Risk Perception, Benefit Perception, General Trust and Sceptical Trust in the Government Based on Interview Analysis

Subject	**Risk**	**Benefit**	**Trust – General**	**Trust - Sceptical**	**Risk Acceptance**
1	High risk	Low benefit	Disagree	Unsure	Not accept
2	High risk	No benefit	Unsure	Unsure	Not accept
3	No risk	Some benefit	Agree	Disagree	Accepts
4	High risk	Some benefit	Unsure	Unsure	Not accept
5	Some risk	Low benefit	Unsure	Agree	Not accept
6	No risk	No benefit	Agree	Unsure	Not accept

7	Some risk	No benefit	Unsure	Unsure	Not accept
8	High risk	Some benefit	Unsure	Unsure	Not accept
9	Some risk	No benefit	Agree	Disagree	Accepts
10	High risk	Some benefit	Disagree	Agree	Not accept

Source: Analysis of Interviews Transcripts

Table D-18 Overall Relationship Between Risk Perception, Benefit Perception, General Trust and Sceptical Trust in the Insurance Industry Based on Interview Analysis

Subject	**Risk**	**Benefit**	**Trust – General**	**Trust - Sceptical**	**Risk Acceptance**
1	High risk	Low benefit	Disagree	Unsure	Not accept
2	High risk	No benefit	Unsure	Agree	Not accept
3	No risk	Some benefit	Disagree	Agree	Accepts
4	High risk	Some benefit	Unsure	Unsure	Not accept
5	Some risk	Low benefit	Unsure	Unsure	Not accept
6	No risk	No benefit	Disagree	Unsure	Not accept
7	Some risk	No benefit	Unsure	Unsure	Not accept
8	High risk	Some benefit	Unsure	Unsure	Not accept
9	Some risk	No benefit	Disagree	Disagree	Accepts
10	High risk	Some benefit	Disagree	Agree	Not accept

Source: Analysis of Interviews Transcripts

Table D-19 Analysis of Risk Management

Subject	Interview – Risk Management
1	**Regulator:** • "I think the Government should certainly legislate to provide some restrictions at least where it makes sense to insurance companies using genetic test results, yes" • "an independent third party should be involved to regulate their use" "Perhaps a Government appointed agency" • "I think that such a party would need to be created and given the appropriate powers to hold the Government accountable, I guess that's one of the key things for me, is that in some ways, such as this, a third party can be more powerful than Government itself, in that it can hold Government accountable as well as the insurance industry, and, but be representative of the people" **Members of Third Party:** • "I think that both" (experts and laypersons) "would be the ideal, certainly as concerning, I think sometimes some ordinary people are also needed. Sometimes experts don't always see the wood for the trees and ordinary people without such focused outlooks may have different opinions that are very important" • "in my view, the best practical solution would be to have an independent body made up of members of the public along with experts as you suggested and in that way, you know using an appropriate sample from the public have decisions made. I think that a nationwide referendum wouldn't be, for every decision that had to be made, wouldn't be a practical solution. But there are lots of ways to poll public opinion, text message voting, websites, you know things like this, gather opinions from radio listeners, although of course, it's open to a certain level of bias, but basically, yes, I think that public opinion should be factored in" **Public Participation:** • "I do think so yeah" (think that many people would be interested in providing their opinions) **Personal Participation:** • "No" (I do not believe that I have a say on the content of decisions regarding insurers use of genetic test results) • "No" (I do not have a say in how these decisions are made) • "I personally, I would, but I don't feel I'd be able to articulate my opinions strongly enough to work in that kind of role"

2	**Regulator:** • "I suppose they'd" (Government) "have to appoint who they would think would be best sort of like, to be able to manage it, and once it would be made into law, like you can't break the law, the insurance companies can't" • (do you think that the Government should completely stop the companies from using this information?) "Yes, because I mean, the Government is the law of the land, rather than the insurance companies" **Personal Participation:** • (do you believe you have a say on the content of these decisions?) "I don't think so", "No" • (do you think that you should have a say?) "Yeah, because I think everybody has the right to vote like, and everybody has the right to have their say, sort of like, and I know most people wouldn't like, you know, doing anything in public, but at least if it was a ballot, you have a ballot for this and a right to vote for whatever, and I don't know whether we go into the EU or what like, but I think it's just the same for insurance companies because most people have policies…" **Public Participation:** • "if there was going to be a vote on it or any, or a say from the public on it, I think the information would be freely available on it first, but it's just people don't know about it, I didn't know"
3	**Regulator:** • "I think that we need a control, there needs to be controls, I don't think they should be controlling it themselves, probably a watchdog" • "I think in the initial stages it should be the Government, whether you would then have an ombudsman or a controller, independent of the insurance bodies, but, but initially who, the Government" • "over a period of time there could be misuse, and it's better if consumers have an independent body" • "The politicians make the law, they decide, they get the best possible advice, right, now you can go to ten experts and get ten different focuses, or directions on that advice, at the end of the day, they decipher that, the civil servants do it, not the politicians, and the civil servants, they are, not necessarily scientific, but expert in their own right, interpret that, advise the ministers, the ministers draft the laws and the Government passes them. So I, I think the system is there to work properly"

	• "if they are only operating one test on the basis of perameters laid down in the UK, a lot of laws is based on legislation first drafted in the UK, it's a control, so they seem to be controlling it" • "I am glad to hear there are constraints and they are working to a control system even if it's not our own system, but we should have our own system in place, our own Government's" **Personal Participation:** • (regarding the decision-making by the Government and by the insurance industry, do you believe that you have a say on the content of these decisions?) "Removed, but a say" (with regard to the Government) "with the insurance industry, very little say, unless there's a control involved. In the Government, yes a say, but quite a bit removed, you have, you have one vote but your vote counts, and you should never forget that, and, and again that takes you back to, if you don't like the decision your Government makes, or, the local politician as part of your Government makes then you change direction the next time round, you punish them for a lack of decision, or a lack of doing what you want them to do. So, I think you have a say, but it is well removed" • (Have a say in how these decisions are made) "Only by putting in the Government into place" "quite removed, but that's, but that's the trust, we go, we're in a democracy, that's what they're there for" • "We can form our own opinions and make our own decisions, but we won't be making, individually we won't be making decisions pertaining to industry, Government, it is necessary that Government makes those decisions, and they do it by the best sources available, the best scientific sources available, the politician doesn't make the decision, we just see it as the minister makes the decision and drafts the law and the Government decide on that too. But it is on solid scientific advice. The same way they might approach environmental issues" • "we vote our Government in, people to make the decision for us, we make the decision to put them in there, if you don't like what they do, you take them out"
4	**Regulator:** • "I don't know what the Government have to do with it" • "I'd definitely say that it'd be Government based decisions and even perhaps petitions and stuff" **Public Participation:** • "I just don't think it'd come to a public vote, really. That, that's the way I imagined, that's the way I interpreted the question, would

	the public have a say in whether or not insurers do do this and I thought perhaps not, I thought maybe more so it'd be to do with Government. Maybe the Government might decide on it without putting, I, I don't know if things like that would come to a public vote. It should, by, in, in my mind, but, anyway, in my opinion it should, but I wouldn't be sure so I just thought that it wouldn't, I wasn't a 100% sure basically, but I thought that it wouldn't come to a, I thought I wouldn't have a say because it wouldn't come to a public vote. But, it might, I could be wrong!" **Personal Participation:** • "Yes, definitely" (would like to have a say) **Public Participation:** • "I think it's definitely a matter of, in the public interest, it's definitely for the public to decide or for me myself in anyway, I'd definitely like a say in it. • "it's definitely in the public interest to make decisions, the public should get information to make the decisions on this rather than just the Government"
5	**Regulator:** • "the Government should really be on top of the situation" • (Do you think that the insurance industry is capable of managing its own use of genetic test results?) "yeah they are, but maybe shouldn't the Government…" "make sure that the information is being used appropriately" • (Do you think that it's necessary for the Irish Government to create legislation that prohibits life assurance companies from using genetic test results?) "Yeah, yeah, I do" • (Do you believe that it's the Government that manages the situation or do you believe that the insurance industry should manage itself?) "Both" • "you know like the way banks they were looked on as completely trustable organisation and then, oh my God, there's been this awful thing going on you need to set up another body to manage this body and we're like, you always have to have a police system the system policed by some kind of…" "Authority, like maybe the Government should set up that authority to make sure that things are managed better" • "you can't really predict, like, you say, like, that question there like, whether I think it's going to have a bad effect on future generations or not, I can't really answer that like, we won't really know" "I wouldn't know whether the Government would be able to…" (make decisions now?) "Yeah, like, yeah"

	• (Do you think that there should be different regulations for different countries?) "Well, I think Europe should be mostly the same" **Personal Participation:** • "of course you should have a say, cos otherwise they're (Government) "just going to rule, yeah, no, we just want to rule this" • "I don't think I'd go off campaigning against it, cos you know because, I wouldn't know enough about what I'd be trying to say about this and this and this. It makes you feel a bit sceptical about how it works"
6	**Regulator:** • "at the moment I suppose it isn't" (insurance industry capable of managing its own use of genetic test results) "'cos I take that the little bit I know about genetic testing even in the theoretical and even in the scientific side it is very limited in what it's capable of, so I imagine that industry or the insurance industry would be even less capable of managing it, if scientists can't do it well then…. It's not a proven science yet. It's not proven to enough of a degree to make it fixed enough to be able to use in industry" • "I believe it should be through Government regulation that the public should, it should be a regulated system, whereby it's open, openly known how much, how much access the insurance companies have to genetic information. Perhaps if there was a regulator in the market or something like that to provide this information, it should be there" • "there should be some sort of representation over the system, it should be there" • "I believe it should be a panel of experts in the area, but I believe that all findings from such a meeting should be made public and should be open to the public, but the people actually on the board should be experts in the area" **Personal Participation:** • "at the moment, no, I suppose I don't have a say" • (would you like to have a say?) "Yes, most definitely"
7	**Regulator:** • (Do you think that the insurance industry is capable of managing it's own use of genetic test results, of creating it's own policies?) "No, not to regulate" • "An independent, definitely an independent regulator should be

set up"

- "A mixture of, do you know experts in the field and people who have experience of regulation as well, and people like that" (should be involved) "Try people out, with the politicians or people with an interest really I suppose"
- "you have to regulate it really, they're gonna, they want to, all they're worried about is their profits, so they're going to use, they're going to do everything they can to get those figures, so they can not insure those people or charge them a lot of money"

Personal Participation:

- "Individually you can write letters to your TD's and get them to pass it on, in order to express your concerns, but, I don't know you imaging they'd have some kind of independent, laypeople on the committee expressing general concerns"
- "I probably wouldn't be interested, but, I'd prefer someone with more knowledge. I don't have much knowledge about the whole thing."
- (Do you believe you have a say in how these decisions are made?) "Not really, no. I don't know if there's an ombudsman for this kind of thing, but it's really by writing letters to TDs that really all you can do"

Public Participation:

- (Do you think that if people had more knowledge about insurers use of genetic test results that they might be more willing to participate in any discussions regarding whether or not insurers have any access to genetic test results?) "Definitely, yeah. But you don't read about it in the paper and I read the paper every day so"
- (Do you believe that the public should participate more in any decisions made regarding this matter?) "Yes, because it affects everybody that needs insurance like, it's a major deal, and all these decisions are being made without most people knowing about them"

8	**Regulator:** • (Do you think that the Government can manage insurers use of genetic test results successfully?) "I don't, unless they have something like a specific task force, you know like people who are, experts themselves in that area, I don't know if they have, if they would employ a group of people, you know go to that kind of trouble and expanse of employing, the like of a committee or a task force to, to overview this kind of, the genetic testing. I, I suppose of it's, if it's, insurance companies, I mean I feel like insurance companies benefit more , they benefit more than society, or the person who's getting the insurance, so I think that, I kind of feel that the Government, I can't imagine that they would have a huge role in it or that they would really care whether or not, you know, that they, I kind of feel that they wouldn't put, be a hundred percent in putting effort into making sure that it's running properly and also that, I suppose they feel, you know, the whole expertise side of it as well, whether they would know enough about it or employ people" • (The Government) "have to react in some way, but like that it wouldn't be off their own bat, I think it'd be a matter of somebody pushing the matter on them and having to maybe set up a committee or something" • "I would hope that if it did go to the fact that you would have to have a test done or you would have to expose information, I think that I would like to see a third party involved in it" • (Which Government body do you think should be involved or can you think of any body that should be acting as controller?) "I suppose the logical one would be the Department of Health, I suppose you'd think that you'd have hopefully expertise in that kind of department, but also maybe independent people outside Government as well" • "maybe a group, a consulting group who are not, politically driven or that, that it's also, maybe a bit of both maybe, a bit of a kind of private but also a kind of public sector people" • Insurance companies are "just going to be there to make as much money as possible so I feel that's why the Government should be regulating it" **Personal Participation:** • "I imagine that I wouldn't have a lot of say • "the fact that I don't agree with it, I suppose I would like a say, you know if you had a choice to vote or if you had a choice to, like as a person who, like at the moment I don't, but as a person who would pay for insurance, or life assurance, that you would have a say in, in the way it's going to develop in the future or what kind of testing is going to come in"

	• "if I did have to make a choice I would hope that it would be informed and that you would be, you know, get your fact sheets get your for and against and why you would vote yes or no, so I'd hope that I should be informed before making a decision" • (do you think that maybe people who do have a genetic disorder within their own families, within their own history, medical history, should be involved with the decisions made regarding this matter or do you think that that's necessary?) "I don't think it's necessary, like you'd hope that if there was, you know a third party, and that they're there you know for the benefit of you know the ordinary person that they would you know kind of represent you to the best of their ability and that you wouldn't have to personally become involved or be on a panel but you'd hope that those kind of people would be able to do the job without, you know, I, I, I would trust those people to do it themselves, maybe a third party to come in and assess it" **Public Participation:** • "you know I can't imagine you know there being a referendum on, on whether this information should be given out or not, I think that it will be very much between the Government and the insurance companies and that the people wouldn't have a say, I can't imagine how I, could influence a decision like that. I couldn't see it"
9	**Personal Participation:** • (Do you believe that you have a say on the content of these decisions?) "not as an individual, no" • "Yeah, it would be good but I don't think something like that would be put to a vote" • "if I had concerns I don't think I'd speak to anyone" • "I suppose if I did have concerns I'd possibly write a letter to somebody in the Government, but I don't think it's the Government, you'd go to you local politicians with your concern. There's a hierarchy" **Public Participation:** • "I think that yeah they probably should contribute" (the public)
10	**Public Participation:** • (Do you believe you have a say on the content of these decisions with regard to insurers use if genetic test results?) "I think so, if enough people say, no, hold on a second know, this is information is now rather private, crucial and everything else, it really has to be dealt with in a balanced way, and get on to the Government

	ministers, get on to the Government media, whatever, making their voice, making their point saying no, let's see how this thing is going to work through the system and really the less we do, it's not going to be dealt with accurately because the vested interest will deal with it in their own way, and then the Government tends to be incompetent, lacking in information, too busy doing other things, there's not a specific group to deal with this sort of way and it just gets lost in the maelstrom while or the insurance companies tend to be very much for their own interest"

Table D-20 Percentage Extra on Regular Premium

Subject	Percentage Extra on Regular Premium
1	• "around something like 10%, not a lot, obviously because, yeah, not a lot"
6	• "I suppose I would be willing to, I would be willing to pay, I suppose if it was 15 or 20%, I would not mind having to pay if it had to be, even if it had to be quite a substantial amount" • (would you go above 20%?) "no I probably wouldn't go too far up, no"
7	• "I don't know, if you think like that, a few percent" • "Yeah, 5%" • (would you go over maybe 10%) "If I had to make it fair for everyone, I suppose" • "you wouldn't want to be going over about 25 or so"
9	• "Well speaking in terms of figures rights say for a mortgage, life assurance would you know on average you would probably pay 20 euros a month, I think now I could be wrong, I don't think 10% is going to break anyone" • (Would you pay more than 20%?) "Well, you're probably just going, 20 maybe, but…" • (Would you pay more than 20%?) "Well, I wouldn't see a need for it" • "this is an awful thing to say, but people who smoke or should be penalised before others"
10	• "That's a very good question, again if I logically follow my own line of thought, I've have to see what the market could bare, if a company said to me it would cost you around 10 or 20%, I might say yeah to 10%, I might say no to 20%, I'm not going to live with that"

Source: Analysis of Interviews Transcripts

Table D-21 Accountability

Subject	Accountability
1	**Government** "me as an individual? No" (the public have the means) "I guess in general perhaps more so these days, I still think that, the public by themselves need to have a representative committee that is recognised at Government level in order to have their voices heard on issues like this. Public outcry on the radio for example, it can work, but I think that independent regulators would probably be the most effective way" **Insurance Industry** "My answer would be similar" to that of the Government
3	**Government** "Only with our vote, and you can vote them out, but again far removed" **Insurance Industry** "Only if the Government enact a consumer watch dog"
5	**Government** "No" means to hold the Government accountable for their decisions on this matter **Insurance Industry** "Yeah, I mean yeah, if they're the ones making all the decisions about it, I think we could, yeah, if they're making bad decisions, we could say well you're the one making all the decisions, so, we can point a finger at them"
6	**Government** "I believe at the moment, yes, the public do, yes, I believe that as a member of the public I would have some power over it, yes" **Insurance Industry** "To some degree, yes"
8	**Government** "I think so because they are you know they would speak for the people and for such, for something that if it was given the go-ahead you know for, for tests and God only knows what would lead from one test and then you could have half a dozen tests and so on, I kind of feel that yeah, they would be responsible, fully responsible unless the people had actually had a choice in the matter, but I think it would be very much, I can imagine that it would be very much a decision between the Government and the insurance companies themselves rather than anybody else" **Insurance Industry** "well I suppose they're just going to be there to make as much money as possible so I feel that's why the Government should be regulating it

	and, if the people can't make a choice on it then the Government should be there to make sure you're not taken advantage of and that, they, I think they can be held accountable but at the end of the day, they're in for the business and it's, it's, they're going to make as much money as possible"
9	**Government** "If there was strong enough public feeling, yeah I would" **Insurance Industry** "No" means to hold the insurance industry accountable for their decisions
10	**Government** "Ultimately in the ballot box, ultimately we're citizens of the state, yes, yes, yes, but you have to work very hard to keep that up to spec, up to speed up to the mark, you know you have to continually be watchful, because you have to get out there and have your voice heard, it isn't easy but you've got to do it" **Insurance Industry** "Probably less so because it's commercial entities which tend to be the offerors of or the providers of in a lot of cases mandatory insurance, perhaps in the case of you're buying a house because the insurance company don't want their assets to burn down or if it's life assurance it could be because you want to protect the family or it could be that it could be car insurance, you know so whatever they have all their cards, they're continually always in the driving seat, they have all the information. You know I wouldn't, no, I wouldn't necessarily trust them or I wouldn't consider that they would, because they are vested interest, bottom-line profit, bottom-line shareholder, they are going to deal with the information in a way that I suppose secures their best interest and not necessarily the clients. Therefore, the client has to be watchful of the companies, and how they deal with the information and why they're looking for the information. I don't think we as easily have a lead on them as we would have on the Government. For the commercial entities, they are very hard to get at, not like the ballot box, you tend to have to go to court to get them to do this that or the other or and that's going to cost most people, but they have vast, huge deep pockets, they can do with it relatively as they will or they can, you know go a long way with these pockets you know paying lots of legal people or to protect their interests"

Source: Analysis of Interviews Transcripts

Table D-22 Personal and General Family History Statements

Subject	Family History
1	**Personal and Family History** • "I myself, have a form of heamachromatosis, which I know was inherited from one of my parents" **General - Family History** • "I think I was" asked to provide family history information. "I personally didn't mind, the information that I had to provide was okay in my circumstance, as I mentioned earlier I have heamachromatosis and I had no reservations in sharing that with the insurance company, however, I could see why other families may wish to keep that or other information like that to themselves for privacy reasons. And I know that other member s of my family would be a lot less inclined to share the same information" • (Do you think that insurance companies should be allowed to use family history information when underwriting individual applicants?) "I guess it's relevant because, something you mentioned earlier in your questions that didn't spring to mind to me, was that, I as an individual with no, tendencies towards certain genetic disorders may be, may end up paying a higher premium to compensate for other who do have a high likelihood of developing a genetic illness, if such disclosure isn't required, however, I think my answer is still tending towards no, I think that family history, of course it does indicate that there is an increased likelihood of a certain illness but it's not positive indicator by any means. For example in my family three out of nine children of my mother's siblings got, or carried the heamachromatosis while the other six did not, whereas I would assume that an insurance company would take the approach that all of these nine children would be equally likely to carry heamachromatosis, and increase their premiums accordingly, that I wouldn't be in favour of, so I think basically no" • "in my disclosing that my family has a history of heamachromatosis, I wasn't asked for a genetic test to verify whether I had it or not, so I guess if disclosing family history there is a case for genetic testing for the benefit of those that don't actually carry genetic disorders" • (it's okay to give a genetic test if it proves that you don't have a disorder, but if you do have a disorder then you don't have to give the test result) • "I guess that would be a win-win situation for the, people seeking insurance and obviously not the… it would be weighted in their favour and obviously not in the insurance company's. So, yeah,

	ideally I would like that but guess that's not realistic" • (Insurance companies should be allowed to use family history information and genetic test information?) "I would prefer neither"
2	**General - Family History** • "I think that it's (family history) private information, and, and I mean it's only, again should be used for medical reasons and I mean, I mean everybody has a right, whether, you know sort of like, because none of us know but we could have something genetic , but it just hasn't come about yet" • "On all forms that you go to fill out, like, they always and ever ask you, like background" • (Genetic testing is different to family history information) "Yes, I do, because, like I said earlier, just because your mother or father died of a certain thing which could be hereditary, it doesn't necessarily mean that you're going to go at the same time. And because medicine is improving and improving all of the time, like if you were to take my father for instance, he had malignant hypertension, which is extremely high blood pressure, by 53, but nowadays there'd be no such thing as dying at 53 years of age with that, I mean that's how much medicine has come on, and do you know, I think it would be very wrong to punish an individual who because of their family history" • "I don't think so, no" (think that insurance companies should be allowed to use family history) • "family history information doesn't necessarily mean, sort of like, like your parents could have been killed, they could have been, you know, anything else sort of, but genetic testing is very much an invasion of… just too invasive…"
3	**General - Family History** • "family history information is purely just information, it's, it's what's been collected over the years, it's, it's, it's history, whereas a genetic test will actually show if there's a vulnerability, it's, it's, it's proof" • "Put the two of them together and you will gain more accuracy"
4	**Personal and Family History** • "my Dad would have high blood pressure, I don't know if you'd consider that a genetic risk, I don't know if that would be considered a high enough risk, I suppose risk of heart attack, so yeah there would be, and I know that other than my immediate family, I know that there is, I would have Aunts and Uncles with high blood pressure as well" • "I could because of cells that they found when I went to a

	coloscopy clinic, so that could come against me because I'd have more risk toward ovarian cancer now than other females" **General - Family History** • "I would see it as unfair for them to use that (family history information above) in anyway, because there's no saying that I'm definitely going to suffer, I could lead, I could decide from now on, I'm going to lead an extremely healthy life and put myself in the least possible risk to getting high cholesterol and high blood pressure, and therefore I could say that's extremely unfair, 'cos I've done everything in my power to stop myself from getting high blood pressure" • "No, I wouldn't" (think it's fair that life assurance companies should be allowed to use this information) "because there's no definite answer that I would ever get ovarian cancer, I'm just at a more slight risk than another person, but then again another person, another woman could develop this in a year's time who took out an insurance policy and in a year's time she could actually get ovarian cancer, you know as in it could start a lot quicker, or anything like that. No, I don't think it would be fair at all"
5	**Personal and Family History** • "Diabetes, definitely, that's a possibility, because, I don't know what way it works, it's been in my father's family" **General - Family History** • (Is it fair that insurance companies should be allowed to use family history information to decide what price to charge you for insurance?) "If I owned an insurance company, I'd be like, yes, but as a person, or as a consumer, no, I think it's cheating"
6	**Personal and Family History** • "I can think of only one genetic disease which I have picked up, I have a low thyroid level that is genetic, I picked it up from my Grandmother" **General - Family History** • "no, I don't" (think that insurance companies should be allowed to use this information when deciding what premium to charge you for life assurance) • "The other person should be, should be, in general I believe that the other person should not have to pay more insurance premium for, for factors that is outside their control" therefore insurance companies should not be allowed to us family history information

7	**Personal and Family History** • "No" hereditary disease in family • **Family History versus Genetic Testing** • "No" insurers should not be allowed to use family history information
8	**Personal and Family History** • "my mother died of cancer, so that would be, for me, that would be an issue, but I don't think that it was hereditary, her, her, type, and also with my skin cancer, which is something which I know is caused by the sun, I know it's not, it's not a hereditary condition, but on my Father's side there is a heart, you know there would be, you know, he would have had a brother and Father who have died of heart attacks and also another brother who would have had a triple bypass so there would be a problem with, the heart, you know that side of it, you know his side of the family there would be issues that like with heart disease and things like that, so, I suppose then for me, I, you know, I suppose yeah, you could see that side of it in the sense that I could have heart trouble or I could get cancer or you know, but I suppose then it is another generation, so I don't know if that means I'm you know predisposed to get those things, so…" **General - Family History** • "I suppose I can see like, based on information, that's one thing filling up a form but then I think testing is a completely different issue, as in actually going and saying are you going to be predisposed to such a thing that you may never get, but, or you could die of getting knocked down, you know you never know what's going to happen, but I think, you know it's a fair question to ask what has happened in your family but I think a matter of you going into a hospital, getting blood tests and basing facts on that, I don't agree with that side of it" • "genetic testing just delves too much into you know something that might not happen and basing everything on, on a test that, you know, I mean I think it's one thing saying that my, such and such a person died and such and such a person died, which doesn't mean, you know, which means you could get these, but I think the fact that you, you know go to the stage of having blood tests and having them analysed and then basing that on a premium for you know, for your premium, your charge that, I think they're two different things, I consider too serious, too personal, too private"

10	**Personal and Family History** • "my mother, my grandmother, my great mother had severe ostoperosis" **General - Family History** • "if I'm following a logical line of thought in relation to genetic testing, no, I don't think that they should be allowed to use family history either, because it's one and the same basis. You've come from somewhere. You're genetic history is your family history, and eventually you're coming from the same source, your mother came from somewhere, your grandmother came from somewhere, the merged I presume, with another individual who was a male and who equally had a history, a family genetic history, so those two come together to produce you or me, you know, it's the same genetic history combined, egg, sperm, female egg, x and y chromosomes, you eventually come up with a physical make up with a genetic history first of all, and then over time a family history, a personal history, you have a medical history"

Source: Analysis of Interviews Transcripts

Table D-23 Case Selection Based on Survey Analysis

Case	Subject.	Knowledge	Risk	Benefit	Gvt. General	Gvt. Sceptical	Ins. General	Ins. Sceptical	Acceptance	Sex	Age	Employment.	Ins. Own
Represent-ative	4	Some	Some	Some	Disagree	Don't know	Disagree	Don't know	Not accept	F	26.35	Employed	No
Represent-ative	5	Some	Some	Low	Don't know	Agree	Don't know	Don't know	Not accept	M	18-25	Self-employed	No
Represent-ative	10	None	Some	Low	Disagree	Agree	Disagree	Agree	Not accept	M	56-65	Employed	Yes
Extremely Varied	1	Some	Some	Low	Agree	Don't know	Agree	Agree	Don't know	M	26.35	Employed	Yes
Extremely Varied	6	None	Low	Low	Agree	Don't know	Disagree	Don't know	Strongly not accept	M	18-25	Student	Yes
Extremely Varied	9	None	Low	Low	Agree	Disagree	Disagree	Disagree	Not accept	M	26-35	Employed	Yes
Critical	2	None	High	Low	Don't know	Don't know	Agree	Agree	Strongly accepts	F	46.55	Employed	No
Critical	7	Some	Some	Low	Agree	Don't know	Disagree	Don't know	Not accept	M	26.35	Student	No
Critical	8	None	Low	Low	Don't know	Don't know	Don't know	Don't know	Not accept	F	26.35	Employed	No
Extreme	3	Some	Low	Some	Agree	Disagree	Disagree	Agree	Accepts	M	46-55	Employed	No

Table D-24 Case Selection Based on Interview Analysis

Case	Subject.	Know.ledge	Risk	Benefit	Gvt. General	Gvt. Sceptical	Ins. General	Ins. Sceptical	Acceptance	Sex	Age	Employment.	Ins. Own
Represent-ative	1	Some	High	Low	Disagree	Don't know	Disagree	Don't know	Not accept	M	26.35	Employed	Yes
Represent-ative	2	Some	High	None	Don't know	Don't know	Don't know	Agree	Not accept	F	46.55	Employed	No
Represent-ative	4	Some	High	Some	Don't know	Don't know	Don't know	Don't know	Not accept	F	26.35	Employed	No
Represent-ative	5	Some	Some	Low	Don't know	Agree	Don't know	Don't know	Not accept	M	18-25	Self-employed	No
Represent-ative	7	Some	Some	None	Don't know	Don't know	Don't know	Don't know	Not accept	M	26.35	Student	No
Represent-ative	8	Some	High	Some	Don't know	Don't know	Don't know	Don't know	Not accept	F	26.35	Employed	No
Represent-ative	10	Good	High	Some	Disagree	Agree	Disagree	Agree	Not accept	M	56-65	Employed	Yes
Extremely Varied	6	Some	None	None	Agree	Don't know	Disagree	Don't know	Not accept	M	18-25	Student	Yes
Extremely Varied	9	Some	Some	None	Agree	Disagree	Disagree	Disagree	Accepts	M	26-35	Employed	Yes
Extreme	3	Some	None	Some	Agree	Disagree	Disagree	Agree	Accepts	M	46-55	Employed	No

APPENDIX E

E.1 Case Selection

Table E-1 Case Selection Based on Survey Analysis

Case	Subject	Knowledge	Risk	Benefit	Gvt. General	Gvt. Sceptical	Ins. General	Ins. Sceptical	Acceptance	Sex	Age	Employment.	Ins. Own
Represent-ative	4	Some	Some	Some	Disagree	Don't know	Disagree	Don't know	Not accept	F	26.35	Employed	No
Represent-ative	5	Some	Some	Low	Don't know	Agree	Don't know	Don't know	Not accept	M	18-25	Self-employed	No
Represent-ative	10	None	Some	Low	Disagree	Agree	Disagree	Agree	Not accept	M	56-65	Employed	Yes
Extremely Varied	1	Some	Some	Low	Agree	Don't know	Agree	Agree	Don't know	M	26.35	Employed	Yes
Extremely Varied	6	None	Low	Low	Agree	Don't know	Disagree	Don't know	Strongly not accept	M	18-25	Student	Yes
Extremely Varied	9	None	Low	Low	Agree	Disagree	Disagree	Disagree	Not accept	M	26-35	Employed	Yes
Critical	2	None	High	Low	Don't know	Don't know	Agree	Agree	Strongly accepts	F	46.55	Employed	No

Critical	7	Some	Some	Low	Agree	Don't know	Disagree	Don't know	Not accept	M	26.35	Student	No
Critical	8	None	Low	Low	Don't know	Don't know	Don't know	Don't know	Not accept	F	26.35	Employed	No
Extreme	3	Some	Low	Some	Agree	Disagree	Disagree	Agree	Accepts	M	46-55	Employed	No

Table E-2 Case Selection Based on Interview Analysis

Case	Subject	Knowledge	Risk	Benefit	Gvt. General	Gvt. Sceptical	Ins. General	Ins. Sceptical	Acceptance	Sex	Age	Employment	Ins. Own
Represent-ative	1	Some	High	Low	Disagree	Don't know	Disagree	Don't know	Not accept	M	26.35	Employed	Yes
Represent-ative	2	Some	High	None	Don't know	Don't know	Don't know	Agree	Not accept	F	46.55	Employed	No
Represent-ative	4	Some	High	Some	Don't know	Don't know	Don't know	Don't know	Not accept	F	26.35	Employed	No
Represent-ative	5	Some	Some	Low	Don't know	Agree	Don't know	Don't know	Not accept	M	18-25	Self-employed	No
Represent-ative	7	Some	Some	None	Don't know	Don't know	Don't know	Don't know	Not accept	M	26.35	Student	No
Represent-ative	8	Some	High	Some	Don't know	Don't know	Don't know	Don't know	Not accept	F	26.35	Employed	No
Represent-ative	10	Good	High	Some	Disagree	Agree	Disagree	Agree	Not accept	M	56-65	Employed	Yes
Extremely Varied	6	Some	None	None	Agree	Don't know	Disagree	Don't know	Not accept	M	18-25	Student	Yes
Extremely Varied	9	Some	Some	None	Agree	Disagree	Disagree	Disagree	Accepts	M	26-35	Employed	Yes
Extreme	3	Some	None	Some	Agree	Disagree	Disagree	Agree	Accepts	M	46-55	Employed	No

E.2 Descriptive Statistics

Table E-3 Distribution of Knowledge Percentage Population

	Knowledge of genetic testing	Knowledge of use of genetic test results by life insurers	Knowledge of policies governing use of genetic test results
No Knowledge	36.7	70.0	81.1
Some knowledge	32.8	22.2	18.9
General knowledge	19.4	7.8	0
Good knowledge	8.9	0	0
Perfect knowledge	2.2	0	0
Total	**100.0**	**100.0**	**100.0**

(n = 180)

Table E-4 Distribution of Risk Perception Percentage Population

	Risks to Society	Risks to Family
Strongly disagree	5.6	6.1
Disagree	20.0	18.9
Don't know	11.7	10.6
Agree	47.2	47.8
Strongly Agree	15.6	16.7
Total	**100.0**	**100.0**

(n = 180)

Table E-5 Distribution of Benefit Perception
Percentage Population

	Benefits to Society	Benefits to Family	Overall Benefit Perception
Strongly disagree	19.4	21.7	16.7
Disagree	50.6	47.8	48.9
Don't know	9.4	10.0	17.2
Agree	20.6	20.6	17.2
Strongly Agree	0	0	0
Total	**100.0**	**100.0**	**100.0**

(n = 180)

Table E-6 Distribution of Risk Acceptance
Percentage Population

	Risk Acceptance
Strongly disagree	32.8
Disagree	40.6
Don't know	5.6
Agree	18.3
Strongly Agree	2.8
Total	**100.0**

(n = 180)

Table E-7 Distribution of Beliefs in Adequate Regulation
Percentage Population

	Strongly Disagree	Disagree	Don't Know	Agree	Strongly Agree	Total
Government adequately regulates	8.3	31.1	53.9	5	1.7	**100**
IIF adequately regulates	10.6	28.9	53.3	4.4	2.8	**100**

(n = 180)

Table E-8 Distribution of Trust in Irish Government Percentage Population

	Strongly Disagree	Disagree	Don't Know	Agree	Strongly Agree	Total
Government - does good job	7.8	25.6	53.9	12.8	0	**100**
Government – competent	11.1	37.2	15	33.9	2.8	**100**
Government – skilled	12.2	33.9	14.4	35	4.4	**100**
Government - distorts facts	0	23.9	37.2	29.4	9.4	**100**
Government - changes policies without good reason	2.8	22.8	37.8	28.9	7.8	**100**
Government - influenced by life insurance companies	0	22.2	34.4	38.3	5	**100**
Government - acts in public interest	5.6	26.1	35	33.3	0	**100**
Government - listens to public concerns	7.8	28.9	23.3	38.9	1.1	**100**
Government - listens to ordinary people	8.9	35.6	22.8	31.7	1.1	**100**
Government - fair decision-making	7.2	29.4	38.3	23.9	1.1	**100**
Government - provides information	35	33.3	22.2	8.3	1.1	**100**

(n = 180)

Table E-9 Distribution of Trust in Irish Insurance Industry Percentage Population

	Strongly Disagree	Disagree	Don't Know	Agree	Strongly Agree	Total
Insurance Industry - does good job	10.6	25	53.3	11.1	0	**100**
Insurance Industry – competent	15.6	41.7	18.9	23.9	0	**100**
Insurance Industry – skilled	13.9	28.3	22.8	34.4	0.6	**100**
Insurance Industry - distorts facts	0	13.9	27.8	33.9	24.4	**100**
Insurance Industry - changes policies without good reason	0	18.3	31.7	35	15	**100**
Insurance Industry - influenced by life insurance companies	0	9.4	24.4	43.9	22.2	**100**
Insurance Industry - acts in public interest	20	45	21.1	13.9	0	**100**
Insurance Industry - listens to public concerns	19.4	45.6	21.7	13.3	0	**100**
Insurance Industry - listens to ordinary people	20.6	46.7	21.1	11.7	0	**100**
Insurance Industry - fair decision-making	15.6	42.2	31.1	11.1	0	**100**
Insurance Industry - provides information	38.3	35.6	20	4.4	1.7	**100**

(n = 180)

Figure E-1 Graphical Distribution Regarding Beliefs of Lack of Credibility, Reliability and Integrity

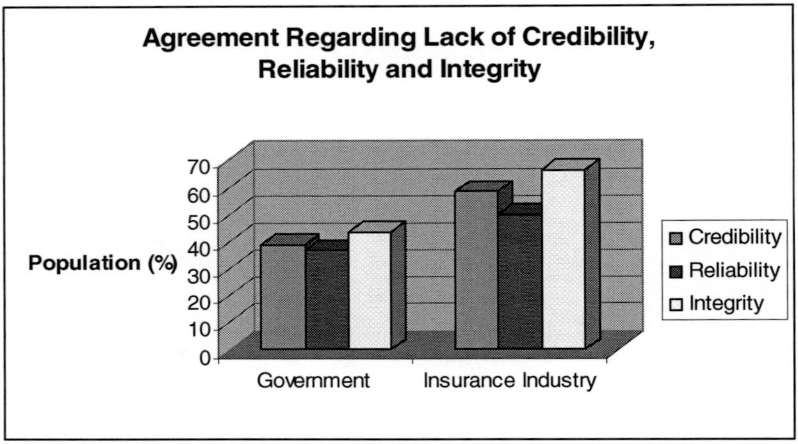

(n = 180)

Figure E-2 Graphical Distribution Regarding Beliefs of Lack of Credibility, Reliability and Integrity

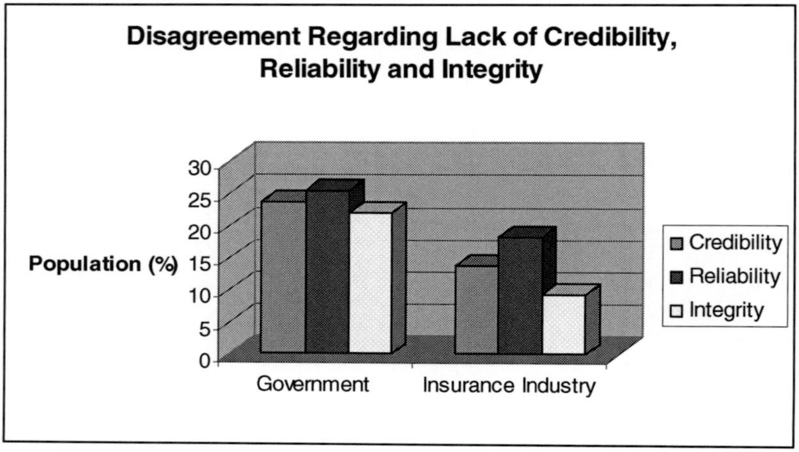

(n = 180)

Figure E-3 Graphical Distribution Regarding Evaluation of Care

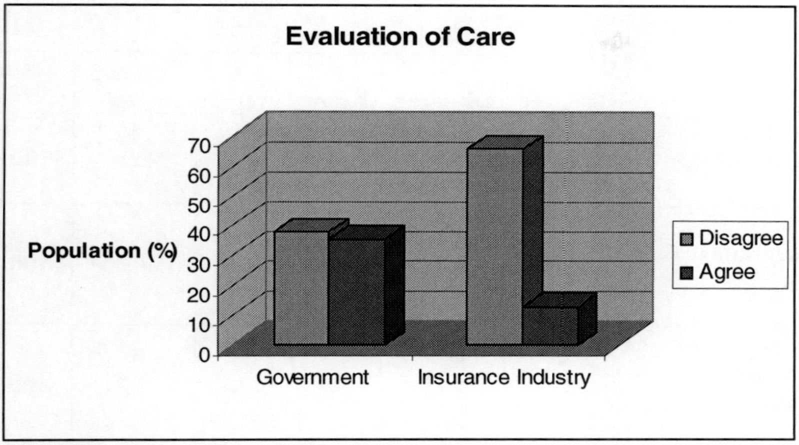

(n = 180)

Table E-10 Distribution of Satisfaction with Current Rules and Regulations and Need for Legislation Percentage Population

	Strongly Disagree	Disagree	Don't Know	Agree	Strongly Agree	Total
Sufficiency of current rules and regulations	7.8	31.1	50.6	8.9	1.7	**100**
Need for legislation	0	15.6	10.6	52.8	21.1	**100**

(n = 180)

Table E-11 Distribution of Evaluation of Public Participation and Involvement in Risk Management Process
Percentage Population

	Strongly Disagree	Disagree	Don't Know	Agree	Strongly Agree	Total
Public participation	0	2.8	5.6	46.1	45.6	**100**
Say on decision content	28.3	43.3	10	16.7	1.7	**100**
Say on how decisions are made	30.6	40.6	8.9	17.8	2.2	**100**

(n = 180)

Table E-12 Percentage Agreement: Say on Decision Content

	Frequency	Percent
Strongly disagree	51	28.3
Disagree	78	43.3
Don't know	18	10.0
Agree	30	16.7
Strongly Agree	3	1.7

(n = 180)

Table E-13 Percentage Agreement: Say on How Decisions are Made

	Frequency	Percent
Strongly disagree	55	30.6
Disagree	73	40.6
Don't know	16	8.9
Agree	32	17.2.8
Strongly Agree	4	2.2

(n = 180)

Table E-14 Correlations: Risk Perception and Say on Decision Content

		Risks to society	Say on decision content
Risk perception	Pearson Correlation	1	-.165
	Sig. (2-tailed)	.	.027
Say on decision content	Pearson Correlation	-.165	1
	Sig. (2-tailed)	.027	.

(n = 180)
* Correlation is significant at the 0.05 level (2-tailed).

Table E-15 Correlations: Risk Perception and Say on How Decisions are Made

		Risks to society	Say on how decisions are made
Risks to society	Pearson Correlation	1	-.245
	Sig. (2-tailed)	.	.001
Say on how decisions are made	Pearson Correlation	-.245	1
	Sig. (2-tailed)	.	.001

(n = 180)
** Correlation is significant at the 0.01 level (2-tailed).

Table E-16 Means to hold Government Accountable

	Frequency	Percent
Strongly disagree	30	16.7
Disagree	55	30.6
Don't know	23	12.8
Agree	63	35.0
Strongly Agree	8	4.4

(n = 180)

Table E-17 Means to Hold Insurance Industry Accountable

	Frequency	Percent
Strongly disagree	45	25.0
Disagree	55	30.6
Don't know	24	13.3
Agree	48	26.7
Strongly Agree	8	4.4

(n = 180)

Table E-18 Correlations: Risk Perception and Means to Hold Government Accountable

		Risk Perception	Means to hold government accountable
Risk perception	Pearson Correlation	1	-.085
	Sig. (2-tailed)	.	.255
Means to hold government accountable	Pearson Correlation	-.085	1
	Sig. (2-tailed)	.255	.

(n = 180)

Table E-19 Correlations: Risk Perception and Means to hold Insurance Industry Accountable

		Risks to family	Means to hold insurance industry accountable
Risks to family	Pearson Correlation	1	-.093
	Sig. (2-tailed)	.	.213
Means to hold insurance industry accountable	Pearson Correlation	-.093	1
	Sig. (2-tailed)	.213	.

(n = 180)

Table E-20 Descriptive Statistics: Public Participation

	Frequency	Percent
Disagree	5	2.8
Don't know	10	5.6
Agree	83	46.1
Strongly Agree	82	45.6

(n = 180)

Table E-21 Age Group – Risk Perception

	Risk Perception					
	N	Strongly disagree	Disagree	Don't know	Agree	Strongly Agree
18-25	15		26.7	13.3	33.3	26.7
26-35	50		20.0	18.0	42.0	20.0
36-45	47	6.4	23.4	17.0	36.2	17.0
46-55	36	5.6	8.3	8.3	44.4	33.3
>56	32	3.1	15.6	18.8	43.8	18.8

(n = 180)

Table E-22 Age Group – Benefit Perception

	Benefit Perception				
	N	Strongly disagree	Disagree	Don't know	Agree
18-25	15	6.7	46.7	20.0	26.7
26-35	50	16.0	60.0	12.0	12.0
36-45	47	14.9	51.1	19.1	14.9
46-55	36	22.2	44.4	16.7	16.7
>56	32	18.8	34.4	21.9	25.0

(n = 180)

Table E-23 Current occupational status

	Frequency	Percent
Employed	101	56.1
Self-employed	15	8.3
Unemployed	30	16.7
Retired	19	10.6
Student	15	8.3
Total	**180**	**100.0**

(n = 180)

Table E-24 Economic Status and Risk Acceptance

	Agreement with life insurers' use of genetic test results				
	Strongly disagree	Disagree	Don't know	Agree	Strongly Agree
Employed	27.7	46.5	5.0	19.8	1.0
Self-employed	46.7	26.7		20.0	6.7
Unemployed	46.7	33.3	10.0	10.0	
Retired	26.3	31.6	5.3	26.3	10.5
Student	33.3	40.0	6.7	13.3	6.7

(n = 180)

Table E-25 Economic Status and Benefit Perception

	Benefit Perception			
	Strongly disagree	Disagree	Don't know	Agree
Employed	15.8	56.4	17.8	9.9
Self-employed	26.7	40.0	20.0	13.3
Unemployed	20.0	36.7	13.3	30.0
Retired	15.8	36.8	10.5	36.8
Student	6.7	46.7	26.7	20.0

(n = 180)

Table E-26 Economic Status and Risk Perception

	Risk Perception				
	Strongly disagree	Disagree	Don't know	Agree	Strongly Agree
Employed	5.0	15.8	11.9	42.6	24.8
Self-employed		26.7	26.7	20.0	26.7
Unemployed	3.3	20.0	13.3	46.7	16.7
Retired		5.3	21.1	47.4	26.3
Student		40.0	26.7	26.7	6.7

(n = 180)

Table E-27 Economic Status and Trust in the Government

	Government - Overall Trust			
	Strongly disagree	Disagree	Don't know	Agree
Employed		5.9	84.2	9.9
Self-employed	6.7	6.7	73.3	13.3
Unemployed		6.7	80.0	13.3
Retired		15.8	52.6	31.6
Student		6.7	86.7	6.7

(n = 180)

Table E-28 Economic Status and Trust in the Insurance Industry

	Insurance industry - Overall Trust		
	Disagree	Don't know	Agree
Employed	5.9	84.2	9.9
Self-employed		86.7	13.3
Unemployed	3.3	83.3	13.3
Retired	15.8	57.9	26.3
Student	20.0	73.3	6.7

(n = 180)

Table E-29 Ownership of Life Insurance and Risk Acceptance

		Agreement with life insurers' use of genetic test results				
		Strongly disagree	Disagree	Don't know	Agree	Strongly Agree
Currently own life insurance	Yes	35.8	36.7	5.0	18.3	4.2
	No	26.7	48.3	6.7	18.3	0

(n = 180)

Table E-30 Future Purchase of Life Insurance and Risk Acceptance

		Agreement with life insurers' use of genetic test results			
		Strongly disagree	Disagree	Don't know	Agree
Future purchase of life insurance	Yes	30.0	50.0	6.7	13.3
	No	28.0	40.0	4.0	28.0

(n = 180)

Table E-31 Age Group – Trust in Government

	Government - Overall Trust			
	Strongly disagree	Disagree	Don't know	Agree
18-25		6.7	93.3	
26-35	2.0	10.0	78.0	10.0
36-45		6.4	80.9	12.8
46-55			88.9	11.1
56-65		11.1	72.2	16.7
66-75			55.6	44.4
76-85		50.0	50.0	
86-95				100.0

(n = 180)

Table E-32 Age Group – Trust in Insurance Industry

	Insurance Industry - Overall Trust		
	Disagree	Don't know	Agree
18-25	13.3	80.0	6.7
26-35	6.0	82.0	12.0
36-45	8.5	78.7	12.8
46-55		91.7	8.3
56-65	16.7	72.2	11.1
66-75		66.7	33.3
76-85	25.0	75.0	
86-95			100.0

(n = 180)

Table E-33 Age Group and Risk Acceptance

	Agreement with life insurers' use of genetic test results					
	N	Strongly disagree	Disagree	Don't know	Agree	Strongly Agree
18-25	15	20.0	46.7	13.3	20.0	
26-35	50	38.0	46.0	2.0	12.0	2.0
36-45	47	29.8	46.8	4.3	17.0	2.1
46-55	36	36.1	33.3	8.3	16.7	5.6
>56	32	31.3	28.1	6.3	31.3	3.1

(n = 180)